INTRODUCTION TO
APPLIED SOLID STATE PHYSICS

TOPICS IN THE APPLICATIONS OF
SEMICONDUCTORS, SUPERCONDUCTORS,
FERROMAGNETISM, AND THE NONLINEAR
OPTICAL PROPERTIES OF SOLIDS

SECOND EDITION

INTRODUCTION TO
APPLIED SOLID
STATE PHYSICS

TOPICS IN THE APPLICATIONS OF
SEMICONDUCTORS, SUPERCONDUCTORS,
FERROMAGNETISM, AND THE NONLINEAR
OPTICAL PROPERTIES OF SOLIDS

SECOND EDITION

RICHARD DALVEN

Department of Physics
University of California
Berkeley, California

PLENUM PRESS · NEW YORK AND LONDON

Library of Congress Cataloging-in-Publication Data

Dalven, Richard.
    Introduction to applied solid state physics : topics in the
  applications of semiconductors, superconductors, ferromagnetism, and
  the nonlinear optical properties of solids / Richard Dalven. -- 2nd
  ed.
        p.   cm.
    Includes bibliographical references.
    ISBN 0-306-43434-2
    1. Solid state physics. 2. Semiconductors.   I. Title.
  QC176.D24   1990
  530.4'1--dc20
                                                              89-72108
                                                                   CIP

© 1990, 1980 Plenum Press, New York
A Division of Plenum Publishing Corporation
233 Spring Street, New York, N.Y. 10013

Printed in the United States of America

To my father, JOSEPH DALVEN
and to the memory of my mother,
RUTH NEWTON DALVEN

Preface to the Second Edition

In addition to the topics discussed in the First Edition, this Second Edition contains introductory treatments of superconducting materials and of ferromagnetism. I think the book is now more balanced because it is divided perhaps 60% – 40% between devices (of all kinds) and materials (of all kinds). For the physicist interested in solid state applications, I suggest that this ratio is reasonable. I have also rewritten a number of sections in the interest of (hopefully) increased clarity.

The aims remain those stated in the Preface to the First Edition; the book is a survey of the physics of a number of solid state devices and materials. Since my object is a discussion of the basic ideas in a number of fields, I have not tried to present the "state of the art," especially in semiconductor devices. Applied solid state physics is too vast and rapidly changing to cover completely, and there are many references available to recent developments. For these reasons, I have not treated a number of interesting areas. Among the lacunae are superlattices, heterostructures, compound semiconductor devices, ballistic transistors, integrated optics, and light wave communications. (Suggested references to those subjects are given in an appendix.) I have tried to cover some of the recent revolutionary developments in superconducting materials. However, as of this writing, this field is still in ferment and the story is far from complete, so my presentation may well be obsolete in a short time. I have also elected to discuss only the physics of individual discrete devices, rather than the functions of the integrated circuits which serve as digital and analog building blocks. It seems to me that an applied physicist would probably be more concerned with, say, how field-effect transistors work than how they are used in an integrated circuit.

The prerequisites for reading this book remain the same as those for the First Edition, with the addition of a knowledge of ferromagnetism equivalent to that presented in an introductory solid state physics course.

Many people have commented on the manuscript or have helped me in other ways. My debts from the First Edition remain outstanding, and I have new debts to P. Berdahl, G. Y. Chin, J. Clarke, M. L. Cohen, L. M. Falicov, T. H. Geballe, E. L. Hahn, J. D. Jackson, T. W. Kenny, C. Kittel, R. U. Martinelli, A. M. Portis, A. C. Rose-Innes, M. Tinkham, and R. M. White. All have helped improve the book, but the responsibility for errors and obscurities remains mine alone. Thanks are also due the AAPT for permission to use some of the material in Chapter 8 which appeared in preliminary form in the *American Journal of Physics*. The text was typed by Rita Jones and Claudia Madison, both of whom were exceptionally skillful and patient. John Clarke's kind hospitality at the Lawrence Berkeley Laboratory was, as always, most helpful.

Last, I must thank D. for making this book possible, and G. for keeping it going.

RICHARD DALVEN

Berkeley, California

Preface to the First Edition

The aim of this book is a discussion, at the introductory level, of some applications of solid state physics.

The book evolved from notes written for a course offered three times in the Department of Physics of the University of California at Berkeley. The objects of the course were (a) to broaden the knowledge of graduate students in physics, especially those in solid state physics; (b) to provide a useful course covering the physics of a variety of solid state devices for students in several areas of physics; (c) to indicate some areas of research in applied solid state physics.

To achieve these ends, this book is designed to be a *survey* of the *physics* of a number of solid state devices. As the italics indicate, the key words in this description are physics and survey. Physics is a key word because the book stresses the basic qualitative physics of the applications, in enough depth to explain the essentials of how a device works but not deeply enough to allow the reader to design one. The question emphasized is how the solid state physics of the application results in the basic useful property of the device. An example is how the physics of the tunnel diode results in a negative dynamic resistance. Specific circuit applications of devices are mentioned, but not emphasized, since expositions are available in the electrical engineering textbooks given as references. To summarize, the aim of the book is the physics underlying the applications, rather than the applications themselves.

The second key word is survey. The book is designed to be broad rather than deep. Although the survey approach is not to everyone's taste, it has proved popular with the approximately 120 Berkeley graduate students (mostly in physics) who took or audited the course in 1973, 1974, and 1977. They seemed to want to learn something, but not everything, about the applications of the solid state physics they already knew. As a survey, the selection of topics is a compromise between recognition of the overwhelming technological importance of semiconductor devices and a desire to have some breadth of coverage. To this end, about 70% of the material covers applications of semiconductors, and the remainder is divided about evenly between nonlinear optical devices and superconductive materials and devices. Since the physics of the applications is the central interest of the book, no special effort was made to select the latest devices or to indicate the present "state of the art."

The book is a textbook ("A textbook explains, a treatise expounds"— J. M. Ziman) in that its aim is frankly tutorial. The book is essentially a collection of material from a number of sources, ranging from introductory textbooks to research journals, organized and presented with the intent of emphasizing the basic physics involved. There is no original work included. More advanced treatments and discussions of fine points are left to the literature. However, the reader is provided with references where fuller and/or more advanced treatments may be found. Further, a special effort has been made to give very specific references, telling where values of parameters, etc., were obtained. A selection of problems can be found at the end of each chapter. These are derivations, illustrative calculations, or invitations to explore the physics of some application. It is believed that these points harmonize with the attempt to provide a broad selection of the applications of solid state physics, while telling the reader where further information may be found.

The order of the first seven chapters is more or less linear. After a first chapter that is partly review and partly new material that will be useful, Chapter 2 treats the semiconductor p–n junction in some detail. The third chapter exploits this treatment in a discussion of several device applications. Chapter 4 treats the physics of metal–semiconductor and metal–insulator–semiconductor junctions, and the results are used in Chapter 5 to explore a few applications. In Chapter 6, a potpourri of "other" devices is discussed; they were chosen principally on the basis of my own interests. The seventh chapter treats a number of detectors and generators (principally semiconductors) of electromagnetic radiation. Chapter 8 is mostly concerned with the physics of Josephson junction devices, but concludes with a short discussion of the transition temperature in superconductors. Finally, Chapter 9 covers the interaction of electromagnetic waves in nonlinear solids, and concludes with a few applications.

In teaching a course on these topics, I have found that this book contains too much material for a one-quarter course. One semester would seem about right, particularly if appropriate review material were included. The notation used is standard, except perhaps that I have used $\mathscr{E}$ for the electric field vector to avoid confusion with energy E, particularly in band diagrams. The chapter on nonlinear optics reverts to the more common **E** for electric field because there seemed little possibility of ambiguity.

The presentation relies on a number of standard sources. Charles Kittel's classic introductory text on solid state physics is constantly quoted and used as a reference. Other books on which I have drawn particularly are *Solid State Electronic Devices* by B. G. Streetman; "Optical Second Harmonic Generation and Parametric Oscillation" by A. Yariv, in *Topics in Solid State and Quantum Electronics*, W. D. Hershberger (editor); *The Feynman Lectures on Physics* by R. P. Feynman, R. B. Leighton, and M. Sands; and *Long-Range Order in Solids* by R. M. White and T. H. Geballe.

This book is at the introductory level in that no particular prior knowledge of solid state device physics is assumed. However, the introductory level is not the same for all topics. For example, the treatment of the applications of nonlinear optical effects in solids is more complex than the treatment of the $p–n$ junction. As for prerequisites, it is assumed that the reader has had an introductory course in solid state physics at the level of Kittel's *Introduction to Solid State Physics*, Fifth Edition. In particular, it is assumed that the reader has a knowledge of energy bands, semiconductors, and superconductivity equivalent to that covered in Chapters 7, 8, and 12 of Kittel's book. In addition, this book assumes a knowledge of electromagnetic theory at the level of Reitz and Milford's *Foundations of Electromagnetic Theory*, of optics at the level of Stone's *Radiation and Optics* or Fowles's *Modern Optics*, and of quantum mechanics at the level of Bohm's *Quantum Theory*.

Many people have shared their expertise with me and have commented on the manuscript at various stages. I would like to thank N. Amer, T. Andrade, B. Black, R. W. Boyd, J. Clarke, M. L. Cohen, L. M. Falicov, L. T. Greenberg, E. L. Hahn, G. I. Hoffer, M. B. Ketchen, A. F. Kip, R. U. Martinelli, R. S. Muller, W. G. Oldham, and P. L. Richards for helping me improve the book. However, the responsibility for errors and misconceptions is mine alone. Special thanks are due M. L. Cohen and C. Kittel for their encouragement during the development of the course. T. H. Geballe kindly provided me with a prepublication copy of his work. I would like to thank M. L. Cohen, D. Long, G. S. Kino, W. G. Oldham, J. Tauc, D. Adler, E. Gutsche, J. Millman, B. G. Streetman, T. C. Harman, H. Y. Fan, and J. Clarke for permission to use figures from their publications. The hospitality extended by John Clarke was invaluable and is sincerely appreciated. Linda Billard typed part of the manuscript with great skill, and Leslie Hausman typed the first draft. Gloria Pelatowski executed the drawings with exceptional skill and enthusiasm.

Last, but also first, I would like to thank D. and G. for making this book a reality.

RICHARD DALVEN

Contents

1. Review of Semiconductor Physics

2. The Semiconductor *p-n* Junction

3. Semiconductor *p-n* Junction Devices

4. Physics of Metal-Semiconductor and Metal-Insulator-Semiconductor Junctions

5. Metal-Semiconductor and Metal-Insulator-Semiconductor Devices

6. Other Semiconductor Devices

7. Detectors and Generators of Electromagnetic Radiation

8. Superconductive Devices and Materials

9. Physics and Applications of the Nonlinear Optical Properties of Solids

10. Ferromagnetic Materials

INTRODUCTION TO
APPLIED SOLID STATE PHYSICS

1

Review of Semiconductor Physics

Introduction

The aim of this chapter is a brief discussion of some topics in semiconductor physics that will be useful in our discussion of applications. Since it is assumed that the reader has had an introductory course in solid state physics at the level of the book by Kittel,[1] some of the chapter will be review material. However, since some of the topics may be new to some readers, references to more complete and/or advanced treatments are given.

Metals, Insulators, and Semiconductors

We recall[2] that there are $2N$ independent states in each energy band of the band structure of a crystal containing N primitive unit cells. Each energy band can therefore hold $2N$ electrons. If the atomic arrangement in the unit cell is such that each primitive cell contains one valence electron (e.g., the $3s$ electron in sodium), there will be a total of N electrons occupying the $2N$ states in the band formed from the $3s$ atomic level. This $3s$ band is therefore only half-full and we expect sodium to be a metal. This situation is illustrated schematically in Figure 1.1, in which the vertical axis is electron energy and the horizontal axis[†] is distance (in real space) within the crystal.

[†] This type of drawing is also used without ascribing any particular meaning to the horizontal axis. We, however, will usually assign the meaning of distance to the horizontal axis.

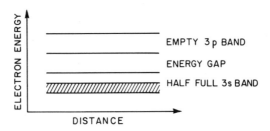

Figure 1.1. Energy bands (schematic) in sodium, showing the half-filled 3s band.

In this figure, we see a half-full 3s valence band separated by an energy gap from the next higher band, the completely empty 3p band.

If the number of valence electrons per primitive unit cell is an even integer, then we will have, if all other things are equal,[†] a filled valence band and the crystal will be an insulator. An example is diamond, which has eight electrons per primitive unit cell. This situation is indicated schematically in Figure 1.2, which shows the energy gap E_g between the highest of the four filled valence bands in diamond and the lowest empty band. We expect diamond to be an insulator, at least at 0 K. If the magnitude of the energy gap E_g in Figure 1.2 is not too large, then, at temperatures greater than 0 K, we would expect a few electrons in the highest filled valence band to be thermally excited into the lowest empty band, as indicated schematically in Figure 1.3. The presence of a few mobile electrons in the lowest empty band gives the crystal a weak electrical conductivity at temperatures above 0 K. For this reason, the lowest empty band, above the highest valence band, is called the conduction band. In such a case,

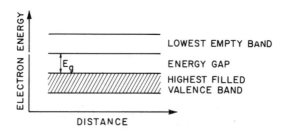

Figure 1.2. Diagram (schematic) of the highest filled valence band and lowest empty band separated by an energy gap E_g in an insulator (e.g., diamond).

[†] In this case, "all other things equal" means that there are no bands that overlap in energy making the crystal a metal[2] rather than an insulator.

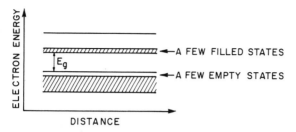

Figure 1.3. Energy bands (schematic) in a crystal in which a few electrons have been thermally excited from the filled band to the empty band.

the crystal is a semiconductor. A semiconductor differs only quantitatively from an insulator. The magnitude of the energy gap is smaller (about 1 eV) in a semiconductor than in an insulator (several eV). At temperatures above 0 K, a semiconductor will, because of its smaller energy gap, have more electrons thermally excited into its conduction band than will an insulator. The semiconductor will thus have a greater electrical conductivity than does the insulator.

Band Structure Diagrams

Consider a perfect crystal of diamond. The carbon atom has the electronic configuration $(1s)^2(2s)^2(2p)^2$, in which the $2s$ and $2p$ valence electrons are four in number. There are two atoms, and hence eight electrons, per primitive unit cell in the diamond crystal. We therefore expect diamond to have completely filled bands and to be an insulator. In an approximate way, we may think of the one $2s$ level and the three $2p$ levels in the carbon atom as each giving rise to a band in the crystal.[3,4] This gives a total of four bands arising from the atomic energy levels containing the valence electrons of carbon. These four bands are called the valence bands of the band structure of diamond. The valence bands of diamond (including spin) are shown[5] in the reduced zone scheme in Figure 1.4. The valence band structure in Figure 1.4 is shown for two directions, [111] and [100], in k space. The right-hand half of the drawing shows the valence bands for the [100] direction; the left-hand half, for the [111] direction. The symbols Γ, X, etc., in Figure 1.4 label certain special points in the Brillouin zone. While they are connected with the group-theoretical description of the symmetry properties of the crystal lattice, we will use them simply as "band labels."

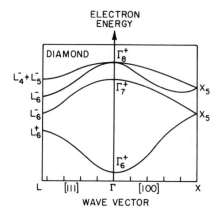

Figure 1.4. Valence band structure of diamond (reduced zone scheme) in the [111] and [100] directions in k-space. (Adapted from Long.[5])

We now consider the allocation of the eight valence electrons of carbon among the four valence bands. Since we consider $2N$ atoms in N primitive unit cells of the diamond lattice, we have a total of $8N$ electrons to put into four valence bands. Considering, for example, the [100] direction, the lowest valence band is the $\Gamma_6^+X_5$ band; it has N energy levels, which can accept $2N$ electrons (including spin). The same is true of the $\Gamma_7^+X_5$ band and the two $\Gamma_8^+X_5$ valence bands. We find, as expected, a grand total of $8N$ electrons in the four valence bands ($\Gamma_6^+X_5$, $\Gamma_7^+X_5$, $\Gamma_8^+X_5$, $\Gamma_8^+X_5$) of diamond.

We see from Figure 1.4 that the point in k space at which a valence electron has the highest energy is the Γ point at the center of the Brillouin zone. We therefore say that the valence band maximum occurs at the center of the Brillouin zone. At 0 K, all of the valence band states shown in Figure 1.4 are occupied by electrons, and all of the higher bands of the diamond band structure (which are not shown in Figure 1.4) are empty. At temperatures above 0 K, a few electrons will be thermally excited from the highest valence band into the vacant next higher band. The next empty band above the highest valence band is called the conduction band, because electrons thermally excited into it will find empty states available for the electrical conduction process.

The element silicon, which is of great technological importance as a semiconductor, has the same crystal structure as diamond. Figure 1.5 shows the band structure[6] (excluding spin) of silicon in the [111] and [100] directions. The upper valence band is doubly degenerate in the absence of spin–orbit coupling. The valence band (abbreviated VB) maximum energy is at the Γ point at the center of the Brillouin zone (abbreviated BZ). The conduction band (abbreviated CB) minimum energy is at a point in the

[100] direction, 0.86 of the distance between the zone center at Γ and the edge of the zone at the point X. The minimum energy gap E_g is the energy difference between the conduction band minimum and the valence band maximum, so the magnitude E_g of the energy gap in silicon is given by the relation

$$E_g = E(\text{CB minimum}) - E(\text{VB maximum}) \tag{1.1}$$

Since the conduction band minimum and the valence band maximum occur at different points of the zone, silicon is said to have an indirect minimum energy gap. If the conduction band minimum and the valence band maximum occur at the same point of the zone, the semiconductor is said to have a direct minimum energy gap. Examples of semiconductors with direct energy gaps are indium antimonide (InSb) and lead sulfide (PbS). Figure 1.6 shows the band structure[7] of InSb, in the [100] and [111] directions, near the center of the Brillouin zone. The zone edge is not shown. The conduction band minimum is at the Γ point at the zone center. The valence band maximum is (almost exactly[7]) at the zone center also, so the energy gap in InSb is a direct gap. (As a point of terminology, the conduction band minimum is often called the conduction band edge, and the valence band maximum is called the valence band edge.) In InSb, there are four valence bands V_1, V_2, V_3 (the V_3 band is twofold degenerate) which contain the eight valence electrons in this III–V semiconductor.

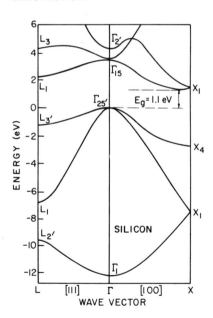

Figure 1.5. Band structure of silicon in the [111] and [100] directions in k-space. The minimum energy gap $E_g = 1.1$ eV at room temperature. (Adapted from Chelikowsky and Cohen.[6])

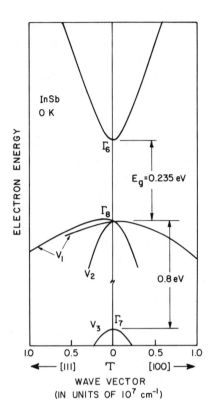

Figure 1.6. Band structure of InSb in the [100] and [111] directions in k-space. The energy separations, including the energy gap E_g, are at 0 K. (Adapted from Long.[7])

Holes in Semiconductors

We recall[8] the picture of a single electron missing from a state at the top ($k = 0$) of an otherwise filled band. If an electric field is applied, then the collective motion of the electrons in the band is equivalent to the motion of the empty state in the direction of decreasing (i.e., more negative) k. If an electron is missing from a state of wave vector k_e in a band, we say that there is a hole in that state. To quote Kittel,[9] "The physical properties of the hole follow from those of the totality of electrons in the band." We recall that, if an electron of effective mass m_e (effective mass is reviewed below) is missing from a state of wave vector k_e and energy $E(k_e)$, then the hole has the following properties:

1. Wave vector $k_h = -k_e$
2. Effective mass $m_h = -m_e$; the hole effective mass is positive
3. Energy $E(k_h) = -E(k_e)$
4. Positive charge $(+e)$, where $(-e)$ is the electron charge

Note that relation (3) means that, if we plot electron energy increasing upward, hole energy increases downward. In other words, electrons tend to "sink"; holes tend to "float," on the usual plots of *electron* energy.

Effective Mass of Carriers in Semiconductors

Consider an electron in the periodic potential of the lattice. If an electric or magnetic field is applied, the electron is accelerated relative to the lattice *as if* it had a mass m^* whose value is different from the free electron mass m_0. The effective mass m^* of an electron in a state in an energy band whose energy–wave-vector relation is $E = E(k)$ is given by the relation

$$m^* = \frac{\hbar^2}{d^2E/dk^2} \qquad (1.2)$$

For an electron in such a band, we see from (1.2) that the effective mass m^* is related to the curvature of the band $E = E(k)$. Consider two cases in which the $E(k)$ relation is quadratic in k and given by

$$E = \frac{\hbar^2}{2m^*} k^2 \qquad (1.3)$$

but for which the effective masses are different. Figures 1.7a and 1.7b show the two cases; in Figure 1.7a, the electron effective mass is m_1^* and in Figure 1.7b it is m_2^*. From the figure, we see that the $E(k)$ function is, qualitatively speaking, more "curved" in Figure 1.7a than it is in Figure 1.7b, so d^2E_1/dk^2 is larger than d^2E_2/dk^2, using $E = E_1(k)$ and $E = E_2(k)$ for the bands in Figures 1.7a and 1.7b, respectively. From equation (1.2), we have the result that the effective mass m_1^* is smaller than the effective mass m_2^*.

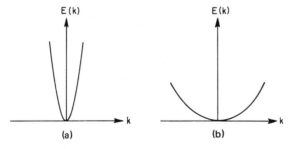

Figure 1.7. (a) Band with a small effective mass $[E_1 = (\hbar^2/2m_1^*)k^2]$. (b) Band with a larger effective mass $[E_2 = (\hbar^2/2m_2^*)k^2]$.

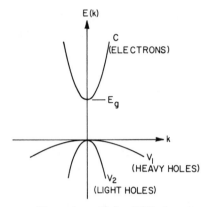

Figure 1.8. Schematic band structure with a direct energy gap E_g and two valence bands V_1 and V_2. The conduction band is denoted by C.

Note that, if the $E(k)$ function is proportional only to k^2 (a parabolic band), then (1.2) shows that the effective mass m^* is independent of k and is a constant over the whole band in question.

As an example of these definitions, consider the semiconductor band structure shown in Figure 1.8. This band structure shows a conduction band (C) and two valence bands (V_1 and V_2). The energy gap is direct at the center of the zone and is of magnitude E_g, since the zero of energy has been taken at the valence band maximum. In the conduction band, the dispersion relation (i.e., the energy–wave-vector relation) is

$$E(k) = E_g + \frac{\hbar^2}{2m_c} k^2 \tag{1.4}$$

where the quantity m_c in (1.4) is seen to be positive. The dispersion relations for the valence bands V_1 and V_2 are

$$E(k) = -\frac{\hbar^2}{2m_{V_1}} k^2 \tag{1.5}$$

and

$$E(k) = -\frac{\hbar^2}{2m_{V_2}} k^2 \tag{1.6}$$

where m_{V_1} and m_{V_2} are positive quantities. We are thus considering a semiconductor in which all of the bands under consideration are parabolic.

To find the effective mass m_e^* of an electron in the conduction band, we apply equation (1.2) to the conduction band dispersion relation (1.4) and find, since $d^2E/dk^2 = \hbar^2/m_c$,

$$m_e^* = m_c \tag{1.7}$$

Equation (1.7) tells us that the effective mass m_e^* of an electron in this conduction band is equal to m_c and is therefore positive. Note also that, as expected, the electron effective mass m_e^* is constant because the dispersion relation (1.4) for the conduction band is quadratic in k.

We now want to find the effective mass $m_{h_1}^*$ of a hole in the valence band V_1. In this band, the dispersion relation is given by (1.5), so the definition (1.2) of the effective mass gives, since $d^2E/dk^2 = -\hbar^2/m_{V_1}$, the result

$$m^* = -m_{V_1} \tag{1.8}$$

However, m^* in equation (1.8) is the mass of an *electron* in the valence band V_1. From relation (2) concerning holes given earlier, we see that the effective mass of a *hole* in a state in the valence band is the *negative* of the effective mass of an electron in the same state. Applying this result to equation (1.8) gives

$$m_{h_1}^* = -m^* = -(-m_{V_1}) = +m_{V_1} \tag{1.9}$$

We note that the hole effective mass $m_{h_1}^*$ is positive because the quantity m_{V_1} is positive and that $m_{h_1}^*$ is independent of k because equation (1.5) tells us that the valence band V_1 is parabolic.

We note also from Figure 1.8 the difference in curvature between the conduction band C and the valence band V_1. For this band structure, then, we have the result that the hole effective mass $m_{h_1}^*$ is larger than the electron effective mass m_e^*. This is usually the case for real semiconductors: the hole effective mass is larger than the electron effective mass.

Note also that we have tacitly been discussing electrons in states near the conduction band minimum and holes in states near the valence band maximum. These states[10] are important because they are where the carriers are located in most semiconductor devices. The results we have obtained are, of course, for parabolic bands. This assumption is at least approximately valid for the band edges of most of the semiconductors of device interest that we will discuss.

Some semiconductors have more than one valence band (like the band structure in Figure 1.8) which may (possibly) contribute holes to the process of electrical conduction. From the curvatures of the valence bands in Figure 1.8, we see that the hole effective mass $m_{h_1}^* = m_{V_1}$ in valence band V_1 will be larger than the hole effective mass $m_{h_2}^* = m_{V_2}$ in band V_2, where m_{V_2} is the quantity in the dispersion relation (1.6) for valence band V_2. For this reason, the holes in band V_1 are called "heavy holes" and those in band V_2 are called "light holes." An example[11] is GaAs, which has a band structure like that in Figure 1.8, and for which (near $k = 0$) $m_{h_1}^* \simeq 0.5m_0$, $m_{h_2}^* \simeq 0.08m_0$, and the conduction band electron effective mass $m_c^* \simeq 0.06m_0$, all at 300 K. Effective mass values for other semiconductors may be found in the literature.[11a]

Conductivity of Semiconductors

Consider a pure semiconductor crystal at 0 K. None of the electrons in the conduction band are thermally excited into the valence band. Suppose one electron is excited from the valence band into the conduction band. The result is that there is a hole (i.e., a missing electron) in the valence band and an electron in the conduction band. The situation is shown schematically in Figure 1.9, which is of the same type as Figure 1.2. If an electric field $\mathscr{E}$ is now applied to the crystal, the electron and hole will have opposite velocities. This is shown in Figure 1.10, in which q is the electric charge of a carrier, $\mathbf{v}$ is its velocity, and $\mathbf{J}$ is the current density. Since

$$\mathbf{J} = nq\mathbf{v} \tag{1.10}$$

where n (or p) is the number density of electrons (or of holes), we have that the current densities $\mathbf{J}_n$ and $\mathbf{J}_p$ of electrons and holes, respectively, are given by

$$\mathbf{J}_n = -ne\mathbf{v}_n, \qquad \mathbf{J}_p = pe\mathbf{v}_p \tag{1.11}$$

This leads to the result that $\mathbf{J}_n$ and $\mathbf{J}_p$ are in the same direction, as shown in Figure 1.10, because the electron velocity $\mathbf{v}_n$ and the hole velocity $\mathbf{v}_p$ are in opposite directions. The electrons and holes in a semiconductor therefore both contribute in the same way to the total current density.

From Ohm's law, we have, for electrons,

$$\mathbf{J}_n = \sigma_n \mathscr{E} = -ne\mathbf{v}_n \tag{1.12}$$

where σ_n is the electrical conductivity due to electrons. For holes,

$$\mathbf{J}_p = \sigma_p \mathscr{E} = pe\mathbf{v}_p \tag{1.13}$$

We define the electron and hole mobilities μ_n and μ_p by the vector equations

$$\mathbf{v}_n \equiv -\mu_n \mathscr{E} \tag{1.14}$$

$$\mathbf{v}_p \equiv \mu_p \mathscr{E} \tag{1.15}$$

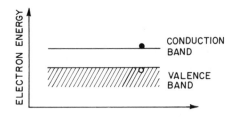

Figure 1.9. Excitation of an electron from the valence band to the conduction band, producing a free electron (●) and a free hole (○).

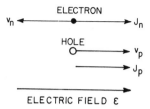

Figure 1.10. Velocity vectors $\mathbf{v}_n$ and $\mathbf{v}_p$, and current density vectors $\mathbf{J}_n$ and $\mathbf{J}_p$ for an electron ($\bullet$) and a hole ($\bigcirc$) in an electric field $\mathscr{E}$. The current density vectors $\mathbf{J}_n$ and $\mathbf{J}_p$ are in the same direction.

where the negative sign in (1.14) means the electron velocity $\mathbf{v}_n$ is opposite to the direction of the electric field $\mathscr{E}$. The mobilities μ_n and μ_p are both positive quantities. Combining (1.12)–(1.15) gives

$$\mathbf{J}_n = ne\mu_n\mathscr{E} = \sigma_n\mathscr{E} \tag{1.16}$$

$$\mathbf{J}_p = pe\mu_p\mathscr{E} = \sigma_p\mathscr{E} \tag{1.17}$$

and the total current density $\mathbf{J} = \mathbf{J}_n + \mathbf{J}_p$ is given by

$$\mathbf{J} = (ne\mu_n + pe\mu_p)\mathscr{E} = \sigma\mathscr{E} \tag{1.18}$$

where the total conductivity σ due to both electrons and holes is

$$\sigma = e(n\mu_n + p\mu_p) = \sigma_n + \sigma_p \tag{1.19}$$

The carrier mobilities are determined by the collisions of the carriers with various imperfections (e.g., phonons) in the crystal lattice. We recall[12] the expression

$$\sigma = ne^2\tau_n/m_0 \tag{1.20}$$

for the DC conductivity σ in the free-electron model of a solid, where n is the free-electron concentration, m_0 is the free-electron mass, and τ_n is the collision time, or relaxation time, for electrons. Comparing (1.20) and (1.19), we have

$$\mu_n = e\tau_n/m_0 \tag{1.21}$$

If the electrons have an effective mass m_e^*, (1.21) becomes

$$\mu_n = e\tau_n/m_e^* \tag{1.22}$$

for electrons, while for holes the analogous relation is

$$\mu_p = e\tau_p/m_h^* \tag{1.23}$$

where m_h^* is the hole effective mass and τ_p is the collision time for holes.

In a semiconductor at 300 K, the dominant carrier scattering process is usually scattering by phonons. It is thus this type of scattering[13] that determines the collision time τ in this case. In some cases, however, scattering by ionized impurities may dominate.

We note from equations (1.22) and (1.23) that the carrier mobility μ is, for a given scattering time τ, inversely proportional to the carrier effective mass m^*. Since the hole effective mass in a given semiconductor is generally larger than the electron effective mass, it is usually true that the electron mobility μ_n is larger than the hole mobility μ_p in a semiconductor. Furthermore, since it is known[14] that the electron effective mass varies directly with the magnitude of the energy gap E_g, semiconductors with small energy gaps will have small values of effective mass, and hence large values of the electron mobility.[15] (This conclusion is often true for holes also, but is not as clear-cut because of the relatively complex structures of semiconductor valence bands, leading to light and heavy holes and other complications.) For silicon,[15] the electron and hole mobilities at room temperature are, respectively, 1350 and 480 cm^2 V^{-1} sec^{-1}. For GaAs,[15] the corresponding values are 8000 and 300 cm^2 V^{-1} sec^{-1}. (The very high electron mobility in GaAs makes it of current interest for fast devices.)

Carrier Density in an Intrinsic Semiconductor

In a pure semiconductor crystal, each electron thermally excited into the conduction band creates a hole in the valence band. One speaks of the creation of electron–hole pairs. In this case the number of electrons is necessarily equal to the number of holes. A semiconductor for which this is true is called an intrinsic semiconductor, defined by the condition

$$n = p \tag{1.24}$$

where n is the electron density and p is the hole density. It is important to note that electrons and holes can, and do, simultaneously coexist in semiconductors. This fact is the physical basis for the action of the bipolar transistor.

We now recall[16] a few standard results concerning the intrinsic carrier density in a pure semiconductor. Consider the Fermi–Dirac distribution function $f(E)$, where

$$f(E) = \{\exp[(E - E_F)/k_B T] + 1\}^{-1} \tag{1.25}$$

and where E is the energy, E_F is the Fermi level, and k_B is Boltzmann's constant. (Strictly speaking,[16a] the quantity E_F in equation (1.25) is the chemical potential, which is usually temperature-dependent. At $T = 0$, the chemical potential is equal to the Fermi energy, which is the energy of the filled state of highest energy. However, it is usual in semiconductor work to refer to the chemical potential at any temperature as the "Fermi level," and we will follow this common usage.) The function $f(E)$ gives the probability that the state of energy E is occupied by a fermion; further, $f(E_F) = 1/2$. If "not too many" electrons[17] have been excited into the conduction band, then it will be true that $(E - E_F) \gg k_B T$, the Fermi function $f(E)$ will be much smaller than unity, and the semiconductor is referred to as nondegenerate. In such a case, the Fermi function may be approximated by

$$f(E) \cong \exp[-(E - E_F)/k_B T] \tag{1.26}$$

meaning that classical statistics furnish a valid description of the carrier densities involved. If the carrier densities are large enough that it is not true that $f(E) \ll 1$, then the approximation (1.26) is not valid, the exact Fermi function (1.25) must be used, and the semiconductor is referred to as degenerate.

Using equation (1.26) for $f(E)$, and a parabolic energy dependence of the densities of states in the valence and conduction bands of the semiconductor, one may calculate[16] the electron and hole densities n and p. The results, on taking the zero of energy at the valence band maximum so that the energy is E_g at the conduction band minimum, are

$$n = N_C \exp[(E_F - E_g)/k_B T] \tag{1.27}$$

$$p = N_V \exp[-E_F/k_B T] \tag{1.28}$$

where E_g is the energy gap and

$$N_C \equiv 2(m_e^* k_B T/2\pi\hbar^2)^{3/2} \tag{1.29}$$

$$N_V \equiv 2(m_h^* k_B T/2\pi\hbar^2)^{3/2} \tag{1.30}$$

If we multiply the values of n and p given by (1.27) and (1.28), we obtain

$$np = 4(k_B T/2\pi\hbar^2)^3 (m_e^* m_h^*)^{3/2} \exp(-E_g/k_B T) \tag{1.31}$$

showing that the np product is independent of the Fermi energy in a nondegenerate semiconductor. The np product is thus a constant at a given temperature. At 300 K the value[16] of np for silicon is 2.10×10^{19} cm^{-6}.

We next invoke the condition (1.24) that the semiconductor is intrinsic, and equate equations (1.27) and (1.28) for n and p, solving for the Fermi energy E_F. The result is

$$E_F = \tfrac{1}{2}E_g + \tfrac{3}{4}k_BT\ln(m_h^*/m_e^*) \tag{1.32}$$

giving the position of the Fermi energy (relative to the energy zero chosen at the valence band maximum) in an intrinsic semiconductor. The Fermi energy is seen to be a function of the carrier effective masses m_e^* and m_h^*. If $m_e^* = m_h^*$, then $E_F = E_g/2$ and the Fermi energy is exactly in the center of the energy gap. If, as is usual in semiconductors, m_h^* is larger than m_e^*, the second term in (1.32) is positive, and the Fermi energy lies somewhat above the center of the energy gap. It is conventional in semiconductor physics to refer to E_F as the Fermi level.

Impurity Conductivity (Extrinsic Conductivity)

In general, for device applications, we will be interested in semiconductors that are not intrinsic, but in which the concentration of electrons or of holes is controlled by deliberately added impurities. Such semiconductors are called extrinsic semiconductors or impurity semiconductors.[18] Certain impurities (e.g., arsenic) added to silicon enter the lattice substitutionally and can give up an electron to the conduction band. Such impurities, called donors, add free electrons to the crystal. The amount of energy E_d necessary to free the electron from the donor atom and liberate it into the conduction band is called the donor ionization energy. Values of E_d are about 0.01 eV and 0.05 eV for various pentavalent impurities in germanium and silicon, respectively. Figure 1.11 shows schematically the energies involved in donor ionization. Figure 1.11a shows the electron (–●–) bound to a donor atom; this energy level is an energy E_d below the energy of the conduction band minimum, so the addition of the donor ionization energy E_d will promote the electron into a conduction band (CB) state, as shown in Figure 1.11b. An empty donor state (—) remains in the energy gap. The process may be summarized by the equation

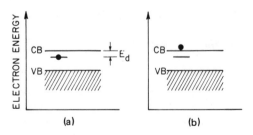

(a)   (b)

Figure 1.11. Ionization (schematic) of electron bound to a donor atom. In (a), the electron (–●–) is in a bound state an energy E_d below the energy of the conduction band (CB) minimum. In (b), the electron (●) has been promoted into a CB state by adding the ionization energy E_d, leaving behind an empty donor state (—) in the energy gap. (The energy spacings are not to scale.)

Figure 1.12. Ionization (schematic) of hole bound to an acceptor atom. In (a), the hole (–O–) is in a bound state at an energy E_a above the energy of the valence band (VB) maximum. In (b), the hole (O) has been raised to a valence band state by adding the ionization energy E_a,

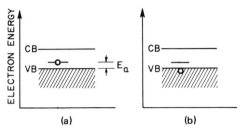

leaving behind an empty acceptor state (—) in the energy gap. (The energy spacings are not to scale.)

$$D^0 + E_d \rightarrow D^+ + e^- \tag{1.33}$$

where D^0 is the neutral donor impurity, E_d is the ionization energy of the electron e^-, and D^+ is the positively charged ionized donor.

In the same way, a trivalent impurity (e.g., boron) in silicon can accept an electron from the valence band, thereby generating a free hole. In this process the impurity atom is called an acceptor, and adding an energy E_a promotes an electron from the valence band (VB) maximum to the energy level of the neutral acceptor, leaving behind a free hole and creating a negatively charged acceptor center. One can write the equation

$$VB + E_a + A^0 \rightarrow A^- + (VB - e^-) \tag{1.34}$$

where A^0 is the neutral acceptor, A^- is the negatively charged acceptor, VB represents the valence band, and $(VB - e^-)$ represents the valence band minus one electron, or, in other words, a hole. It is more usual to write (1.34) as

$$E_a + A^0 \rightarrow A^- + h^+ \tag{1.35}$$

where h^+ is a hole, and the energy E_a is called the ionization energy of the acceptor. Values of E_a for trivalent impurities in silicon range from 0.016 to 0.065 eV; the value of E_a for germanium is about 0.01 eV.

It is also common to describe the process in terms of holes rather than electrons. One says that the hole is bound to the acceptor forming the neutral acceptor A^0. Adding the ionization energy E_a frees the hole into the valence band, leaving a negatively charged acceptor A^-. This description is shown in Figure 1.12, which is the analog for holes of Figure 1.11. Figure 1.12a shows the hole (–O–), bound to the acceptor atom, at an energy E_a above the valence band maximum, and where we recall that hole energy increases in the downward direction on a plot of electron energy like Figure 1.12. The addition of the acceptor ionization energy E_a raises the hole (O) into the valence band, as shown in Figure 1.12b. An empty hole state (—) remains behind in the energy gap.

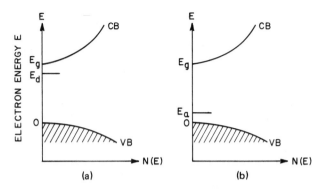

Figure 1.13. Density of states $N(E)$ as a function of electron energy E for (a) an n-type, and (b) a p-type semiconductor. In (a), the donor levels are an energy E_d below the conduction band edge, where E_d is the donor ionization energy. In (b), the acceptor levels are an energy E_a above the valence band edge, where E_a is the acceptor ionization energy. (In both drawings, the energy separations are exaggerated for clarity.)

For future reference, we may also display the ideas of Figures 1.11 and 1.12 on a plot of the density of electron states $N(E)$ as a function of energy E. This is shown in Figure 1.13, in which $N(E)$ is plotted horizontally and E is plotted vertically. Figure 1.13a shows an n-type semiconductor, in which the donor states are an energy E_d below the conduction band, and Figure 1.13b shows a p-type semiconductor with acceptor levels an energy E_a above the valence band. (The energy spacings of the impurity levels from the bands are greatly exaggerated for clarity.)

The addition of donors to a pure semiconductor adds electrons to the conduction band. Such a semiconductor is called n type, meaning that its electrical conduction is due primarily to electrons. Similarly, the addition of acceptors adds holes to the valence band, producing a semiconductor, called p type, in which electrical conduction is mainly due to holes. Equation (1.31) tells us that the np product n_i^2 is constant at a given temperature. This means that an increase in the density of electrons in a semiconductor results in a decrease in the density of holes, and vice versa.

Suppose that the donor and acceptor ionization energies in a semiconductor are E_d and E_a, respectively, and that only one or the other type of impurity is present. If, in the n-type semiconductor, the donor atom density is N_d, and if the temperature is low enough that the condition $E_d \gg k_B T$ is satisfied, then the electron density n due to the ionization of donors is given by[18]

$$n \cong (n_0 N_d)^{1/2} \exp[-E_d/2k_B T] \qquad (1.36)$$

where the quantity n_0 is equal to $[2(m_e^* k_B T/2\pi\hbar^2)^{3/2}]$. A similar equation holds for the hole density p in a p-type semiconductor in which the acceptor atom density is N_a. (The case in which both donors and acceptors are present in comparable concentrations is more complicated.[19])

Most semiconductor devices are designed to operate at 300 K, at which temperature the usual donor or acceptor impurities in germanium or silicon are essentially completely ionized. We will therefore usually be interested in intrinsic n-type or p-type semiconductors in which $n = N_d$ or $p = N_a$, i.e., the electron concentration n will be taken as equal to the donor atom concentration N_d, and similarly for holes.

Suppose we have a semiconductor containing N_d ionized donors and N_a ionized acceptors per unit volume, where N_d is larger than N_a. Then this semiconductor would have a net extrinsic electron concentration equal to $N_d - N_a$. We say that the N_a acceptors have neutralized or compensated N_a of the donors, leaving $N_d - N_a$ donors uncompensated and able to supply free electrons. This semiconductor would be referred to as a compensated n-type semiconductor.

Fermi Level Position in Extrinsic Semiconductors

We are next interested in considering the Fermi level in an impurity or extrinsic semiconductor[19-23] since these are the semiconductors of greatest interest in applications. Equations (1.27) and (1.28) give the electron and hole concentrations n and p in terms of the Fermi level E_F and the various semiconductor parameters. These equations apply to any nondegenerate semiconductor with parabolic conduction and valence bands, and hence hold for an extrinsic semiconductor with those properties. We may therefore use these equations to find out how the position of the Fermi level will vary with the carrier concentration in an extrinsic semiconductor. We repeat the equations for convenience:

$$n = N_C \exp[(E_F - E_g)/k_B T] \qquad (1.27)$$

$$p = N_V \exp[-E_F/k_B T] \qquad (1.28)$$

and consider an n-type semiconductor containing N_d ionized donors per unit volume, where the electron density n is equal to N_d. Consider two semiconductor samples with ionized donor densities N_d and N_d', where N_d' is larger than N_d; the electron densities are $n = N_d$ and $n' = N_d'$, where

n' is larger than n. Equation (1.27) must hold, so it must be true that

$$\exp[(E_F{}' - E_g)/k_B T] > \exp[(E_F - E_g)/k_B T] \qquad (1.37)$$

where $E_F{}'$ and E_F are, respectively, the Fermi level position in the samples of electron density n' and n. Equation (1.37) requires that

$$E_F{}' - E_g > E_F - E_g \qquad (1.38)$$

so that it is also true that

$$E_g - E_F{}' < E_g - E_F \qquad (1.39)$$

Equation (1.39) tells us that the energy separation between the energy E_g of the conduction band and the energy E_F of the Fermi level is smaller in the sample with the larger electron density n'. Put another way, the Fermi level will be closer to the conduction band in the sample with the larger electron density. For example, if we compare an n-type extrinsic semi-conductor with an intrinsic semiconductor, the extrinsic sample will have the larger electron density. We expect, therefore, that the Fermi level will be closer to the conduction band in the n-type extrinsic sample, as shown schematically in Figure 1.14. In that figure, the electron energy zero is taken at the valence band maximum, the conduction band minimum is at the forbidden gap energy E_g, and the Fermi level energy E_F is close to that of the conduction band. If we compare Fermi level positions in two n-type samples, we expect the Fermi level to be closer to the conduction band for the sample with the larger donor density and thus with the larger electron concentration.

We may discuss a p-type semiconductor in the same way. From equa-tion (1.28), as the hole density p becomes larger, $\exp[-E_F/k_B T]$ becomes larger, and thus $\exp[E_F/k_B T]$ becomes smaller. This results in the Fermi

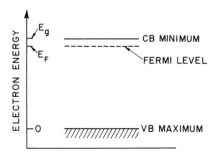

Figure 1.14. Position (schematic) of the Fermi level in an n-type semiconductor of energy gap E_g. The Fermi level is the dashed line.

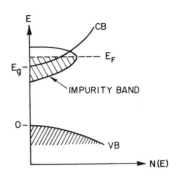

Figure 1.15. Plot of density of electron states $N(E)$ as a function of electron energy E for a high donor density and impurity band formation. States occupied by electrons are shown shaded. The impurity band of donor states overlaps the conduction band, and the position of the Fermi level E_F is above the conduction band minimum energy E_g. (The width of the impurity band is exaggerated for clarity.)

energy E_F becoming smaller as p increases, meaning that the position E_F of the Fermi level (relative to the energy zero at the valence band maximum) becomes closer to the valence band. If we compare the Fermi level positions in two p-type samples, we expect the Fermi level to be closer to the valence band in the sample with the larger acceptor density and thus with the larger hole concentration.

As the carrier density is increased further, the position of the Fermi level moves closer to the relevant band edge. Eventually the assumption that $E - E_F \gg k_B T$, made in the derivation of equations (1.27) and (1.28), fails and the classical approximation (1.26) is no longer valid. In that case, use of the exact Fermi function (1.25) is required, and the semiconductor is referred to as degenerate. Such semiconductors with very high carrier densities are of considerable interest for applications, and we will discuss, qualitatively, their physical situation, considering donors in an n-type semiconductor for concreteness. As the donor and electron densities are increased, the donor centers are closer together, and their wave functions overlap, forming a band called an impurity band.[24-26] The donor states are no longer discrete, with a well-defined ionization energy; the ionization energy approaches zero[24] with increasing donor, and electron, density. The impurity band may, at high donor concentrations, be wide enough to overlap the conduction band, as shown schematically in Figure 1.15. This figure plots density of electron states $N(E)$ as a function of electron energy E. The donor electrons are free carriers in conduction band states, and there are electrons in the conduction band up to the energy E_F of the Fermi level. From the point of view of applications, the important results are that, in a degenerate heavily doped n-type semiconductor, the Fermi level is above the conduction band minimum energy, as shown in Figure 1.16, and that there is a density of free electrons in the conduction band.

Similar statements may be made about impurity banding in heavily doped p-type semiconductors. The analogous results are that, for such a

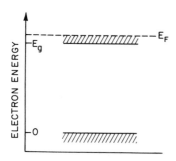

Figure 1.16. Position (schematic) of the Fermi level E_F in a degenerate n-type semiconductor. The states occupied by electrons are shown shaded, and the impurity states are not shown.

degenerate p-type semiconductor, the Fermi level is below the valence band maximum, and there is a density of free holes in the valence band.

We may also discuss the temperature dependence of the Fermi level in a nondegenerate extrinsic semiconductor,[21,23] using an n-type semiconductor as an example. We consider first such a semiconductor at 0 K. The situation is shown in Figure 1.17. At 0 K, all of the donors are un-ionized so the probability that a donor level *is* occupied is unity. Since there are no electrons in the conduction band at 0 K, the probability that a conduction band state is occupied is zero. Since, at 0 K, the Fermi function $f(E)$ has the shape[27] shown in Figure 1.17b, and since $f(E_F) = \frac{1}{2}$, the Fermi energy E_F must lie between the donor level energy and the conduction band edge at 0 K. It can be shown[28] that, at 0 K, the Fermi level E_F lies exactly half-way between the donor level energy and the conduction band edge in an uncompensated n-type semiconductor.

We now consider the same semiconductor at a temperature greater than 0 K, so some of the donor are ionized, as shown in Figure 1.18. Since the donors are no longer all un-ionized, the probability that a donor energy level is occupied is no longer exactly unity. Similarly, the probability that a conduction band state is occupied is no longer zero. This results in the energy E_F (at which the probability of occupancy of a state is exactly $\frac{1}{2}$) being lower (i.e., further from the conduction band edge) at a temperature above 0 K

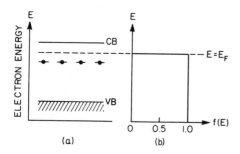

Figure 1.17. (a) Band diagram for an n-type nondegenerate semiconductor at 0 K. The energy separations are exaggerated for clarity. The symbol $(-\bullet-)$ represents an un-ionized donor. (b) Plot of the Fermi function $f(E)$, at 0 K, as a function of electron energy E.

Figure 1.18. (a) Band diagram for an *n*-type nondegenerate semiconductor at a temperature above 0 K. The symbols (–●–), (——), and (●) refer, respectively, to an un-ionized donor, an ionized donor, and a free electron. The energy separations are exaggerated for clarity. (b) Plot of the Fermi function $f(E)$, at a temperature above 0 K, as a function of electron energy E.

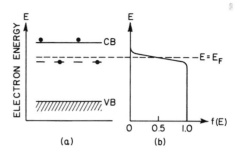

than it is at 0 K. This result is exhibited in Figure 1.18b. As the temperature is increased further, the Fermi level E_F continues to "drop" toward the position it would occupy if the semiconductor were intrinsic.[21,23]

Analogous results may be obtained for the temperature dependence of the Fermi level in a *p*-type extrinsic semiconductor.[23]

Carrier Lifetime in Semiconductors

If a carrier is excited into the conduction band, it will eventually recombine and disappear after a time called its lifetime.[29,30] The lifetime of a carrier in a given semiconductor will depend on the kinetics of its mode of recombination, either by direct recombination, or by recombination via an intermediate state or "trap." While the subject of carrier lifetimes in semiconductors is a complex one, a few qualitative observations which will be useful in discussing applications may be made here. In semiconductors with direct energy gaps (e.g., InSb, GaAs) direct recombination of an electron and a hole, both with the same value of wave vector, to produce a photon, is a process of high probability and is often the dominant process determining carrier lifetime. However, in a semiconductor with an indirect energy gap (e.g., germanium or silicon), direct recombination of an electron and hole, with different values of wave vector, must involve a phonon in the process. Because this is essentially a three-body process, direct recombination of an electron and hole, in an indirect-gap semiconductor, with the emission of a photon and the participation of a phonon to conserve wave vector, is a process of low probability. For this reason, the dominant recombination process in indirect-gap semiconductors usually proceeds through an intermediate "trap" state. The kinetics of these recombination processes taking place via traps can be very complex.[29,30]

However, for our purposes, an important qualitative conclusion concerning carrier lifetime may be reached. We expect carrier lifetimes to be longer in semiconductors with indirect energy gaps. Further, since electron–hole recombination in indirect-gap semiconductors generally proceeds through trap states, which may be impurity levels, one would expect a decreased trap density to favor longer lifetimes. For this reason, a smaller doping impurity concentration in a semiconductor with an indirect energy gap will favor a longer carrier lifetime. To summarize, we expect that an indirect band gap and a low impurity density in a semiconductor will favor longer carrier lifetimes.

The complexity of the recombination kinetics involved makes detailed statements about values of carrier lifetime difficult. However, typical values[31] in pure semiconductors range from 10^{-3} to 10^{-8} sec. For example, the electron lifetime in pure silicon[32] is of the order of 10^{-3} sec; for silicon doped with gold (which acts as a recombination center), the electron lifetime[33] is of the order of nanoseconds. Tables of values of carrier lifetimes in silicon and in GaAs are given by Hovel.[34]

Problems

1.1. *Band Structure of GaAs.* (a) Using any convenient reference, sketch the band structure of GaAs in the reduced zone scheme, indicating the highest valence band and the lowest conduction band. (b) Allocate the valence electrons in GaAs to the valence bands. Is GaAs a metal or a semiconductor? (c) Where in the Brillouin zone is the minimum energy gap located? Is the gap direct or indirect? (d) Is there a light hole band in GaAs? (e) In which valence band would you expect to find holes in intrinsic GaAs at room temperature?

1.2. *Intrinsic Carriers in Silicon.* (a) Using results given in the text, calculate the concentration of intrinsic electrons in pure silicon at 300 K. (b) What is the hole concentration in pure silicon at 300 K? Why?

1.3. *Fermi Level Position in Extrinsic Silicon.* (a) Consider a sample of silicon in which the density of ionized arsenic atoms in substitutional lattice sites is 10^{17} cm^{-3}. Using the carrier effective mass values given in Problem 1.2 above, find the position of the Fermi level (relative to the appropriate band edge) at 300 K. (b) Sketch the band diagram, taking the impurity ionization energy of arsenic in silicon as 0.049 eV.

1.4. *Band Diagrams for Degenerate p-Type Semiconductor.* For a heavily doped, *p*-type semiconductor, make drawings that are analogous to Figures 1.15 and 1.16 for the *n*-type case.

References and Comments

1. C. Kittel, *Introduction to Solid State Physics*, Sixth Edition, John Wiley, New York (1986).
2. C. Kittel, Reference 1, pages 176–178.
3. See, for example, L. Pincherle, *Electron Energy Bands in Solids*, Macdonald, London (1971), Section 6.3, page 172.
4. M. Tinkham, *Group Theory and Quantum Mechanics*, McGraw-Hill, New York (1964), Section 8.3, page 277.
5. Adapted from D. Long, *Energy Bands in Semiconductors*, John Wiley, New York (1968), Figure 2.7(a), page 39.
6. Adapted from J. R. Chelikowsky and M. L. Cohen, *Physical Review B*, 10, 5095 (1974), Figure 2 (using the nonlocal pseudopotential calculation).
7. Adapted from D. Long, Reference 5, Figure 6.1, page 101.
8. C. Kittel, Reference 1, pages 214–217.
9. C. Kittel, *Introduction to Solid State Physics*, Fourth Edition, John Wiley, New York (1971), page 328.
10. A. J. Dekker, *Solid State Physics*, Prentice-Hall, New York (1957), Section 10.4, gives a brief discussion of electron motion at higher energies in the bands.
11. J. S. Blakemore, "Semiconducting and Other Major Properties of GaAs", in *Gallium Arsenide* (J. S. Blakemore, editor), American Institute of Physics, New York (1987), Table XII, page 39 and Figure 45, page 34.
11a. See, for example, M. L. Cohen and J. R. Chelikowsky, *Electronic Structure and Optical Properties of Semiconductors*, Springer-Verlag, Berlin (1988), Table 5.2, page 49.
12. C. Kittel, Reference 1, page 142.
13. See, for example, A. J. Dekker, Reference 10, pages 329–331.
14. See, for example, C. Kittel, *Quantum Theory of Solids*, John Wiley, New York (1963), page 187.
15. C. Kittel, Reference 1, Chapter 8, Tables 1 and 3, pages 185 and 205, give values of energy gaps and carrier mobilities for a number of semiconductors.
16. C. Kittel, Reference 1, pages 200–204.
16a. C. Kittel and H. Kroemer, *Thermal Physics*, Second Edition, John Wiley, New York (1980), pages 154–156.
17. See, for example, R. A. Smith, *Semiconductors*, Second Edition, Cambridge University Press (1978), pages 81–82.
18. C. Kittel, Reference 1, pages 206–211.
19. R. A. Smith, Reference 17, Section 4.3, pages 86–96.
20. A. S. Grove, *Physics and Technology of Semiconductor Devices*, John Wiley, New York (1967), Section 4.4, pages 100–106, especially Figure 4.7, page 104.
21. S. Wang, *Solid State Electronics*, McGraw-Hill, New York (1966), Section 3.5, pages 146–152.
22. S. M. Sze, *Physics of Semiconductor Devices*, Second Edition, John Wiley, New York (1981), pages 22–27.
23. A. J. Dekker, Reference 10, Section 12.4, pages 310–314.
24. J. S. Blakemore, *Semiconductor Statistics*, Pergamon Press, New York (1962), Section 3.5, pages 166–169.
25. N. W. Ashcroft and N. D. Mermin, *Solid State Physics*, Holt, Rinehart and Winston, New York (1976), page 584.

26. H. M. Rosenberg, *Low Temperature Solid State Physics*, Oxford University Press, Oxford (1963), pages 237–240.
27. See, for example, C. Kittel and H. Kroemer, Reference 16a, Chapters 6 and 7.
28. S. Wang, Reference 21, page 149, equation (3.57).
29. See, for example, W. R. Beam, *Electronics of Solids*, McGraw-Hill, New York (1965), Section 4.6, pages 190–200.
30. S. Wang, Reference 21, Section 5.5, pages 275–282.
31. N. W. Ashcroft and N. D. Mermin, Reference 25, page 604.
32. W. R. Beam, Reference 29, page 165.
33. A. S. Grove, Reference 20, page 142.
34. H. J. Hovel, *Semiconductors and Semimetals*, R. K. Willardson and A. C. Beer (editors), Academic Press, New York (1975), Volume 11, pages 11, 12, and 14.

Suggested Reading

C. KITTEL, *Introduction to Solid State Physics*, Sixth Edition, John Wiley, New York (1986). This modern classic is our basic background reference on solid state physics. Chapter 8 provides an introduction to semiconductor physics and also contains tables of values of semiconductor parameters. (Chapter 1 includes tables of values of crystallographic quantities.) Earlier editions of this book will sometimes be referred to.

A. J. DEKKER, *Solid State Physics*, Prentice-Hall, New York (1957). This text is now somewhat out of date but is clearly written. It also contains some interesting material not readily found elsewhere at the introductory level. Chapters 12 and 13 discuss semiconductors.

R. A. SMITH, *Semiconductors*, Second Edition, Cambridge University Press (1978). This book provides detailed discussions of many topics (especially transport properties) in semiconductor physics.

D. LONG, *Energy Bands in Semiconductors*, John Wiley, New York (1968). This short book provides, among other things, a compendium of band structures and other data on semiconductors. Even though more than twenty years old, it is still quite useful.

B. G. STREETMAN, *Solid State Electronic Devices*, Second Edition, Prentice-Hall, New York (1980). Chapter 3 of this fine textbook provides an introduction to semiconductor physics, written with applications in mind.

N. W. ASHCROFT and N. D. MERMIN, *Solid State Physics*, Holt, Rinehart, and Winston, New York (1976). Chapter 28 of this advanced-level textbook discusses the physics of semiconductors.

T. S. MOSS (series editor), *Handbook on Semiconductors*, North-Holland (1981). This four-volume treatise contains review articles, written by different experts, on just about every aspect of semiconductor physics, including devices. The treatments are advanced and authoritative.

P. N. BUTCHER, N. H. MARCH, and M. P. TOSI (editors), *Crystalline Semiconducting Materials and Devices*, Plenum Press, New York (1986). A shorter and more recent treatise than the series edited by Moss, this single volume is a group of rather theoretically oriented review articles. While these papers discuss a number of subjects of interest for applications, the presentations are definitely not of an introductory nature.

J. S. BLAKEMORE, editor, *Gallium Arsenide*, American Institute of Physics (1987); W. T. LINDLEY, editor, *Gallium Arsenide and Related Compounds*, 1986, Institute of Physics (1986). These collections of review and research papers discuss GaAs, which is of increasing importance for semiconductor devices.

M. L. COHEN and J. R. CHELIKOWSKY, *Electronic Structure and Optical Properties of Semiconductors*, Springer-Verlag, Berlin (1988). This monograph includes band structures, optical properties, and charge densities of semiconductors and an extensive list of references.

C. M. WOLFE, N. HOLONYAK, and G. E. STILLMAN, *Physical Properties of Semiconductors*, Prentice-Hall, Englewood Cliffs (1989). This recent text at the graduate level presupposes a knowledte of semiconductors equivalent to the treatment in Streetman's book listed above. The discussion covers the basic topics in semiconductor physics, and also includes chapters on surfaces and on heterostructures.

2

The Semiconductor *p–n* Junction

Introduction

The aim of this chapter is a discussion of the physics of a semiconductor *p–n* junction, i.e., a semiconductor structure in which there is a change from *n* type to *p* type over some region of space. A simple qualitative picture is used first to obtain the energy band diagram of a *p–n* junction; a quantitative treatment follows. The important ideas underlying the effect of an applied potential on the junction are then discussed, both qualitatively and quantitatively. These are followed by a calculation of the current through the junction, culminating in the celebrated Shockley equation. Finally, the majority and minority carrier components of the current are discussed.

Qualitative Discussion of the *p–n* Junction in Equilibrium

The main applications of solid state physics are in the area of solid state electronics. The semiconductor *p–n* junction is not only of interest technologically but, more important for our purposes, illustrates a wide range of interesting phenomena of importance in applied solid state physics.

We begin with a qualitative physical discussion of the electronic processes that take place during the formation of a semiconductor *p–n* junction. Consider a semiconductor in which there is a change from *n* type to *p* type over a very small distance, as shown schematically in Figure 2.1. Let there be N_a ionized acceptors per unit volume in the *p*-type region and N_d ionized

p-TYPE	n-TYPE
$p = N_a$	$n = N_d$

Figure 2.1. Schematic representation of a semiconductor with an abrupt change from *p* type (with N_a ionized acceptors per unit volume) to *n* type (with N_d ionized donors per unit volume).

donors per unit volume in the *n*-type region, so the hole concentration *p* equals N_a in the *p*-type region and the electron concentration *n* equals N_d in the *n*-type region. The structure shown in Figure 2.1 is thus an idealized semiconductor *p–n* junction in which the change of conductivity type takes place abruptly at a certain point in space. This structure is called an ideal abrupt *p–n* junction.

We will discuss the ideal abrupt junction by considering the process of bringing together a piece of *n*-type semiconductor and a piece of *p*-type semiconductor to form the junction. The band diagrams of the *n*- and *p*-type semiconductors, when still separated in space, are shown in Figure 2.2. In Figure 2.2, the vacuum energy E_{VAC} is defined[1] as the energy of an electron at rest (and hence with zero kinetic energy) just outside the crystal surface and not interacting (via image forces) with the crystal. With this definition, the energy E_{VAC} has the same value for the *n*- and *p*-type semiconductors.

If the two semiconductors are brought together in space, the band diagram in Figure 2.3 results because the two semiconductors have the same vacuum energy E_{VAC}. The junction between the *n*- and *p*-type regions is the vertical line between the two regions. From Figure 2.3 we see that, just after contact, the *n*-type side has an excess of electrons relative to the *p*-type side. Similarly, the *p*-type side has an excess of holes relative to the *n*-type side. There thus exists a concentration gradient of both electrons and holes at the junction between the *n*- and *p*-type sides.

In order to attain equilibrium, a diffusion of electrons and holes,

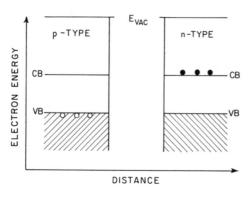

Figure 2.2. Band diagrams of *p*- and *n*-type semiconductors when separated in space. Shaded regions indicate filled electron states; dots (●) indicate free electrons, and circles (○) indicate free holes; CB and VB mean conduction band and valence band, respectively. Fermi levels are not shown and E_{VAC} is the vacuum energy.

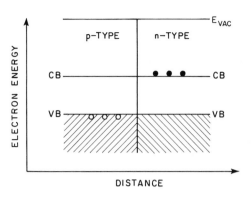

Figure 2.3. Band diagram of *n*- and *p*-type semiconductors just after contact and before any flow of holes or electrons. The junction is the vertical line separating the *n* and *p* regions, and $E_{\rm VAC}$ is the vacuum energy.

driven by their respective concentration gradients, takes place. Holes diffuse across the junction and into the *n*-type side, and electrons diffuse into the *p*-type side. Considering the diffusion of holes, the diffusing holes enter the *n*-type side, where they combine with the free electrons present there. Recalling that the recombination of a free electron and a free hole "annihilates" both carriers, the result of the diffusion of holes across the junction is to remove free electrons from the *n*-type side. A region in the *n*-type side, near the junction, thus becomes deficient in free electrons. This region in the *n*-type side therefore has an excess of positively charged ionized donor centers. Since these ionized donors are at fixed positions in the semiconductor crystal lattice, a region of positive space charge (with a concentration of N_d positive charges per unit volume) is created in the *n*-type region near the junction.

In a similar fashion, electrons diffusing from the *n*-type region into the *p*-type region create a region of fixed negative charge (of concentration N_a negative charges per unit volume) in the *p*-type region in the vicinity of the junction. The sum of these two regions, near the junction, containing fixed space charges is called the space charge region. The space charge region is also called the depletion layer because it is depleted of mobile charges due to electron–hole recombination.

Figure 2.4 shows a band diagram of the situation at this stage, i.e., after carrier diffusion is complete and equilibrium has been attained. In Figure 2.4, the plus and minus signs represent fixed ionized donors and acceptors, respectively, and the space charge region or depletion layer is the region between the vertical dashed lines on both sides of the junction. Outside the space charge region, further from the junction, there has been no recombination of electrons on the *n*-type side with holes diffusing from the *p* side, or of holes on the *p*-type side with electrons diffusing from the

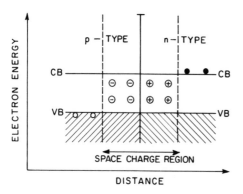

Figure 2.4. Band diagram (electron energy plotted as a function of distance) of an ideal abrupt *p–n* junction after equilibrium is attained and the space charge region has formed. The plus and minus signs represent fixed ionized donors and acceptors, respectively.

n side. We see that, outside the space charge region, electrical neutrality continues to be maintained between electrons and positively charged ionized donors (on the *n*-type side), and between holes and negatively charged ionized acceptors (on the *p*-type side). The regions outside the space-charge layer are called the *n*- and *p*-type neutral regions; these regions are the parts of Figure 2.4 to the right and left, respectively, of the vertical dashed lines which delineate the space charge region. From Figure 2.4 we see also that the space charge region extends a certain distance into the *n*-and *p*-type material on either side of the junction between the *n*- and *p*-type sides. (The *n* and *p* sides of the space charge layer are drawn as equal in extent in Figure 2.4; this need not be the case, as we shall see later.) Finally, we note that the entire semiconductor structure remains electrically neutral *overall* since no net charge has been created or destroyed.

We note also that the creation of these two regions of space charge of opposite sign sets up an electric field $\mathscr{E}$ which is known as the "built-in" electric field. The situation is shown schematically in Figure 2.5, which includes the built-in electric field extending over the space charge region between the vertical dashed lines on either side of the junction. The field $\mathscr{E}$ is directed from the *n*-type space charge region toward the *p*-type space-charge region. Figure 2.5 also shows the forces F_e and F_h which the electric field $\mathscr{E}$ exerts on electrons and holes, respectively. It is seen that the force exerted by the built-in electric field opposes the diffusion of electrons out

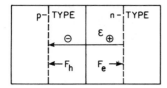

Figure 2.5. Built-in electric field $\mathscr{E}$ in the space charge region.

of the *n*-type side and of holes out of the *p*-type side. In this way, the development of the built-in field brings about a condition of equilibrium in the *p–n* junction.

The built-in electric field $\mathscr{E}$ corresponds to a gradient $-dV/dx$ of the electrostatic potential $V(x)$, where $V(x)$ is a function of the distance x in the semiconductor. For simplicity, we will make the assumption[1] that the built-in electric field is confined to the space charge region of the ideal abrupt junction we are considering. This in turn means that we are assuming that the electrostatic potential $V(x)$ is constant in the neutral regions outside the space charge layer. If we denote the electrostatic potential in the neutral *n*- and *p*-type regions by V_n and V_p, respectively, then Figure 2.6 shows a plot of $V(x)$ as a function of distance x in the semiconductor. Since the vector $\mathscr{E} = -\nabla V$, and $\mathscr{E} = -\mathscr{E}\hat{\mathbf{i}}$ (where $\mathscr{E}$ is positive and $\hat{\mathbf{i}}$ is a unit vector in the x direction), we have, as shown in Figure 2.6, the result that

$$dV/dx > 0 \tag{2.1}$$

meaning that the electrostatic potential V experienced by an electron in a *p–n* junction increases on going from the *p* region to the *n* region. The existence of the built-in electric field therefore increases the electrostatic potential of an electron in the neutral *n* side of the junction by an amount

$$V_n - V_p \equiv V_0 \tag{2.2}$$

where V_n and V_v are the constant electrostatic potentials in the neutral

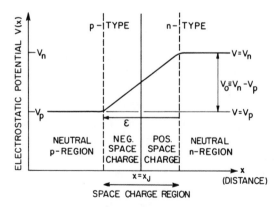

Figure 2.6. Electrostatic potential $V(x)$ shown schematically as a function of distance x in an ideal abrupt *p–n* junction. The built-in electric field $\mathscr{E}$ is confined to the space charge region and the point $x = x_J$ is the location of the junction between the *p*- and *n*-type sides denoted by the vertical solid line.

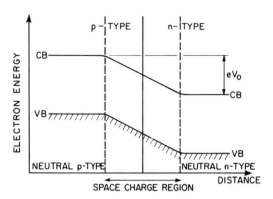

Figure 2.7. Electron energy shown schematically as a function of distance in the ideal abrupt *p–n* junction. The shading indicates filled valence band states.

n and *p* regions, respectively. The quantity V_0 is called the contact potential or the diffusion potential.

We can now interpret these results in terms of electron energy. The potential energy of an electron of electrostatic potential V is $-eV$. We see that the result of the built-in electric field is to change the potential energy of the electrons in the neutral *n* region by an amount

$$-e(V_n - V_p) \equiv -eV_0 \qquad (2.3)$$

relative to the electrons in the neutral *p* region. Since the band diagrams we use are plots of electron energy, the electron energy levels in the neutral *n* region are lowered by an amount eV_0. Figure 2.7 shows electron energy plotted as a function of distance in the junction after equilibrium has been attained. This figure shows the valence band levels and conduction band levels as a function of distance in the junction, and we see that electron energies, in both valence and conduction bands, are lower by an amount eV_0 in the neutral *n* region than they are in the neutral *p* region. There is thus an energy difference eV_0 between valence band states in the neutral *n* region and valence band states in the neutral *p* region, and an energy difference eV_0 between conduction band states in the neutral *n* region and conduction band states in the neutral *p* region. The situation may be described physically by saying that the flow of electrons due to diffusion produces an electric field that opposes that flow by lowering the electron energy in the *n*-type side of the junction. There is thus created an energy barrier of height eV_0 to electron flow from the *n* region to the *p* region.

At equilibrium, there is also a barrier of height of eV_0 to the flow of holes from the *p* region to the *n* region. This can be seen from Figure 2.7

if we recall that, on an electron energy diagram, holes tend to "float," so holes in the valence band in the p side "see" an energy barrier of height eV_0 between them and the valence band in the neutral n-type region. (Another way of seeing the energy barrier to the flow of holes is to turn Figure 2.7 upside down, which effectively converts it into a plot of hole energy as a function of distance.)

A very useful way of discussing the attainment of equilibrium between n- and p-type semiconductors forming a p–n junction is to consider the Fermi level or chemical potential. An important result[2] of statistical mechanics is that "two systems that can exchange energy and particles are in equilibrium when the temperatures and the chemical potentials are equal." We consider again n- and p-type semiconductors which are in contact, before the attainment of equilibrium, by redrawing Figure 2.3 to show the Fermi levels on both sides. The result is Figure 2.8, which shows the Fermi level positions in the n- and p-type semiconductors. In the p-type semiconductor, the Fermi level is close to the valence band, while in the n-type semiconductor, it is close to the conduction band.

The energy band diagrams for a p–n junction at equilibrium may be obtained from Figure 2.8 by making the Fermi energies on the two sides equal. This result is shown in Figure 2.9. This figure shows the change in energy between the n- and p-type sides taking place discontinuously over a zero length. However, we know physically that the change takes place over a region of nonzero length, so the drawing in Figure 2.10 is more realistic. Figure 2.10 shows the result of equalizing the Fermi energies on the two sides of the junction when equilibrium is attained. This process is equivalent to decreasing the electron energies on the n side of the junction. This is, of course, just what we described earlier as the change in the electrostatic potential of the electrons brought about by the built-in electric field.

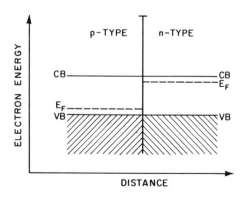

Figure 2.8. Band diagram of a p–n junction before the attainment of equilibrium. The Fermi levels E_F are shown as dashed lines.

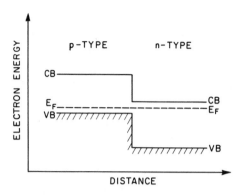

Figure 2.9. Band diagram of a *p–n* junction showing equalization of the Fermi levels.

From Figures 2.8 and 2.9, we see that the shift in energy between *n*- and *p* sides, when their Fermi levels are equalized, is just the difference in the Fermi energies on the two sides. If we call $E_F(n)$ the Fermi energy on the *n* side, and $E_F(p)$ the Fermi energy on the *p* side, then this energy shift is just $E_F(n) - E_F(p)$. However, on comparing Figure 2.9 with Figure 2.8, we see graphically that the energy shift between the two sides is just the height eV_0 of the energy barrier between the two sides. We therefore have the relation that

$$eV_0 = E_F(n) - E_F(p) \tag{2.4}$$

where V_0 is the contact potential defined by equation (2.2). Equation (2.4) shows that the contact potential V_0 is the difference $E_F(n) - E_F(p)$ between the Fermi levels on the two sides, divided by the charge e on the proton.

Finally, we can calculate the width of the space charge region of the *p–n* junction. If Q_+ is the number of positive charges on the *n* side of the space charge region, and Q_- is the number of negative charges on the *p*

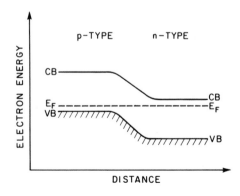

Figure 2.10. Band diagram of a *p–n* junction at equilibrium, showing equalized Fermi levels and the space charge region.

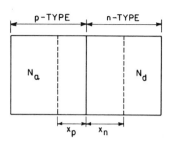

Figure 2.11. Widths x_p and x_n of the space charge regions in the *p*- and *n*-type sides of an ideal abrupt *p–n* junction (shown schematically). The donor and acceptor densities are N_d and N_a, respectively, and the cross-section area of the sample is assumed to be a constant.

side, then electrical neutrality requires that

$$Q_+ = Q_-$$ (2.5)

If there are N_d ionized donors per unit volume on the *n* side, and N_a ionized acceptors on the *p* side, then (2.5) becomes

$$N_a x_p = N_d x_n$$ (2.6)

where x_p is the distance the space-charge region extends into the *p*-type side of the junction, and x_n is the distance the space charge region extends into the *n*-type side of the junction, and assuming that the cross-section area is the same throughout the sample. This result, and its consequence that x_p and x_n are unequal if N_a and N_d are unequal, are shown in Figure 2.11.

Quantitative Treatment of the *p–n* Junction in Equilibrium

We now have a physical picture of the semiconductor *p–n* junction in equilibrium and have obtained, in Figure 2.7, a qualitative idea of the electron energy level diagram of the junction. We now consider a quantitative discussion[3,4] of the *p–n* junction in equilibrium.

The equilibrium described in the previous section is a dynamic state, in which there are equal and opposite fluxes of both carrier types. For electrons, we have

$\mathbf{J}_{n1}$ = current density of electrons from *n* to *p* due to diffusion

$\mathbf{J}_{n2}$ = current density of electrons from *p* to *n* due to the built-in electric field

The current $\mathbf{J}_{n1}$ is called the diffusion current of electrons; $\mathbf{J}_{n2}$ is called the generation or drift current of electrons. At equilibrium, these two current densities are equal in magnitude and opposite in direction, so the net electron current density is equal to zero. Similarly, we have the hole current densities

$\mathbf{J}_{p1}$ = current density of holes from p to n due to diffusion

$\mathbf{J}_{p2}$ = current density of holes from n to p due to the built-in electric field

At equilibrium, the diffusion current density of holes $\mathbf{J}_{p1}$ is equal in magnitude and opposite in direction to the generation current of holes $\mathbf{J}_{p2}$, so the net hole current density vanishes. The conditions for equilibrium are expressed by the equations

$$\mathbf{J}_{n1} + \mathbf{J}_{n2} = 0 \tag{2.7}$$

and

$$\mathbf{J}_{p1} + \mathbf{J}_{p2} = 0 \tag{2.8}$$

The diffusion of carriers across the junction is due to the concentration gradient between the n and p sides of the junction. We recall[5] that the flux $\mathbf{F}$ (the number of particles crossing unit area in unit time) is given by

$$\mathbf{F} = -D\nabla C$$

where C is the concentration of particles and D is the diffusion coefficient or diffusivity. The current density $\mathbf{J}$ of particles of charge q due to diffusion is then

$$\mathbf{J} = q\mathbf{F} = -qD\nabla C \tag{2.9}$$

We first apply equation (2.9) to the diffusion of electrons of charge $-e$, yielding

$$\mathbf{J}_{n1} = eD_n\nabla n \tag{2.10}$$

where D_n is the diffusion coefficient for electrons and n is the spatially varying electron density.

To obtain the current density $\mathbf{J}_{n2}$ of electrons due to the built-in electric field $\mathscr{E}$, we recall that the current density of charges q of concentration n and drift velocity $\mathbf{v}$ is equal to $nq\mathbf{v}$. Further, we have

$$\mathbf{v}_p = \mu_p\mathscr{E} \tag{2.11}$$

for holes and

$$\mathbf{v}_n = -\mu_n\mathscr{E} \tag{2.12}$$

for electrons, where μ_p and μ_n are the hole and electron mobilities, both of which are defined as positive quantities. The minus sign in equation (2.13) shows that the electron drift velocity is in the direction opposite to that of the electric field. Then the current density $\mathbf{J}_{n2}$ is given by

$$\mathbf{J}_{n2} = nq\mathbf{v}_n = n(-e)(-\mu_n \mathscr{E}) = en\mu_n \mathscr{E} \qquad (2.13)$$

showing that the drift (or generation) current density $\mathbf{J}_{n2}$ is parallel to the built-in electric field $\mathscr{E}$.

In a similar manner, we may obtain the current density

$$\mathbf{J}_{p1} = -eD_p \nabla p \qquad (2.14)$$

of holes due to diffusion, and the current density

$$\mathbf{J}_{p2} = pq\mathbf{v}_p = ep\mu_p \mathscr{E} \qquad (2.15)$$

of holes due to the built-in electric field. The conditions (2.7) and (2.8) then become

$$eD_n \nabla n + en\mu_n \mathscr{E} = 0 \qquad (2.16)$$

$$-eD_p \nabla p + ep\mu_p \mathscr{E} = 0 \qquad (2.17)$$

as the equations that must hold at equilibrium. From equation (2.16), we can see that the vector ∇n is opposite to the direction of the built-in electric field $\mathscr{E}$ because D_n, μ_n, and n are all positive quantities. This is as it should be because the vector $\mathscr{E}$ points from the n side to the p side of the junction and the vector ∇n points in the spatial direction of increasing electron concentration n, i.e., from the p side to the n side.

If we now specialize the equilibrium conditions (2.16) and (2.17) to our one-dimensional case, they become

$$eD_n \frac{dn(x)}{dx} + en(x)\mu_n \mathscr{E}(x) = 0 \qquad (2.18a)$$

$$-eD_p \frac{dp(x)}{dx} + ep(x)\mu_p \mathscr{E}(x) = 0 \qquad (2.18b)$$

where $n(x)$, $p(x)$, and $\mathscr{E}(x)$ are the spatially varying electron concentration, hole concentration and built-in electric field, respectively. Physically, these last two equations are the conditions for equilibrium in our one-dimensional p–n junction, and state the equality of two opposing current densities for both electrons and holes. The first term in each equation represents the diffusion current density and the second term is the current density due to the action of the built-in electric field. This latter current is often called the drift current density. We may use the Einstein relation[6]

$$eD = \mu k_B T$$

where k_B is Boltzmann's constant, relating the mobility μ of a carrier at temperature T to its diffusion coefficient D, to eliminate the mobility from equations (2.18). The results[7] are

$$\frac{dn(x)}{dx} + \left(\frac{e}{k_B T}\right) n(x) \mathscr{E}(x) = 0 \qquad (2.19a)$$

and

$$\frac{dp(x)}{dx} - \left(\frac{e}{k_B T}\right) p(x) \mathscr{E}(x) = 0 \qquad (2.19b)$$

as the differential equations relating the electron and hole concentrations to the built-in electric field $\mathscr{E}(x)$ of the junction. Since $\mathscr{E}(x)$ is equal to the gradient $-dV/dx$ of the electrostatic potential $V(x)$ in the junction, we obtain, on rearranging terms,

$$\frac{1}{n}\frac{dn}{dx} - \left(\frac{e}{k_B T}\right)\frac{dV}{dx} = 0 \qquad (2.20)$$

and

$$\frac{1}{p}\frac{dp}{dx} + \left(\frac{e}{k_B T}\right)\frac{dV}{dx} = 0 \qquad (2.21)$$

as the differential equations relating the spatially varying electrostatic potential $V(x)$ to the spatially varying hole and electron concentrations $p(x)$ and $n(x)$ in the junction.

We next consider the boundary conditions subject to which the differential equations (2.20) and (2.21) will be solved. To do so, we must recall that we have been discussing the idealized case[1,8] of an abrupt junction in which the semiconductor changes from n type to p type at a given point in space. This means that we are assuming that the space charge region has abrupt boundaries and that the built-in electric field is confined to the space charge region. Further, this also means that the semiconductor is assumed to be neutral outside the boundaries of the space charge region. The model of an ideal abrupt junction is clearly an approximation but is amenable to calculation, and more realistic models[8,9] (e.g., graded junctions) do not further illuminate the physics significantly.

We set up the geometry of the junction as shown in Figure 2.12, in which the junction is located at $x = 0$, the region of positive x is n type

Figure 2.12. Idealized abrupt *p-n* junction located at $x = 0$, with $-x_p \leq x \leq 0$ as the negative space charge layer on the *p*-type side, and $0 \leq x \leq x_n$ as the positive space charge layer on the *n*-type side. The ionized donor and acceptor concentrations are, respectively, N_d and N_a.

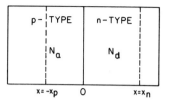

with donor concentration N_d, and the region of negative x is p type with acceptor concentration N_a. The space charge layer in the *n*-type side extends from $x = 0$ to $x = x_n$, and the space charge layer on the *p*-type side extends from $x = 0$ to $x = -x_p$. We note that equation (2.6) implies that the magnitude $|x_n|$ will be unequal to the magnitude $|-x_p|$ if the donor concentration N_d is unequal to the acceptor concentration N_a, so the space charge region may extend different distances into the n and p sides of the junction.

We consider next the differential equation (2.21) for the hole concentration $p(x)$ and set the following boundary conditions on the variables $p(x)$ and $V(x)$. First, that the hole concentration p has its equilibrium values for *n*- and *p*-type material in the neutral regions outside the space charge region. If we use the symbols p_p and p_n for the equilibrium hole concentrations in *p*- and *n*-type material respectively, this boundary condition is

$$p = p_p \qquad \text{for } -x_p \geq x \tag{2.22}$$

and

$$p = p_n \qquad \text{for } x \geq x_n \tag{2.23}$$

Our second boundary condition is that the electrostatic potential $V(x)$ has particular values in the neutral regions outside the space charge region. This condition is expressed as

$$V = V_p \qquad \text{for } -x_p \geq x \tag{2.24}$$

and

$$V = V_n \qquad \text{for } x \geq x_n \tag{2.25}$$

Equations (2.22)–(2.25) tell us that the hole concentration varies from $p = p_p$ to $p = p_n$, and the electrostatic potential varies from $V = V_p$ to $V = V_n$, across the space charge region which extends from $x = -x_p$ to $x = x_n$. Rearranging (2.21) gives

$$\frac{1}{p}\, dp = -\left(\frac{e}{k_B T}\right) dV \tag{2.26}$$

which can be integrated over the space charge region as

$$- \frac{e}{k_B T} \int_{V_p}^{V_n} dV = \int_{p_p}^{p_n} \frac{1}{p} \, dp \tag{2.27}$$

to give

$$\ln\left(\frac{p_n}{p_p}\right) = \frac{-e}{k_B T}(V_n - V_p) = \frac{-eV_0}{k_B T} \tag{2.28}$$

where the definition (2.2) of the contact potential V_0 has been used. Rewriting (2.28) gives

$$p_n/p_p = \exp(-eV_0/k_B T) \tag{2.29}$$

We note in passing that equation (2.4) gives V_0 in terms of the Fermi levels in the neutral n and p regions; the latter are (as discussed in Chapter 1) determined by the carrier concentrations N_d and N_a. The contact potential is thus determined by the impurity densities on the two sides of the junction.

Equation (2.29) is important because it gives the ratio of the hole concentrations in the neutral n and p regions. Since $V_0 \equiv V_n - V_p$ is a positive quantity, (2.29) tells us that p_n is smaller than p_p. This is as it should be; there are fewer holes in n-type material than there are in p-type material. Further, we recall that, at equilibrium at a temperature T, the product

$$np \equiv n_i^2 \tag{2.30}$$

is a constant, so the np product is a constant in both the neutral n and p regions of the junction. Since this is the case, it is true that

$$n_n p_n = n_p p_p = n_i^2 \tag{2.31}$$

where n_n is the electron concentration in the neutral n region and n_p is the electron concentration in the neutral p region. Combining (2.31) with (2.29), we get

$$\frac{p_n}{p_p} = \frac{n_p}{n_n} = \exp\left(\frac{-eV_0}{k_B T}\right) \tag{2.32}$$

for the two ratios p_n/p_p and n_p/n_n. We also know that, since $p_p = N_a$, we have from equation (2.30) that

$$n_p = n_i^2/N_a \tag{2.33}$$

in the neutral *n*-type region. Since $n_n = N_d$, (2.30) gives

$$p_n = n_i^2/N_d \tag{2.34}$$

in the neutral *n*-type region. Equations (2.32)–(2.34) give the carrier concentrations in the neutral regions as functions of V_0 and the impurity concentrations. These equations will be important in our study of current flow through the junction.

We now turn our attention to the space-charge region of the junction. From the geometry of Figure 2.12, we see that electrical neutrality requires that

$$N_d x_n = N_a x_p \tag{2.35}$$

where N_d and N_a are the densities of ionized donors and acceptors, respectively. At any point x, the space charge density $\varrho(x)$ is given by

$$\varrho(x) = -en(x) + ep(x) - eN_a + eN_d \tag{2.36}$$

since ionized donors and acceptors have charges $+e$ and $-e$, respectively. Equation (2.36) gives us the space charge density $\varrho(x)$ at any point x of the space charge region. At any point x, Poisson's equation

$$\frac{d^2 V}{dx^2} = -\frac{4\pi}{\varepsilon} \varrho(x) \tag{2.37}$$

for the electrostatic potential $V(x)$ must be satisfied; in (2.37), ε is the dielectric constant. Substituting the expression (2.36) for the space charge density into Poisson's equation gives

$$\frac{d^2 V}{dx^2} = -\frac{4\pi}{\varepsilon} [-en(x) + ep(x) - eN_a + eN_d] \tag{2.38}$$

solution of which will give us $V(x)$ in the space charge region. We next make the reasonable approximation[10] that the *net* number $p(x) - n(x)$ of mobile free carriers in the space charge region is small compared to either N_a or N_d, the numbers, respectively, of fixed negative and positive charges. With this approximation, the space charge density $\varrho(x)$ becomes

$$\varrho(x) \cong -eN_a \qquad \text{for } -x_p \leq x \leq 0 \tag{2.39}$$

and

$$\varrho(x) \cong +eN_d \qquad \text{for } 0 \leq x \leq x_n \tag{2.40}$$

so Poisson's equation (2.38) reduces to

$$\frac{d^2V}{dx^2} = \frac{4\pi e}{\varepsilon} N_a \qquad \text{for } -x_p \leq x \leq 0 \qquad (2.41)$$

and

$$\frac{d^2V}{dx^2} = \frac{-4\pi e}{\varepsilon} N_d \qquad \text{for } 0 \leq x \leq x_n \qquad (2.42)$$

for the two parts of the space charge region. The charge distribution represented by (2.39) and (2.40) is shown in Figure 2.13. Since it is true that

$$N_a x_p = N_d x_n \qquad (2.43)$$

the two rectangular areas in Figure 2.13 are equal, thus demonstrating the required overall charge neutrality. We keep in mind that we are dealing with the ideal abrupt junction whose charge density distribution is given in Figure 2.13.

We now integrate Poisson's equation (2.41) and (2.42) for the electrostatic potential $V(x)$ and electric field $\mathscr{E}(x) = -dV/dx$ in the space charge region $-x_p \leq x \leq x_n$. Considering first the n-type side $0 \leq x \leq x_n$ of the space charge region, (2.42) can be written as

$$\frac{d^2V}{dx^2} = -\frac{d\mathscr{E}}{dx} = \frac{-4\pi e}{\varepsilon} N_d \qquad (2.44)$$

where $\mathscr{E} = \mathscr{E}(x)$ is the built-in electric field. Equation (2.44) can be integrated immediately to give

$$\mathscr{E}(x) = \frac{4\pi e}{\varepsilon} N_d x + C_1 \qquad (2.45)$$

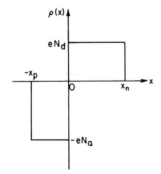

Figure 2.13. Space charge density $\varrho(x)$ given by equations (2.39) and (2.40) shown as a function of x.

The constant of integration C_1 is determined by the boundary condition, appropriate for the ideal junction model we are considering, that the built-in electric field is confined to the space charge region and vanishes outside of it. This condition is expressed as

$$\mathscr{E}(x_n) = 0 \tag{2.46}$$

which, when applied to (2.45), leads to the value

$$C_1 = (-4\pi e/\varepsilon)N_d x_n \tag{2.47}$$

for the constant of integration C_1. We thus obtain the expression

$$\mathscr{E}(x) = \frac{4\pi e}{\varepsilon} N_d(x - x_n) \tag{2.48}$$

for the electric field in the *n*-type side $0 \leq x \leq x_n$ of the space-charge region, in which the space charge is positive. In a similar manner, considering the region $-x_p \leq x \leq 0$ of negative space charge on the *p*-type side of the junction, (2.41) becomes

$$\frac{d^2V}{dx^2} = -\frac{d\mathscr{E}}{dx} = \frac{4\pi e}{\varepsilon} N_a \tag{2.49}$$

which, on integrating, gives

$$\mathscr{E}(x) = \frac{-4\pi e}{\varepsilon} N_a x + C_2 \tag{2.50}$$

for the built-in electric field in the negative space charge region. The constant C_2 is determined by the boundary condition

$$\mathscr{E}(-x_p) = 0 \tag{2.51}$$

a condition restricting the electric field to the space charge region and analogous to (2.46). The resulting value of $C_2 = (-4\pi e/\varepsilon)N_a x_p$ leads to

$$\mathscr{E}(x) = \frac{-4\pi e}{\varepsilon} N_a(x + x_p) \tag{2.52}$$

for the built-in electric field in the region $-x_p \leq x \leq 0$ of negative space charge. We note from (2.48) and (2.52) that $\mathscr{E}(x)$ is negative for all x in the region $-x_p \leq x \leq x_n$, meaning that the built-in electric field is directed in the $-x$ direction, i.e., from the n side of the junction to the p

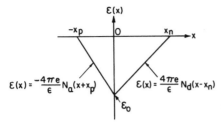

Figure 2.14. Electric field $\mathscr{E}(x)$ in the space charge region $-x_p \leq x \leq x_n$ shown as a function of x for the ideal abrupt *p-n* junction whose charge density distribution is that in Figure 2.13.

side. We see also that the electric field must be continuous at $x = 0$, so, using (2.48) and (2.52), we have

$$\mathscr{E}(0) = \frac{-4\pi e}{\varepsilon} N_d x_n = \frac{-4\pi e}{\varepsilon} N_a x_p \qquad (2.53)$$

in agreement with the condition (2.43) expressing the overall electrical neutrality of the space charge region. Figure 2.14 shows a plot of $\mathscr{E}(x)$ given by equations (2.48) and (2.52) as a function of distance x in the junction. From this figure, we see that the built-in electric field has its maximum magnitude $\mathscr{E}_0 \equiv |\mathscr{E}(0)|$, where

$$\mathscr{E}_0 = \frac{4\pi e}{\varepsilon} N_a x_p = \frac{4\pi e}{\varepsilon} N_d x_n \qquad (2.54)$$

at the point $x = 0$, i.e., at the position of the junction between the *n*-type and *p*-type regions. From the figure, we see that the magnitude of the electric field increases linearly from zero at $x = x_n$, reaches its maximum magnitude $\mathscr{E}_0$ at the junction, and then decreases linearly to zero at $x = -x_p$.

Since the electric field $\mathscr{E} = -dV/dx$, we now integrate again to obtain the electrostatic potential $V(x)$ in the space charge region. We will integrate (2.48) and (2.52) subject to the following boundary conditions which must be obeyed by $V(x)$. In the neutral *n*-type region, the electrostatic potential is equal to a constant, denoted as before by V_n, so we have

$$V(x_n) = V_n \qquad (2.55)$$

Similarly, in the neutral *p*-type region, the electrostatic potential is a constant, denoted by V_p, so

$$V(-x_p) = V_p \qquad (2.56)$$

Further, $V(x)$ must be continuous at the junction located at $x = 0$.

Integrating (2.48) written as $-dV/dx = (4\pi e/\varepsilon)N_d(x - x_n)$ gives

$$V(x) = (-4\pi e/2\varepsilon)N_d(x_n - x)^2 + C_3 \qquad (2.57)$$

and (2.55) yields the result that the constant $C_3 = V_n$, so we have

$$V(x) = (-2\pi e/\varepsilon)N_d(x_n - x)^2 + V_n \tag{2.58}$$

for the electrostatic potential in the region $0 \le x \le x_n$ of positive space charge. Similarly, integrating (2.52) written in the form

$$-dV/dx = (-4\pi e/\varepsilon)N_a(x + x_p)$$

and using the boundary condition (2.56), gives

$$V(x) = (2\pi e/\varepsilon)N_a(x + x_p)^2 + V_p \tag{2.59}$$

for the electrostatic potential $V(x)$ in the region $-x_p \le x \le 0$ of negative space charge on the p-type side of the junction. Finally, since the electrostatic potential must be continuous at $x = 0$, the two solutions (2.58) and (2.59) must have the same value at $x = 0$. This requirement leads to the condition

$$-\frac{2\pi e}{\varepsilon} N_d x_n^2 + V_n = \frac{2\pi e}{\varepsilon} N_a x_p^2 + V_p$$

which yields the equation

$$\frac{2\pi e}{\varepsilon} [N_d x_n^2 + N_a x_p^2] = V_n - V_p \equiv V_0 \tag{2.60}$$

where V_0 is the contact potential defined by equation (2.2). Equation (2.60) is a condition that must be obeyed by N_d, N_a, x_p, and x_n, and it, together with the requirement (2.6)

$$N_a x_p = N_d x_n \tag{2.6}$$

of electrical neutrality in the space charge region, constitutes a system of two equations in four unknowns. Equations (2.60) and (2.6) thus determine x_n and x_p as functions of N_a and N_d, so the total width $x_n + x_p$ of the space charge region is determined by the impurity densities N_a and N_d. These results are exhibited in Figure 2.15, which shows the electrostatic potential $V(x)$ given by equations (2.58) and (2.59) plotted as a function of distance x. We note that the features of the curve $V(x)$ in Figure 2.15 are qualitatively similar to the curve in Figure 2.6, but the variation of $V(x)$ in the space-charge region given by (2.58) and (2.59) is, of course, more detailed than that shown qualitatively in Figure 2.6. However, we note that both figures show the important difference V_0 in the electrostatic potential between the n and p sides of the junction.

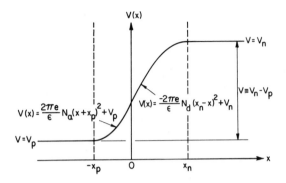

Figure 2.15. Electrostatic potential $V(x)$ given by equations (2.58) and (2.59) shown as a function of x. The figure is drawn for a donor–acceptor ratio $N_d/N_a = \frac{2}{3}$.

Just as we obtained the qualitative variation of electron energy as a function of distance shown in Figure 2.7, we may use Figure 2.15 to obtain the plot shown in Figure 2.16. This figure shows electron energy plotted as a function of distance. If we multiply the curve of $V(x)$ in Figure 2.15 by $-e$, we obtain $-eV(x)$, the potential energy of an electron in the junction. This quantity $-eV(x)$ corresponds, at any point x of the junction, to the valence band energy at that point. In this way, the variation with distance of the valence band electron energy shown in Figure 2.16 is obtained. Since the conduction band energy, at any point, is just the valence band energy plus the magnitude of the energy gap, we obtain also the variation with distance of the conduction band energies. While the variation of electron energy as a function of distance is only schematically shown in Figure 2.16, we could of course obtain analytical expressions for the valence band edge and conduction band edge energies as a function of distance

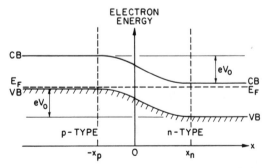

Figure 2.16. Electron energy as a function of distance x in the ideal abrupt p–n junction. (The variation in the space charge region is shown only schematically.)

from (2.58) and (2.59) by multiplying by $-e$, etc., as described above. However, the schematic curve shown in Figure 2.16 will be sufficient for our purposes. We note the similarity between Figure 2.16 and Figures 2.7 and 2.10, which we obtained in our earlier qualitative discussion. Figure 2.16 exhibits clearly the energy barrier of height eV_0 between conduction electrons on the n side and conduction electron states on the p side; holes in the valence band also encounter a barrier of height eV_0 between the p and n sides. We will see that the changes in the height of this energy barrier effected by an applied electric potential are the basis for the useful properties of the p–n junction. Figure 2.16 also indicates the position of the Fermi level as a function of distance through the junction in the equilibrium state in which no net currents flow.

We may now calculate the total width $W = x_n + x_p$ of the space charge region as a function of the impurity concentrations N_d and N_a. Considering equations (2.6) and (2.60), which must be true simultaneously, we can rewrite equation (2.60) as

$$V_0 = \frac{2\pi e}{\varepsilon}\left[N_d x_n\left(x_n + \frac{N_a}{N_d}\,x_p^2 x_n^{-1}\right)\right] \tag{2.61}$$

Since, from (2.6), we have $x_p = (N_d/N_a)x_n$,

$$x_p^2 = (N_d/N_a)x_n x_p \tag{2.62}$$

Substituting (2.62) into (2.61) yields

$$V_0 = \frac{2\pi e}{\varepsilon}\,N_d x_n(x_n + x_p) = \frac{2\pi e}{\varepsilon}\,N_d x_n W \tag{2.63}$$

an equation expressing the space charge width W in terms of the contact potential V_0. Using (2.6), we have

$$W = (x_n + x_p) = x_n + \left(\frac{N_d}{N_a}\right)x_n = x_n\left(\frac{N_a + N_d}{N_a}\right)$$

so we obtain the result that

$$x_n = \left(\frac{N_a}{N_a + N_d}\right)W \tag{2.64}$$

When this result is substituted into (2.63), we obtain

$$V_0 = \frac{2\pi e}{\varepsilon}\left(\frac{N_a N_d}{N_a + N_d}\right)W^2$$

which may be rewritten

$$W = \left[\frac{\varepsilon}{2\pi e} \left(\frac{N_a + N_d}{N_a N_d} \right) V_0 \right]^{1/2}$$ (2.65)

an equation giving the width of the space-charge region in terms of V_0, N_a, and N_d.

We may now express the contact potential V_0 in terms of N_a and N_d in order to have an expression for W in terms of the impurity concentrations alone. From equation (2.28), we have

$$V_0 = (k_B T/e) \ln(p_p/p_n)$$ (2.66)

where p_p and p_n are the hole concentrations in the neutral p and n regions, respectively. From (2.34),

$$p_n = n_i^2/N_d$$ (2.34)

and we know also that the hole concentration in the neutral p region is equal to the ionized acceptor density N_a, so

$$p_p = N_a$$ (2.67)

Combining (2.66), (2.67), and (2.34) yields

$$V_0 = \frac{k_B T}{e} \ln\left(\frac{N_a N_d}{n_i^2} \right)$$ (2.68)

which, when substituted in (2.65) gives the expression

$$W = \left[\frac{\varepsilon k_B T}{2\pi e^2} \left(\frac{N_a + N_d}{N_a N_d} \right) \ln\left(\frac{N_a N_d}{n_i^2} \right) \right]^{1/2}$$ (2.69)

for the total width W of the space charge region in terms of N_d and N_a. From this equation, W can be calculated for typical values of the parameters. Using $N_a = N_d = 10^{15}$ cm^{-3}, and the dielectric constant of silicon ($\varepsilon = 11.7$), one finds that W is about 1.3×10^{-4} cm, a typical value for a silicon *p-n* junction.

It is often of interest to know the penetration distances x_n and x_p of the space charge region into the n and p sides of the junction. We can write, using (2.6),

$$W = x_p + x_n = x_p + \frac{N_a}{N_d} x_p = x_p \left(\frac{N_a + N_d}{N_d} \right) = x_p \left(1 + \frac{N_a}{N_d} \right)$$

leading to the relation

$$x_p = \frac{W}{1 + N_a/N_d} \qquad (2.70)$$

Similarly, one can show that

$$x_n = \frac{W}{1 + N_d/N_a} \qquad (2.71)$$

Equations (2.70) and (2.71) show that, the larger the ionized impurity density on a given side of the junction, the smaller the penetration of the space charge region into that side of the junction. In other words, lighter "doping" means that the space charge region extends further into that side of the junction. Heavily doped junctions have thin space charge layers, and vice versa. We note also, from (2.65), that the width of the space charge layer varies as the square root of the difference in electrostatic potential across the junction. In the equilibrium case we are considering, the electrostatic potential is the contact potential V_0, but, as we will see, applying a voltage across the junction changes this electrostatic potential difference and hence changes the space charge region width. Other useful equations for the width of the space charge region may be found in the literature.[11]

In summary, we see that our quantitative treatment of the abrupt *p-n* junction gave the same basic results as did the qualitative treatment, but with additional detail. It is useful at this point to reiterate that our model of a *p-n* junction is an idealized one in which the space charge region has abrupt boundaries, outside of which the semiconductor is assumed to be neutral. Further, we are tacitly assuming that we may use the results obtained earlier for carrier densities, Fermi levels, etc. in a nondegenerate semiconductor. (For details of these assumed approximations, the interested reader may consult the more advanced literature.[1,8]) Our model is a fairly good description of some alloyed and epitaxial junctions[12] but is often inadequate for discussing diffused junctions.[12] However, this idealized model provides enough physical insight for us to understand the action of a *p-n* junction under an applied potential, the subject which makes the junction of such great technological interest and use.

Before proceeding to do so, however, we digress briefly to discuss two topics which we will need in discussing the behavior of a *p-n* junction under an applied potential. These are the effect of an applied potential on the energy bands in a semiconductor and the diffusion and recombination of excess (i.e., nonequilibrium) carriers introduced into a semiconductor.

Effect of an Applied Potential on Electron Energy Bands

Consider a homogeneous semiconductor bar of length L and constant cross-section area. Let one end of the bar be at $x = 0$ and the other at $x = L$, and apply a constant electrostatic potential V across the sample. If we take the zero of potential, $V = 0$, at $x = 0$, then the applied potential $V(x)$ at point x (where $0 \leq x \leq L$) is

$$V(x) = Vx/L \qquad (2.72)$$

At any point x, the effect of the applied potential $V(x)$ is to change the electron energy at that point by an amount

$$-eVx/L \qquad (2.73)$$

If we look at electron energy as a function of distance x in the sample under the applied potential V, it will be as shown in Figure 2.17, which shows the conduction band (CB) and valence band (VB) of the semiconductor sample as a function of distance x. The application of the potential, corresponding in this case to a constant applied electric field of magnitude (V/L), lowers the electron energy levels, relative to the zero of potential at $x = 0$, as shown in Figure 2.17. At the end $x = L$ of the sample, the decrease in electric energy is equal to (eV), as shown in the figure. This effect is the basic effect of an applied electric field, and is sometimes called "tilting the bands."

We note specifically that, for our model of an abrupt *p-n* junction, "band tilting" will occur only at points in the space charge region. This is because our model of a *p-n* junction is one in which only the space charge region is depleted of free carriers and is of high electrical resistivity compared to the neutral *n-* and *p*-type regions. In our junction model, then, all of the applied potential will appear across the space charge region, so "band tilting" will occur only in that region.

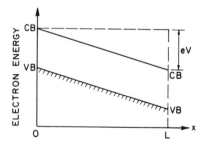

Figure 2.17. Electron energy bands as a function of distance x in a homogeneous semiconductor sample of length L with an applied electrostatic potential V such that, at point x, $V(x) = Vx/L$. The abbreviations CB and VB refer to the conduction and valence band edges, respectively.

Diffusion and Recombination of Excess Carriers[13,14]

We next want to consider excess carriers in a semiconductor. By "excess" is meant carriers introduced in some manner so that the density of such carriers has, at least temporarily, a value different from the equilibrium value. This is not an equilibrium situation, and the system will approach equilibrium as the excess carriers recombine and disappear and the carrier density approaches its equilibrium value.

We will discuss this situation by using the equation of continuity,[15] which, for electric charge, is

$$\frac{\partial \varrho}{\partial t} + \mathbf{\nabla} \cdot \mathbf{J} = G - R \tag{2.74}$$

where ϱ is the electric charge density, $\mathbf{J}$ is the current density, and G and R are the rates of generation and destruction, respectively, of electric charge per unit volume. We consider holes for concreteness. The current density $\mathbf{J}_p$ of holes will have, in general, a drift component given by equation (2.15) and a diffusion component given by (2.14), so

$$\mathbf{J}_p = -eD_p\mathbf{\nabla}p + ep\mu_p\mathscr{E} \tag{2.75}$$

where $\mathscr{E}$ is the electric field producing the drift current density. The electric charge density due to holes is

$$\varrho = ep \tag{2.76}$$

where p is the time- and space-dependent density of holes. Since we will be interested later in a situation in which drift currents are negligible,[16] we set $\mathscr{E} = 0$ in equation (2.75), which becomes, in one space dimension with $p = p(x, t)$,

$$J_p = -eD_p\frac{\partial p}{\partial x} \tag{2.77}$$

In one dimension, the equation of continuity (2.74) becomes, using (2.76),

$$e\frac{\partial p}{\partial t} - eD_p\frac{\partial^2 p}{\partial t^2} = G_p - R_p \tag{2.78}$$

where the subscript p on G_p and R_p refers to holes. In equation (2.78), G_p is the rate of thermal generation of electric charge (due to holes) per unit volume, and R_p is the rate of destruction of charge (holes) due to recombination. (We assume there is no generation of charge due to external factors, such as absorbed photons.) Then $(G_p - R_p)$ is the *net* rate of thermal generation[17] of charge (holes) per unit volume.

We now want to consider an expression for the net rate $(G_p - R_p)$. Let p_0 be the equilibrium value of the hole density, so $(p - p_0)$ is the excess hole density in a nonequilibrium situation. For small densities of excess carriers, the net rate $(G_p - R_p)$ is proportional[18] to the excess density, so

$$G_p - R_p = - \frac{e(p - p_0)}{\tau_p} \qquad (2.79)$$

where τ_p is a constant called the hole lifetime.[13,19] From (2.79), we can see that, for a nonequilibrium hole concentration p larger than the equilibrium value p_0, the net rate of generation $(G_p - R_p)$ of holes is negative, meaning, as it should, a decrease of the hole density toward the equilibrium value p_0. For nonequilibrium values of p smaller than p_0, the opposite is the case, and the net rate of generation of holes is positive.

Substituting equation (2.79) into (2.78) gives

$$e \frac{\partial p}{\partial t} - e D_p \frac{\partial^2 p}{\partial x^2} = - e \left(\frac{p - p_0}{\tau_p} \right) \qquad (2.80)$$

as the equation governing the time- and space-dependent hole density $p(x, t)$, which we rewrite as

$$\frac{\partial p}{\partial t} = D_p \frac{\partial^2 p}{\partial x^2} - \frac{1}{\tau_p} (p - p_0) \qquad (2.81)$$

We consider a situation in which holes are supplied to one end of a semi-infinite bar, and we wish to calculate the space dependence of the hole density in the steady state. Suppose in Figure 2.18 holes are available just to the left of the point $x = x_1$, enter the bar at $x = x_1$, and diffuse away to the right, recombining and disappearing as they go. Suppose further that the rate of supply of holes at $x = x_1$ is such that the steady state obtains, meaning that the hole concentration p is constant in time at every point $x \geq x_1$ of the bar. Then $(\partial p / \partial t) = 0$ in the steady state, and equation (2.81) becomes, writing $(d^2 p / dx^2)$ for the space derivative,

$$\frac{d^2 p}{dx^2} - \frac{1}{D_p \tau_p} (p - p_0) = 0 \qquad (2.82)$$

subject to the following boundary conditions. At the left end $x = x_1$ of the bar, the hole density has the constant value $p(x_1)$. As x approaches $+\infty$, the hole density must approach its equilibrium value p_0 as all of the

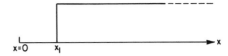

Figure 2.18. Semi-infinite bar of semi-conductor with its left end at the point $x = x_1$.

excess holes recombine and disappear, so we must have $p(+\infty) = p_0$, where p_0 is a constant.

The quantity $(D_p\tau_p)^{1/2}$ has the dimensions of a length, and we define the diffusion length L_p for holes by

$$L_p^2 \equiv D_p\tau_p \tag{2.83}$$

The diffusion length is a measure of the distance a carrier diffuses in its lifetime. It can be shown[20] that the diffusion length is the average distance a carrier diffuses before recombining.

We rewrite (2.82) as

$$\frac{d^2}{dx^2}(p - p_0) = \frac{1}{L_p^2}(p - p_0) \tag{2.84}$$

where $dp_0/dx = 0$ because the equilibrium hole density p_0 is a constant. A solution of (2.84) is

$$p - p_0 = C_1 e^{x/L_p} + C_2 e^{-x/L_p} \tag{2.85}$$

where C_1 and C_2 are constants to be determined by the boundary conditions. Since p must approach p_0 as x approaches $+\infty$, C_1 must equal zero, so

$$p - p_0 = C_2 e^{-x/L_p} \tag{2.86}$$

The second boundary condition is that $p(x_1)$ be a constant, so (2.86) gives us

$$p(x_1) - p_0 = C_2 \exp(-x_1/L_p)$$

and

$$C_2 = [p(x_1) - p_0]e^{x_1/L_p} \tag{2.87}$$

which, substituted into equation (2.86) gives

$$p(x) - p_0 = [p(x_1) - p_0] \exp[-(x - x_1)/L_p] \tag{2.88}$$

as the solution for the spatially dependent hole concentration $p(x)$. If we define the *excess* hole density $\Delta p(x)$ at the point x by the equation

$$\Delta p(x) = p(x) - p_0 \tag{2.89}$$

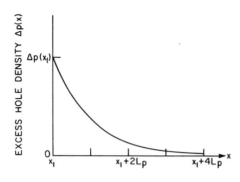

Figure 2.19. Exponential decrease of the excess hole density $\Delta p(x)$ with distance x, from equation (2.90). L_p is the diffusion length for holes and $\Delta p(x_1)$ is the value of $\Delta p(x)$ at $x = x_1$.

then the solution (2.88) becomes

$$\Delta p(x) = [\Delta p(x_1)] \exp[-(x - x_1)/L_p] \tag{2.90}$$

where $\Delta p(x_1)$ is the value of $\Delta p(x)$ at the left end $x = x_1$ of the bar. Equation (2.90) gives the space dependence of the excess hole density $\Delta p(x)$ and tells us that the excess hole density $\Delta p(x_1)$ created at the left end of the bar dies out exponentially with distance x into the bar, as shown in Figure 2.19. From that figure, we see that the *excess* hole density p approaches zero as x gets very large, while the *total* hole density approaches its equilibrium value p_0 in the same limit. As expected from equation (2.90), $\Delta p(x)$ falls to $1/e$ of its initial value $\Delta p(x_1)$ in a distance equal to the diffusion length L_p for holes.

In the same way, we can consider the equation of continuity (2.74) for electrons, and obtain the equation

$$n(x) - n_0 = [n(x_1) - n_0] \exp[-(x - x_1)/L_n] \tag{2.91}$$

describing the diffusion and recombination of excess electrons. In (2.91), $n(x)$ is the electron density at point x, n_0 is the equilibrium electron density, and the electrons are created at $x = x_1$ in a semi-infinite bar. The quantity L_n is the diffusion length for electrons, defined by

$$L_n^2 \equiv D_n \tau_n \tag{2.92}$$

where D_n is the diffusion coefficient for electrons and τ_n is the electron lifetime. Again, defining the excess electron density $\Delta n(x)$ by the relation

$$\Delta n(x) = n(x) - n_0 \tag{2.93}$$

equation (2.91) becomes

$$\Delta n(x) = [\Delta n(x_1)] \exp[-(x - x_1)/L_n] \qquad (2.94)$$

the equation, describing the spatial dependence of the excess electron density $\Delta n(x)$, which is the analog of equation (2.90) for holes.

The process of creating excess carriers in a semiconductor is called *injection* of carriers. It can be done in a number of ways, among which two of the most important are electrical injection and optical injection. We will discuss injection further later. We introduce here some terminology which we will use throughout our study of semiconductor devices. Electrons in an *n*-type semiconductor or holes in a *p*-type semiconductor, are called *majority* carriers. Holes in an *n*-type semiconductor, or electrons in a *p*-type semiconductor, are called *minority* carriers. For example, if we were to create a local excess of concentration of holes in an *n*-type semiconductor, we would speak of the injection of minority carriers in the semiconductor. On the other hand, if we were to remove carriers from some region of the semiconductor, we would speak of carrier *extraction*.

Qualitative Discussion of a Junction under an Applied Potential

At equilibrium, the current densities of electrons and holes due to diffusion (i.e., the effect of concentration gradients) and due to drift (i.e., the effect of the built-in electric field) just cancel each other. This result was expressed in equations (2.7) and (2.8), which can be rewritten as

$$\mathbf{J}_{n1}(\text{diffusion}) + \mathbf{J}_{n2}(\text{drift}) = 0 \qquad (2.95)$$

$$\mathbf{J}_{p1}(\text{diffusion}) + \mathbf{J}_{p2}(\text{drift}) = 0 \qquad (2.96)$$

to emphasize that, for both electrons and holes, there are two different current densities, which move in opposite directions.

If an electrostatic potential V_a is applied across the junction, the situation is changed, equilibrium no longer holds, and net currents flow through the junction. In order to investigate this nonequilibrium situation, we make the assumption that all of the applied potential V_a appears across the space

Figure 2.20. Effect of an applied forward bias potential V_a and applied electric field $\mathscr{E}_a$ on a *p–n* junction; $\mathscr{E}$ is the built-in electric field; W is the width of the space charge region.

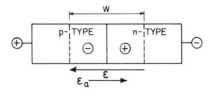

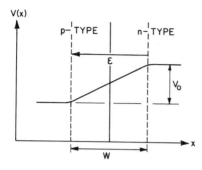

Figure 2.21. Electrostatic potential $V(x)$ (schematic) in the abrupt *p–n* junction, as a function of distance, for the equilibrium case; $\mathscr{E}$ is the built-in electric field, W is the width of the space charge region, and V_0 is the contact potential.

charge region of width W, which is depleted of carriers and hence is a region of higher electrical resistivity than the neutral *n*- and *p*-type regions. Recalling that the built-in electric field $\mathscr{E}$ in the junction at equilibrium is directed from the *n* side of the space charge region toward the *p* side, we see that applying a negative potential to the *n* region will *decrease* the built-in electric field. This is shown in Figure 2.20, in which $\mathscr{E}$ is the built-in electric field and $\mathscr{E}_a$ is the applied electric field, of magnitude V_a/W, due to the negative potential V_a applied to the *n*-type side of the junction. In this case, in which the *n* side is biased negatively, the resulting net electric field in the space charge region is equal to $\mathscr{E} - \mathscr{E}_a$ since $\mathscr{E}$ and $\mathscr{E}_a$ are antiparallel. If the *n* side is biased positively, then the applied electric field $\mathscr{E}_a$ is parallel to the built-in field $\mathscr{E}$, and the net resulting electric field in the space charge region has the increased value $\mathscr{E} + \mathscr{E}_a$.

There is a standard terminology by which the applied potential is described. If the *n*-type region is biased negatively, it is called *forward* bias. If the *n*-type region is biased positively, it is called *reverse* bias.

We need now to consider the effect of an applied potential V_a (or applied electric field $\mathscr{E}_a$) on the width W of the space charge region. Under an applied forward bias, the magnitude of the electric field in the space

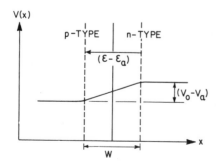

Figure 2.22. Electrostatic potential $V(x)$ (schematic) in the abrupt *p–n* junction, as a function of distance x, for the case of applied forward bias; V_a is the applied potential and $\mathscr{E}_a$ the applied electric field. The net electric field is $\mathscr{E} - \mathscr{E}_a$ and the potential difference is decreased to $V_0 - V_a$. The width W of the space charge layer has decreased.

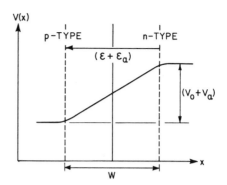

Figure 2.23. Electrostatic potential $V(x)$ (schematic) in the abrupt p–n junction, as a function of distance x, for the case of applied reverse bias; V_a is the applied potential and $\mathscr{E}_a$ the applied electric field. The net electric field is $\mathscr{E} + \mathscr{E}_a$ and the potential difference is increased to $V_0 + V_a$. The width W of the space charge layer has increased.

charge region decreases to $\mathscr{E} - \mathscr{E}_a$. This smaller electric field means that there must be fewer uncompensated ionized impurity atoms in the space charge region. This in turn means that the space charge region must be narrower in extent, so we see that an applied forward bias decreases the width W of the space charge region. Under reverse bias, the electric field increases in magnitude to $\mathscr{E} + \mathscr{E}_a$, so the width W of the space charge region is larger under reverse bias than it is at equilibrium.

To examine next the effect of the applied potential V_a on the difference in electrostatic potential between the n and p sides of the junction, we have seen that both the electric field and the width of the space charge region decrease under forward bias. Since the electrostatic potential $V(x)$ is equal to $W\mathscr{E}(x)$, the difference in electrostatic potential between the n and p regions is smaller under forward bias than it is at equilibrium. This result is shown in Figures 2.21 and 2.22. The sign of the applied potential is defined as follows. For forward bias, the applied potential is equal to V_a, where V_a is a positive quantity. For reverse bias, the applied potential is equal to $-V_a$. We see from these figures that, under forward bias, the potential difference between the n and p sides is decreased to $V_0 - V_a$, where V_0

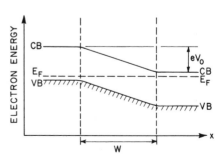

Figure 2.24. Electron energy band diagram (schematic) of the abrupt p–n junction at equilibrium. The energy barrier between n and p sides is eV_0 where V_0 is the contact potential, and W is the width of the space charge layer.

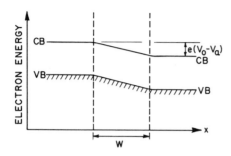

Figure 2.25. Electron energy band diagram (schematic) of the abrupt *p–n* junction under a forward bias potential V_a. The magnitude of the energy barrier has decreased to $e(V_0 - V_a)$, and the width W of the space charge layer has decreased.

is the contact potential at equilibrium. Under reverse bias, the opposite is true, and, as shown in Figure 2.23, the difference in potential between the *n* and *p* sides is increased to $V_0 + V_a$.

We may also plot the electron potential energy $-eV(x)$ as a function of distance *x* through the junction for forward and reverse bias. The energy barrier between the *n* and *p* sides of the junction is eV_0 at equilibrium, as shown in Figure 2.24, which corresponds to the potential diagram Figure 2.21. As seen in Figure 2.25, the application of a forward bias potential equal to V_a *decreases* the height of the energy barrier to $e(V_0 - V_a)$. The application of a reverse bias *increases* the height of the energy barrier to $e(V_0 + V_a)$, as shown in Figure 2.26.

The central, and most important, result above is that a forward bias applied to the junction decreases the magnitude of the energy barrier between the *n*- and *p*-type sides of the junction and a reverse bias increases the magnitude of that barrier. This fact is the basic physics underlying the behavior and operation of a semiconductor *p–n* junction under the influence of an applied electric potential. We shall see that this changing of the magnitude of the energy barrier explains the flow of current in a *p–n* junction under an applied potential, and is the basic physical reason why a semiconductor *p–n* junction acts as a diode rectifier of alternating current.

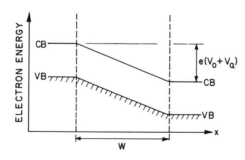

Figure 2.26. Electron energy band diagram (schematic) of the abrupt *p–n* junction under a reverse bias potential V_a. The magnitude of the energy barrier has increased to $e(V_0 + V_a)$, and the width W of the space charge layer has increased.

Qualitative Discussion of Current Flow in the Biased Junction

We have just seen that the application of an applied potential changes the magnitude of the energy barrier between the n and p sides of a p–n junction. We now consider the effect of these changes on the particle currents flowing in the junction.

Considering first the electron currents, we recall that there are two electron currents, which flow in opposite directions. The first, denoted J_{n1}(diffusion) is composed of those majority carrier electrons on the n side which have energy sufficient to surmount the energy barrier (of height eV_0 at equilibrium) between the n and p regions. The high-energy end of the energy distribution of majority electrons on the n side will have enough energy to do this. The diffusion current J_{n1}(diffusion) of majority electrons is indicated schematically in Figure 2.27a. The second current, which we earlier called J_{n2}(drift), is composed of minority electrons on the p side which diffuse into the space charge region and are swept down the energy barrier and into the n-type region. Because these minority electrons are generated by intrinsic thermal excitation across the energy gap, the current J_{n2} is usually called the generation current. This current J_{n2}(generation) is shown schematically in Figure 2.27(b); we will now use this terminology in place of the notation J_{n2}(drift) used earlier. The number of minority electrons comprising the current J_{n2}(generation) will be those created within a diffusion length L_n of the space charge region because electrons generated further into the p side will recombine and disappear before they can reach the space charge region to be swept down the energy barrier. These statements are also true of the diffusion current J_{p1}(diffusion) and the generation current J_{p2}(generation) of holes through the junction.

We consider first the electron currents. The diffusion current of electrons J_{n1} is made up of the small fraction of the majority electrons on the n side which have energy sufficient to surmount the energy barrier. The diffusion current J_{n1} will therefore depend on the magnitude of the energy barrier. The smaller the energy barrier, the larger the fraction of majority

Figure 2.27. (a) Schematic representation of the diffusion current J_{n1} of electrons surmounting the energy barrier and going from the n side to the p side. (b) Schematic representation of the generation current J_{n2} of thermally

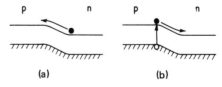

generated electrons on the p side being swept down the energy barrier by the built-in electric field and entering the n-side.

electrons which can go over it and enter into the p side of the junction. Thus, if the height of the energy barrier is decreased, the magnitude of the diffusion current J_{n1} of majority electrons from n to p will increase.

The generation current of electrons J_{n2} from p to n is made up of all of the minority electrons which diffuse into the junction region from the p side. These are relatively few in number but essentially all of them are swept "down the energy hill" into the n side. To a good approximation, then, the generation current J_{n2} of minority electrons is independent of the height of the energy barrier because "it is just as easy for an electron to slide down a high energy hill as a low energy hill," so to speak. Essentially all of the small number of minority electrons that are generated within a diffusion length of the junction will thus enter into the generation current of electrons from p to n.

At equilibrium, the diffusion and generation currents are equal in magnitude and opposite in direction. The diffusion current from n to p is a small fraction of the large number of majority electrons, while the generation current from p to n is a large fraction of the small number of minority electrons.

We can now see how an applied potential affects the electron currents through the junction. Forward bias decreases the height of the energy barrier for electrons between n and p, so the diffusion current J_{n1} increases while the generation current J_{n2} is unchanged. The net electron current $J_{n1} + J_{n2}$ from n to p thus increases on forward bias. In the same way, we can see that an applied forward bias increases the net hole current $J_{p1} + J_{p2}$ moving from p to n. The total net current (composed of both electrons and of holes) moving through the junction increases with increasing forward bias applied to the junction.

To consider the effect of reverse bias, we have seen that reverse bias increases the height of the energy barrier. Thus the diffusion current J_{n1} of electrons from n to p is decreased by an applied reverse bias, while the generation current J_{n2} from p to n is more or less unchanged because the generation current is essentially independent of the barrier height. With increasing reverse bias, the diffusion current J_{n1} decreases until it is negligibly small with respect to the generation current J_{n2}. Eventually, with increasing reverse bias, we reach a state in which the total electron current through the junction is the generation current J_{n2} and is independent of the magnitude of the applied reverse bias. The same conclusions are true of the hole current through the junction. The total current (electrons and holes) through the junction decreases with increasing reverse bias until the current saturates with a magnitude equal to the value of the generation current $J_{n2} + J_{p2}$.

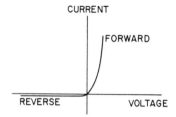

Figure 2.28. Current–voltage plot showing, qualitatively, the behavior expected for forward and reverse biases in a *p–n* junction.

This value of the current is the reverse saturation current of the junction. From these qualitative conclusions, we expect that the total current J (both electrons and holes), where

$$J = J(\text{diffusion}) + J(\text{generation}) \qquad (2.97)$$

will behave approximately as follows. The total current J will be large and increasing with increasing forward bias, and will be small and constant for increasing magnitude of reverse bias; the situation is shown schematically in Figure 2.28.

Quantitative Treatment of Carrier Injection in the Junction

We now treat quantitatively some of the ideas discussed qualitatively about the *p–n* junction under an applied potential. We begin with a treatment of carrier injection and extraction. We have seen that decreasing the energy barrier between the n and p regions allows more majority carriers to diffuse across the junction; these then become minority carriers on the other side of the junction. When a forward bias is applied to the junction, electrons diffuse into the p region and holes diffuse into the n region, thereby increasing the minority carrier densities at the edges of the space charge region. This is the process of minority carrier injection.

On taking the reciprocal of equation (2.32) and making a slight change in notation, we have

$$\frac{p_{p0}}{p_{n0}} = \frac{n_{n0}}{n_{p0}} = \exp\left(\frac{eV_0}{k_B T}\right) \qquad (2.98)$$

where p_{n0} and p_{p0} are the hole concentrations in the neutral n and p regions, respectively, and n_{p0} and n_{n0} are the electron concentrations in the neutral p and n regions, all at *equilibrium*. Thus p_{n0} is the equilibrium minority hole concentration, p_{p0} is the equilibrium majority hole concentration, etc., and V_0 is the contact potential defined by equation (2.2). We saw that V_0, which appears in the exponential in equation (2.98), is the electrostatic potential difference across the junction space charge region at equilibrium. We would expect then, for the nonequilibrium situation in which an external potential V_a is applied across the junction, that the *total* potential difference across the space charge region would appear in the exponential in the equation, analogous to (2.98), appropriate for the *nonequilibrium* case. We would have, then, for an applied potential V_a, that

$$\frac{p_p}{p_n} = \frac{n_n}{n_p} = \exp\left[\frac{e(V_0 - V_a)}{k_B T}\right] \tag{2.99}$$

for the nonequilibrium case of an applied potential. In (2.99), the applied potential V_a is taken as positive for forward bias and is taken as negative for reverse bias. This is because the total potential difference across the junction has magnitude $V_0 - V_a$, with V_a positive, for forward bias, and magnitude $V_0 + V_a$ for reverse bias. Equation (2.99) is true for the carrier densities throughout the neutral n and p regions and is, in particular, true at the edges $x = x_n$ and $x = -x_p$ of the space charge region. We indicate this explicitly by rewriting (2.99) as

$$\frac{n_n(x_n)}{n_p(-x_p)} = \frac{p_p(-x_p)}{p_n(x_n)} = \exp\left[\frac{e(V_0 - V_a)}{k_B T}\right] \tag{2.100}$$

where $n_n(x_n)$ is the majority electron density at $x = x_n$, $n_p(-x_p)$ is the minority electron density at $x = -x_p$, $p_p(-x_p)$ is the majority hole density at $x = -x_p$, and $p_n(x_n)$ is the minority hole density at $x = x_n$, all for the nonequilibrium situation.

We next make the assumption that the *majority* carrier densities $n_n(x_n)$ and $p_p(-x_p)$ are, under the applied potential V_a, essentially unchanged from their equilibrium values. This assumption is reasonable because the majority carrier densities are so large relative to minority carrier densities in extrinsic semiconductors. This assumption is expressed by the equations

$$n_n(x_n) \cong n_{n0}, \qquad p_p(-x_p) \cong p_{p0} \tag{2.101}$$

which, when substituted in (2.100), give

$$\frac{n_{n0}}{n_p(-x_p)} = \frac{p_{p0}}{p_n(x_n)} = \exp\left[\frac{e(V_0 - V_a)}{k_B T}\right] \tag{2.102}$$

Next, we substitute equation (2.98) for the *equilibrium* majority carrier densities n_{no} and p_{po} into equation (2.102). The $\exp[eV_0/k_BT]$ term cancels, leading to the results for the nonequilibrium situation that

$$n_p(-x_p) = n_{po} \exp(eV_a/k_BT) \tag{2.103}$$

and

$$p_n(x_n) = p_{no} \exp(eV_a/k_BT) \tag{2.104}$$

Equations (2.103) and (2.104) give the minority electron density $n_p(-x_p)$ at the edge $x = -x_p$ of the space charge region and the minority hole density $p_n(x_n)$ at the edge $x = x_n$ of the space charge region, both under an applied potential V_a. In (2.103) and (2.104), n_{po} is the *equilibrium* minority electron density in the neutral p region (including the point $x = -x_p$) and p_{no} is the *equilibrium* minority hole density in the neutral n region (including $x = x_n$).

From equations (2.103) and (2.104), we see that, under forward bias, V_a is positive, and the minority carrier densities $n_p(-x_p)$ and $p_n(x_n)$ at the edges of the space charge region both increase to values greater than their equilibrium values n_{no} and p_{po}, respectively, because $\exp(eV_a/k_BT)$ is greater than unity for positive values of V_a. This increase in minority carrier density is the process of minority carrier injection. Similarly, under reverse bias, V_a is negative, and the minority carrier densities $n_p(-x_p)$ and $p_n(x_n)$ at the edges of the space charge region decrease to less than their equilibrium values n_{no} and p_{po}. This decrease in minority carrier density is the process of minority carrier extraction.

If we define the excess minority electron density Δn_p at the edge $x = -x_p$ of the p-type space charge region by

$$\Delta n_p \equiv n(-x_p) - n_{po} \tag{2.105}$$

we see that positive values of Δn_p represent minority carrier injection at $x = -x_p$, and negative values represent minority carrier extraction at that point. Substituting (2.103) into (2.105) we have

$$\Delta n_p = [n_{po} \exp(eV_a/k_BT)] - n_{po} \tag{2.106}$$

$$\Delta n_p = n_{po}[\exp(eV_a/k_BT) - 1] \tag{2.107}$$

Equation (2.107) gives the density Δn_p of excess minority electrons injected or extracted (at the edge of the p-type space charge region) when the external potential V_a is applied to the junction. In a similar way, the excess minority hole density $\Delta p_n \equiv p(x_n) - p_{no}$ at the edge $x = x_n$ of the n-type space charge region is given by

$$\Delta p_n = p_{no}[\exp(eV_a/k_BT) - 1] \tag{2.108}$$

which can be obtained from (2.104). Equation (2.108) gives the density Δp_n of excess minority holes injected or extracted (at the edge of the n-type space charge region) under the external applied potential V_a.

Having obtained the equations (2.107) and (2.108), it is useful at this point to summarize the physical picture of the processes taking place during minority carrier injection or extraction by an applied potential. At equilibrium, the diffusion current of, say, majority electrons from n to p over the energy barrier of magnitude eV_0 is just canceled by the opposing generation current of minority electrons moving from p to n. Under an applied forward bias V_a the magnitude of the energy barrier to the diffusion of majority electrons from n to p is decreased to $e(V_0 - V_a)$, thus increasing the diffusion current of majority electrons into the p region. However, at the same time, the opposing generation current of minority electrons from p to n remains unchanged by the applied potential. The result is an increase Δn_p in the density of minority electrons at the edge of the space charge region on the p-type side of the junction. This increase Δn_p in the minority carrier density on the p side is the process of minority carrier injection. Under an applied reverse bias, exactly the opposite takes place. The generation current of minority electrons from p to n predominates over the decreased diffusion current of electrons of n to p as the magnitude of the energy barrier is increased to $e(V_0 + V_a)$. In this case, Δn_p is negative, meaning that the minority electron density at the edge of the p-type space charge region has decreased. This process is that of minority carrier extraction. While we have used electrons for illustration, the same injection and extraction processes take place for minority holes.

Calculation of the Current through the Junction

We now want to calculate the current flowing through the junction. Denoting the electron current density by $\mathbf{J}_n$ and the hole current density by $\mathbf{J}_p$, we have, from equations (2.10)–(2.15), that

$$\mathbf{J}_n = \mathbf{J}_{n1} + \mathbf{J}_{n2} = eD_n\nabla n + en\mu_n\mathscr{E} \tag{2.109}$$

$$\mathbf{J}_p = \mathbf{J}_{p1} + \mathbf{J}_{p2} = -eD_p\nabla p + ep\mu_p\mathscr{E} \tag{2.110}$$

where the diffusion and drift current densities $\mathbf{J}_{n1}$ and $\mathbf{J}_{n2}$ of electrons and

J_{p1} and J_{p2} of holes, are given by equations (2.10), (2.13), (2.14), and (2.15). In equations (2.109) and (2.110), n and p are the electron and hole concentrations, μ_n and μ_p are the electron and hole mobilities, D_n and D_p are the electron and hole diffusion coefficients, and $\mathscr{E}$ is the electric field. We wish to find the total current density $\mathbf{J}$, where

$$\mathbf{J} = \mathbf{J}_n + \mathbf{J}_p \qquad (2.111)$$

The current densities $\mathbf{J}_n$ and $\mathbf{J}_p$ must both satisfy the equation (2.74) of continuity for charge. Using ep and $(-e)n$ for the charge densities due to holes and electrons, respectively, the equations of continuity are

$$e\frac{dp}{dt} + \boldsymbol{\nabla} \cdot \mathbf{J}_p = -e\left(\frac{p - p_0}{\tau_p}\right) \qquad (2.112)$$

$$(-e)\frac{dn}{dt} + \boldsymbol{\nabla} \cdot \mathbf{J}_n = +e\left(\frac{n - n_0}{\tau_n}\right) \qquad (2.113)$$

In (2.112) and (2.113), the rates of generation G_p and G_n due to external agencies have been set equal to zero, and recombination terms R_p and R_n of the form (2.79) have been used. The symbol p is the hole density, p_0 is the equilibrium hole density, τ_p is the hole lifetime, n is the electron density, n_0 is the equilibrium electron density, and τ_n is the electron lifetime.

Equations (2.109), (2.110), (2.112), (2.113), and Poisson's equation

$$\boldsymbol{\nabla} \cdot \mathscr{E} = 4\pi\varrho/\varepsilon \qquad (2.114)$$

where ε is the dielectric constant and ϱ is the electric charge density, determine the hole and electron densities p and n. However, solution is difficult in most cases and it is necessary[21] to use a model that approximates the situation of interest in order to obtain physically useful information.

The model we shall consider is the idealized model[22] of an abrupt junction we have discussed all along. This junction was shown in Figure 2.12. The depletion layer extends from $x = x_n$ on the n side to $x = -x_p$ on the p side, with the abrupt junction itself located at $x = 0$. The homogeneous p-type neutral region is the region $-\infty \le x \le -x_p$ and the n-type neutral region is $x_n \le x \le +\infty$. We recall that the electrostatic potential was taken as constant in the neutral regions, so, in this model, the neutral regions are free of electric fields. We will thus assume that the electric field $\mathscr{E}$ is negligibly small outside the depletion layer. We assume also that the densities of minority carriers injected into the neutral regions are small compared to the majority carrier densities. This assumption of low-level injection means, along with the assumed small value of $\mathscr{E}$, that the drift currents in (2.109) and (2.110) may be neglected in the neutral regions near the junction. This is done by setting $\mathscr{E} = 0$ in those equations, which become, in one dimension,

$$J_n = eD_n \frac{dn}{dx} \tag{2.115}$$

$$J_p = -eD_p \frac{dp}{dx} \tag{2.116}$$

We now substitute (2.115) and (2.116) into the equations of continuity (2.112) and (2.113), and we consider explicitly minority carrier densities near the edges x_n and $-x_p$ of the depletion layer. We obtain

$$e \frac{dp_n}{dt} - eD_p \frac{d^2 p_n}{dx^2} = -e\left(\frac{p_n - p_{n0}}{\tau_p}\right) \tag{2.117}$$

$$(-e) \frac{dn_p}{dt} + eD_n \frac{d^2 n_p}{dx^2} = +e\left(\frac{n_p - n_{p0}}{\tau_n}\right) \tag{2.118}$$

In (2.117) and (2.118), p_n is the minority hole density in the neutral n region, p_{n0} is the equilibrium minority hole density in the neutral n region, n_p is the minority electron density in the neutral p region, n_{p0} is the equilibrium minority electron density in the neutral p region, and τ_p and τ_n are the minority hole and minority electron lifetimes, respectively.

In the steady state, the minority carrier densities are constant in time, so $dp_n/dt = 0$ and $dn_p/dt = 0$, and (2.117) and (2.118) become

$$\frac{d^2 p_n}{dx^2} - \frac{p_n - p_{n0}}{L_p{}^2} = 0 \tag{2.119}$$

$$\frac{d^2 n_p}{dx^2} - \frac{n_p - n_{p0}}{L_n{}^2} = 0 \tag{2.120}$$

Equations (2.119) and (2.120) describe the time-independent steady state minority carrier densities $p_n(x)$ and $n_p(x)$ which vary with position in the neutral regions. We see that, in this model which neglects drift currents in the neutral regions, the minority carrier currents at the edges of the neutral regions are taken as purely diffusive[16] in nature.

We next make the assumption[23] that there is no generation or recombination of carriers in the depletion layer. We are therefore considering an approximation[22] in which the electron and hole currents are constant across the depletion layer. Hence, if we find the electron or hole current at *any* point of the depletion layer, we know it at every point of the depletion layer. This feature of the model makes the calculation of the total current J through the junction relatively easy.[23] At any point x,

$$J(x) = J_n(x) + J_p(x) \tag{2.121}$$

is the sum of the electron current $J_n(x)$ and the hole current $J_p(x)$ at that point. If we can calculate $J_n(x)$ and $J_p(x)$ at any point of junction, we know $J(x)$ at that point, and hence, at *every* point of the junction because the current through the junction is assumed uniform in space—J does *not* depend on x. In our model, at the edges x_n and $-x_p$ of the depletion layer, the only currents are minority carrier diffusion currents because we are neglecting all drift currents there. The result is that a calculation of the minority carrier diffusion currents at, say $x = x_n$, gives us the total current $J(x_n)$ at $x = x_n$, where

$$J(x_n) = J_n(x_n) + J_p(x_n) \tag{2.122}$$

Then as discussed above, the total current J through the junction is equal to $J(x_n)$, so we have

$$J = J_n(x_n) + J_p(x_n) \tag{2.123}$$

How can we calculate $J_n(x_n)$ and $J_p(x_n)$? We can find $J_p(x_n)$ by solving (2.119) for the minority hole density in the neutral n region, which includes the point $x = x_n$. We recall that equation (2.82), describing the steady state diffusion and recombination of holes from the point $x = x_1$, had the solution

$$p(x) - p_0 = [p(x_1) - p_0] \exp[-(x - x_1)/L_p] \tag{2.88}$$

where $p(x)$ is the hole density at point x, and p_0 is the equilibrium hole density. We apply the solution (2.88) to our problem, the solution of (2.119), simply by substituting p_n for $p(x)$, p_{no} for p_0, and x_n for x_1. We obtain

$$p_n(x) - p_{no} = [p_n(x_n) - p_{no}] \exp[-(x - x_n)/L_p] \tag{2.124}$$

as the equation giving the space-dependent minority hole density $p_n(x)$ in the neutral n region $x_n \le x \le +\infty$ of the junction. The minority hole diffusion current J_p is found from equation (2.116), so

$$J_p(x) = -eD_p \frac{dp_n}{dx}$$

and

$$J_p(x) = -eD_p[p_n(x_n) - p_{n0}](-1/L_p)\exp[-(x - x_n)/L_p] \qquad (2.125)$$

so

$$J_p(x_n) = e(D_p/L_p)[p_n(x_n) - p_{n0}] \qquad (2.126)$$

The quantity $p_n(x_n) - p_{n0}$ in equation (2.126) is just the excess minority hole density at the point x_n. However, from equation (2.104), we have

$$p_n(x_n) = p_{n0}\exp(eV_a/k_BT) \qquad (2.104)$$

where V_a is the potential applied to the junction. Substituting (2.104) into (2.126) gives

$$J_p(x_n) = e(D_p/L_p)p_{n0}[\exp(eV_a/k_BT) - 1] \qquad (2.127)$$

an expression which gives us the minority hole current at x_n in terms of the applied voltage V_a.

Next, equation (2.91) describes the diffusion and recombination of excess electrons created at the point $x = x_1$,

$$n(x) - n_0 = [n(x_1) - n_0]\exp[-(x - x_1)/L_n] \qquad (2.91)$$

where x_1 was assumed positive. If x and x_1 are negative, equation (2.91) becomes

$$n(x) - n_0 = [n(x_1) - n_0]\exp[(x - x_1)/L_n]$$

which, rewritten for the minority electron density $n_p(x)$ created at the edge $x_1 = -x_p$ of the neutral p region, becomes

$$n_p(x) - n_{p0} = [n_p(-x_p) - n_{p0}]\exp[(x + x_p)/L_n] \qquad (2.128)$$

Equation (2.128) describes the spatially dependent steady state minority electron density $n_p(x)$ in the neutral p region $-x_p \geq x \geq -\infty$ of the junction. The minority electron current $J_n(x)$ is given by (2.115),

$$J_n(x) = eD_n(dn_p/dx)$$

so, differentiating (2.128), we find that

$$J_n(x) = eD_n[n_p(-x_p) - n_{p0}](1/L_n)\exp[(x + x_p)/L_n] \qquad (2.129)$$

and, on evaluating $J_n(x)$ at $x = -x_p$, we have

$$J_n(-x_p) = e(D_n/L_n)[n_p(-x_p) - n_{po}] \qquad (2.130)$$

The quantity $n_p(-x_p) - n_{po}$ in equation (2.130) is the excess minority electron density at the point $-x_p$. However, from equation (2.103), we have

$$n_p(-x_p) = n_{po} \exp(eV_a/k_BT) \qquad (2.103)$$

where, again, V_a is the applied potential. Substituting (2.103) into (2.130) gives

$$J_n(-x_p) = e(D_n/L_n)n_{po}[\exp(eV_a/k_BT) - 1] \qquad (2.131)$$

for the minority electron current at the edge $-x_p$ of the neutral p region.

We now have, in equation (2.127), the term $J_p(x_n)$ in the expression (2.123) for the total current J through the junction. How do we find $J_n(x_n)$, the other term in the equation for J, when what (2.131) tells us is $J_n(-x_p)$, the electron current at a different point of the depletion layer? Our assumption that the electron and hole currents are constant across the depletion layer allows us to find $J_n(x_n)$. This is because this assumption says

$$J_n(x_n) = J_n(-x_p) \qquad (2.132)$$

since the electron current at one point $x = x_n$ of the depletion layer must be the same as the electron current at another point $x = -x_p$. We have then

$$J_n(x_n) = e(D_n/L_n)n_{po}[\exp(eV_a/k_BT) - 1] \qquad (2.133)$$

Substituting equations (2.133) for the electron current at x_n and (2.127) for the hole current at x_n into equation (2.123) for the total current J gives

$$J = e\left(\frac{D_n}{L_n}n_{po} + \frac{D_p}{L_p}p_{no}\right)[\exp(eV_a/k_BT) - 1] \qquad (2.134)$$

Equation (2.134) is the celebrated Shockley diode equation which gives the total current density J through the ideal abrupt p–n junction as a function of the applied potential V_a and the various semiconductor parameters of the n and p sides of the junction.

We see from (2.134) that, for magnitudes of reverse bias (V_a negative) such that $|V_a| \gg k_BT/e$, the exponential term becomes very small, and the current density J approaches the value $-J$(generation) defined by

$$J(\text{generation}) \equiv e\left(\frac{D_n}{L_n}n_{po} + \frac{D_p}{L_p}p_{no}\right) \qquad (2.135)$$

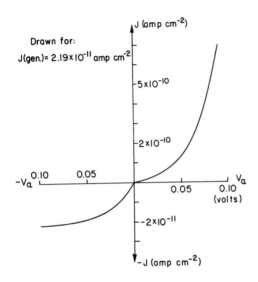

Figure 2.29. Plot of current density J as a function of applied voltage V_a from equation (2.134), drawn for J (generation) = 2.19 $\times 10^{-11}$ A cm^{-2}. Forward bias corresponds to positive values of V_a and the direction of J (generation) is chosen as negative. Note the different scales for positive and negative values of current density J.

The magnitude of J(generation) defined by equation (2.135) is the reverse saturation current density of the p–n junction.

Figure 2.29 shows a plot of equation (2.134) using representative values[†] of the semiconductor parameters for silicon. As discussed above, the applied potential V_a is taken as positive for forward bias and negative for reverse bias. The direction of the current density J(generation), defined by equation (2.135), is chosen as negative. [A typical value of J(generation) is calculated in one of the problems at the end of this chapter.] From the figure, we see that the current density J increases rapidly with increasing magnitude of the forward bias and approaches saturation at a value equal to the reverse saturation current for increasing magnitude of the reverse bias. For forward bias voltages V_a such that $V_a \gg k_B T/e$, the exponential dominates in (2.134) and the current density increases exponentially with increasing magnitude of the forward bias voltage.

From equation (2.135) for the reverse saturation current density J(generation), we see that J(generation) is directly proportional to the

[†] The values used for silicon were $N_a = N_d = 10^{16}$ cm^{-3}, with $n_i^2 = 4.6 \times 10^{19}$ cm^{-6} at 300 K, leading to $p_{n0} = n_{p0} = 4.6 \times 10^3$ cm^{-3}. Values of $D_n = 33.8$ cm^2 sec^{-1} and $D_p = 13.0$ cm^2 sec^{-1} were calculated from the mobility values $\mu_n = 1300$ cm^2 v^{-1} sec^{-1} and $\mu_p = 500$ cm^2 v^{-1} sec^{-1}. Diffusion lengths $L_n = 1.84 \times 10^{-3}$ cm and $L_p = 1.14 \times 10^{-3}$ cm were calculated using an assumed value of 10^{-7} sec for the electron and hole minority carrier lifetimes. The calculated reverse saturation current density of about 2×10^{-11} A/cm^2 is the magnitude of the values observed in silicon diodes at room temperature. (See A. S. Grove, Reference 4, page 178.)

carrier diffusion coefficients D_n and D_p, and also to the equilibrium minority carrier concentrations n_{p0} and p_{n0}. If N_d and N_a are the concentrations of ionized donors and acceptors, respectively, on the two sides of the junction, then we have

$$p_{n0} = n_i^2/N_d, \qquad n_{p0} = n_i^2/N_a \qquad (2.136)$$

From (2.136), we conclude that the larger are N_d and N_a, the smaller is the value of J(generation), all other things being equal. Since J(generation) is inversely proportional to both L_n and L_p, and L_n and L_p vary, respectively, as the square root of the minority carrier lifetimes τ_n and τ_p, the reverse saturation current J(generation) will vary as the $-\frac{1}{2}$ power of the minority carrier lifetimes, again assuming that all other parameters are held constant.

At this point, it is worthwhile to reiterate that the diode equation (2.134) was derived on the basis of an idealized model of a p–n junction. This model assumed (1) an ideal abrupt junction, in which (2) the electric field is confined to the depletion layer, (3) the injected minority carrier densities are small, and (4) there is no generation or recombination of carriers in the depletion layer. This idealized model does not describe real p–n junctions, and for a discussion of the validity of and deviations from this ideal model, the reader is referred to the literature, particularly the books by Moll[8] and by Streetman.[22] However, our discussion of the physics of the p–n junction will be confined to this idealized model.

Majority and Minority Carrier Components of the Junction Current

We calculated the total diode current J given in (2.134) by finding the minority carrier diffusion currents (2.133) and (2.127) at one point of the junction, and adding them. We neglected drift currents of minority carriers due to their small concentrations and neglected drift currents of majority carriers outside the junction because of our assumption that the electric field is confined to the space charge layer of the junction.

As remarked earlier, this assumption about the electric field is not really true because majority carriers in, say, the neutral n region must drift in from the external contact attached to the source of applied potential. These majority electrons fulfill three purposes[24] in the neutral n region of the junction. First, they recombine with the minority holes injected into the n region, thereby giving the familiar decrease in minority hole density with increasing distance into the neutral n region. Second, they provide space charge neutrality for the positive charges of the injected hole distribution.

Third, they supply the electrons for injection into the neutral p region as minority carriers there. Similar statements may be made concerning majority holes in the neutral p region.

We now wish to calculate[25] the majority and minority carrier components of the current in the two neutral regions and in the depletion layer. Consider first the n-type neutral region from $x = x_n$ to $x = +\infty$. The total current J is given by the diode equation (2.134), written in the form

$$J = J_{n0} + J_{p0} \tag{2.137}$$

where the convenient quantities J_{n0} and J_{p0} are defined by

$$J_{n0} = e(D_n/L_n)n_{p0}[\exp(eV_a/k_BT) - 1] \tag{2.138}$$

and

$$J_{p0} = e(D_p/L_p)p_{n0}[\exp(eV_a/k_BT) - 1] \tag{2.139}$$

We know also, from equation (2.125), that the minority hole current density $J_p(x)$ in the neutral n region is

$$J_p(x) = e(D_p/L_p)[p_n(x_n) - p_{n0}]\exp[-(x - x_n)/L_p] \tag{2.125}$$

where also, using equation (2.104),

$$p_n(x_n) = p_{n0}\exp(eV_a/k_BT) \tag{2.104}$$

we obtain

$$J_p(x) = e(D_p/L_p)p_{n0}[\exp(eV_a/k_BT) - 1]\exp[-(x - x_n)/L_p] \tag{2.140}$$

Equation (2.140) may be rewritten, using (2.139), as

$$J_p(x) = J_{p0}\exp[-(x - x_n)/L_p] \tag{2.141}$$

an equation that gives the space dependence of the steady state minority hole density in the neutral n region, where $x \geq x_n$.

With these results, we can find the *majority* electron current density $J_n(x)$ in the neutral n region because it must be true that

$$J_{n0} + J_{p0} = J = J_n(x) + J_p(x) \tag{2.142}$$

because the total current density J must be equal the sum of the minority and majority current densities. Substituting (2.141) for $J_p(x)$ into (2.142) gives

$$J_n(x) = J_{n0} + J_{p0} - J_{p0}\exp[-(x - x_n)/L_p]$$
$$J_n(x) = J_{n0} + J_{p0}\{1 - \exp[-(x - x_n)/L_p]\} \tag{2.143}$$

as the space dependence of the majority electron current density in the neutral n region. Together, equations (2.141) and (2.143) describe the space dependence of the minority and majority current densities in the neutral n-type region of the junction.

In the same way, we can obtain analogous expressions for the minority and majority carrier current densities in the neutral p-type region extending from $x = -x_p$ to $x = -\infty$. The results are that the minority electron current density is

$$J_n(x) = J_{n0} \exp[(x + x_p)/L_n] \tag{2.144}$$

and the majority hole current density is

$$J_p(x) = J_{p0} + J_{n0}\{1 - \exp[(x + x_p)/L_n]\} \tag{2.145}$$

Equations (2.145) and (2.144) are the analogs of equations (2.143) and (2.141), respectively, and give the spatial dependence of the currents in the neutral p region.

To find the electron current density *within* the depletion layer $-x_p \leq x \leq x_n$, we evaluate equation (2.143) at $x = x_n$, obtaining

$$J_n(x_n) = J_{n0} \tag{2.146}$$

and evaluating equation (2.144) at $x = -x_p$, obtaining

$$J_n(-x_p) = J_{n0} \tag{2.147}$$

Equations (2.146) and (2.147) tell us that the electron current density is constant across the depletion layer, an assumption built into the theory when we assumed no carrier generation or recombination in the depletion layer. Similarly, equations (2.141) and (2.145) give us

$$J_p(x_n) = J_{p0} \tag{2.148}$$

$$J_p(-x_p) = J_{p0} \tag{2.149}$$

exhibiting the constancy of the hole current across the depletion layer, also as required by our assumptions.

Equations (2.141), and (2.143)–(2.149) describe the spatial dependence of the electron and hole currents in the entire junction, composed of the depletion layer and two neutral regions. These results are exhibited in Figure 2.30, showing the total current $J = J_n(x) + J_p(x)$ through the junction, and its election component $J_n(x)$ and hole component $J_p(x)$ as func-

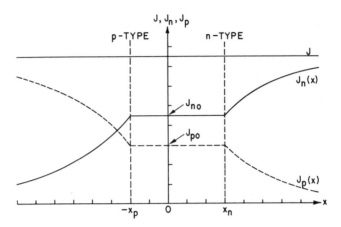

Figure 2.30. The total current $J = J_n(x) + J_p(x)$, where $J_n(x)$ is the electron current given by equations (2.143) and (2.144), and $J_p(x)$ is the hole current given by (2.141) and (2.145). The quantities J_{no} and J_{po} are given by (2.138) and (2.139), and are the values, assumed constant, of the electron and hole currents in the depletion layer of the junction. The figure has been drawn for $J_{no} = 1.5J_{po}$, and $L_n = 1.3L_p$, and where, for clarity and convenience, $L_n = |x_n|$.

tions of distance x, assuming that there is no recombination or generation of carriers in the space charge region. Far from the junction, at large values of $|x|$, the current will be entirely majority carriers, electrons on the n side and holes on the p side, and both will be drift currents driven by the small electric field maintained in the neutral regions. Nearer the edges of the space charge layer, on either side of the junction, there will be a minority carrier diffusion current and, for space charge neutrality, an accompaning diffusion current of majority carriers. At the edges of the space charge layer, majority carriers will be injected across the space charge region and into the opposite side of the junction, where they become minority carriers.

Summary of the Basic Physics of the *p–n* Junction

First, the fact that semiconductors like silicon can exist in both n and p types leads to the idea of a $p–n$ junction. The simultaneous existence of electrons and holes in a semiconductor leads to currents, both diffusion and drift, of the two types of carriers.

Second, the difference in the Fermi level between the two sides of the junction leads to a difference in the electrostatic potential energy of an electron on the two sides of the junction. The fact that the magnitude of this energy barrier can be modulated by an applied potential results in a modulation of the diffusion currents in the junction.

Third, the small (and controllable, through doping) density of minority carriers on each side of the junction leads to a small reverse drift current which is essentially independent of the magnitude of the applied potential.

Fourth, the difference in the effect of an applied potential on the diffusion and drift currents in the junction leads to the great asymmetry of the current–voltage characteristic for forward and reverse bias. It is this asymmetric current–voltage characteristic that makes the *p-n* junction a useful device.

Reverse Breakdown in *p-n* Junctions

When the value of the reverse bias applied reaches a critical value, the magnitude of the reverse current increases sharply, as shown schematically in Figure 2.31. This figure indicates the increase in reverse current at the critical value V_B of the reverse voltage applied to the junction. This effect is called reverse breakdown[26,27] of the *p-n* junction. There are two different mechanisms that produce breakdown.

The first process is called Zener breakdown and is a quantum mechanical tunneling phenomenon. Figure 2.32 shows the band diagram of a *p-n* junction at equilibrium and under an applied reverse bias, showing that reverse bias lowers electron energies on the *n* side relative to electron energies on the *p* side. As shown in Figure 2.32b, there are, under sufficient reverse bias, filled electron states in the valence band on the *p* side at the same energy as empty conduction band states on the *n* side. This situation means that the possibility exists of an electron tunneling from the *p* side to the *n* side. If the width of the energy barrier (through which tunneling takes place) is denoted by *d* in Figure 2.32b, it may be shown[28] that *d*

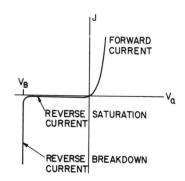

Figure 2.31. Reverse breakdown (schematic) in a *p-n* junction current–voltage characteristic. V_B is the critical breakdown voltage.

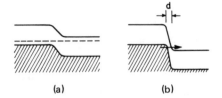

Figure 2.32. Band diagrams for a *p–n* junction (a) at equilibrium, and (b) under reverse bias. In (b), electron tunneling from *p* to *n* is indicated schematically by the arrow, and *d* is the width of the barrier to tunneling.

decreases with increasing magnitude of the applied reverse bias, assuming that the impurity doping is high enough that the width of the space charge region is narrow to begin with. As the magnitude of the reverse bias increases, d decreases and the magnitude of the electric field in the junction increases. The tunnel current of electrons from p to n increases with increasing electric field[29] and with decreasing barrier width, so, as the reverse bias approaches the critical value V_B, a significant tunnel current begins to flow. This flow of electrons from p to n constitutes a reverse current through the junction (just like the reverse current of thermally excited carriers making up the reverse saturation current). At the value V_B of reverse bias, the tunnel current dominates the ordinary J(generation) reverse saturation current, and the reverse breakdown shown in Figure 2.31 takes place.

The value of V_B for a particular junction will depend on the details of the junction type, doping, etc., but the critical value[30] of the electric field in silicon is about 10^6 V cm^{-1} for junction doping values in the 10^{17}–10^{19} cm^{-3} range. This critical field value corresponds to critical breakdown voltages V_B in the neighborhood of a few volts. It should be kept in mind, however, that Zener tunneling will take place only in narrow junctions with high doping.

A second process which can produce reverse breakdown in a *p–n* junction is called avalanche breakdown.[31] In this mechanism conduction electrons are accelerated by the electric field in the junction, thereby gaining kinetic energy. This kinetic energy may be large enough to produce free carriers by impact ionization. In this process, an electron with a kinetic energy at least as large as the energy gap in the semiconductor excites an electron from the valence band to the conduction band, thereby producing an electron–hole pair. These newly generated free carriers can also be accelerated by the electric field, producing more electron–hole pairs by impact ionization. The resulting multiple production of free carriers is called an avalanche and sharply increases the current through the junction, as shown in Figure 2.31.

The avalanche mechanism will produce reverse breakdown in the junction if the Zener tunneling mechanism is not dominant because the junction is not narrow enough for easy tunneling. Generally speaking, if Zener breakdown does not occur with a reverse bias of a few volts, avalanche breakdown will become dominant. Again, the value of the critical voltage for avalanche breakdown depends on the details of the junction.[31]

Other Topics on p–n Junctions

We have touched only on some of the main points of the physics of p–n junctions. Many other interesting topics are to be found in the literature. We have considered current flow in the case of low-level injection of minority carriers; the case of high injection levels, in which the density of minority carriers is comparable to that of the majority carriers, is discussed by Sze[32] and by Shockley.[33] We have discussed only the idealized abrupt junction; other, more realistic types of junction are treated by Moll,[34] Grove,[35] and Streetman.[36] The validity of the approximations made in the idealized model is discussed by Moll.[22] The high-resistivity depletion layer between the more conductive neutral regions has a capacitance associated with it. This topic is treated by, among other authors, Streetman,[37] as are questions of the response of the junction to transient and AC conditions. It is, however, useful to note here an important question. This is the time constant of the response[38–40] of a p–n junction to a sudden change of bias from forward to reverse. If the junction is forward biased, then minority carriers have been injected across the junction, giving the steady state excess carrier density distribution shown in Figure 2.19. If the bias voltage is suddenly switched from forward to reverse, the current through the diode will not immediately fall to its reverse-voltage value. The reason is that it takes a nonzero time interval for the excess minority carrier density to decrease from its large positive value under forward bias to its small negative value under reverse bias. This time interval is called the minority carrier storage time, and limits the speed with which the junction can be switched from a condition of forward bias to one of reverse bias.

Problems

2.1. *Depletion Layer Width.* To get an idea of the magnitudes involved, calculate the width of the depletion layer in an ideal abrupt silicon p–n junction at 300 K. Assume that both the n and p sides contain 10^{15} ionized impurities per cm³. Take $np \equiv n_i^2 = 2.2 \times 10^{18}$ cm⁻⁶ and the dielectric constant equal to 11.7. Check to be sure that your answer has the correct units.

2.2. *Hole Currents in a Junction.* Make a simple band diagram of a p–n junction at equilibrium, showing schematically the diffusion and generation particle fluxes of holes.

2.3. *Currents in a Biased Junction.* Make sketches of the band diagram of a p–n junction under large ($|V_a| \gg k_B T/e$) forward and reverse bias. On each diagram, indicate with arrows the directions and relative magnitudes of all of the particle fluxes and all of the current densities involved. For the generation currents, show where the carriers are generated.

2.4. *Injected Hole Density.* Given a silicon p–n junction at 300 K in which the density of extrinsic carriers is 10^{15} cm^{-3} on either side of the junction, calculate the injected increase in minority holes on the n side for a forward bias of 0.6 V.

2.5. *Junction at Low Temperatures.* Given a silicon p–n junction, for which the total densities of donors and of acceptors are 10^{15} cm^{-3}, maintained at 30 K, discuss what current–voltage characteristic you would expect to observe.

2.6. *Carrier Densities in p Region.* Derive equations (2.144) and (2.145) for the minority electron and majority hole current densities in the neutral p region.

2.7. *Reverse Saturation Current Density.* Calculate the reverse saturation current density of a silicon p–n junction (using the ideal abrupt model). Assume that the doping on both sides of the junction is 10^{16} ionized impurities per cm^3, and that the lifetime of both carriers is 0.1 μsec. Note that carrier diffusion coefficients can be calculated from mobility values, which are $\mu_n = 1300$ cm^2 V^{-1} sec^{-1}, $\mu_p = 500$ cm^2 V^{-1} sec^{-1}. Compare your result with values observed for silicon diodes at room temperature. (See A. S. Grove, Reference 4, page 178.)

References and Comments

1. C. Kittel, *Introduction to Solid State Physics*, Sixth Edition, John Wiley, New York (1986), page 534.
2. See, for example, C. Kittel and H. Kroemer, *Thermal Physics*, Second Edition, Freeman (1980), page 144.
3. B. G. Streetman, *Solid State Electronic Devices*, Second Edition, Prentice-Hall, Englewood Cliffs (1980), Section 5.2.
4. A. S. Grove, *Physics and Technology of Semiconductor Devices*, John Wiley, New York (1967), pages 149–161.
5. C. Kittel, Reference 2, page 399.
6. C. Kittel, *Introduction to Solid State Physics*, Third Edition, John Wiley, New York (1966), page 323.
7. S. M. Sze, *Physics of Semiconductor Devices*, Second Edition, John Wiley, New York (1981), page 51, equations (93a) and (94a).
8. J. L. Moll, *Physics of Semiconductors*, McGraw-Hill, New York (1964), Chapter 7, especially pages 111, 117, 121–123, and 133–140.
9. B. G. Streetman, Reference 3, pages 183–184.
10. R. A. Smith, *Semiconductors*, Second Edition, Cambridge University Press, New York (1978), page 206.

11. B. G. Streetman, Reference 3, page 145.
12. The important subject of the technology of $p-n$ junction fabrication will, unfortunately, not be discussed here, See, for example, S. M. Sze, Reference 7 pages 64–73; A. S. Grove, Reference 4, Chapters 1, 2, 3; B. G. Streetman, Reference 3, pages 126–136.
13. S. Wang, *Solid State Electronics*, McGraw-Hill, New York (1966), pages 272–289.
14. S. M. Sze, Reference 7, pages 50–57.
15. See, for example, J. C. Slater and N. H. Frank, *Introduction to Theoretical Physics*, McGraw-Hill, New York (1933), page 187.
16. W. Shockley, *Electrons and Holes in Semiconductors*, Van Nostrand, New York (1950), page 313, equations (13) and (14).
17. W. Shockley, Reference 16, page 298.
18. N. W. Ashcroft and N. D. Mermin, *Solid State Physics*, Saunders (1976), page 603.
19. W. R. Beam, *Electronics of Solids*, McGraw-Hill, New York (1965), pages 190–200.
20. B. G. Streetman, Reference 3, page 119.
21. J. L. Moll, Reference 8, page 106.
22. J. L. Moll, Reference 8, pages 117, 133–140, discusses the validity of this ideal model, as does B. G. Streetman, Reference 3, pages 176–184.
23. N. W. Ashcroft and N. D. Mermin, Reference 18, Chapter 29, page 609, give a clear discussion.
24. B. G. Streetman, Reference 3, Section 5.3.3.
25. A van der Ziel, *Solid State Physical Electronics*, Second Edition, Prentice-Hall, New York (1968), page 310.
26. S. M. Sze, Reference 7, Section 2.5.
27. B. G. Streetman, Reference 3, Section 5.4.
28. See, for example, B. G. Streetman, Reference 3, page 194, Problem 5-13.
29. S. M. Sze, Reference 7, page 98, equation (68).
30. A. S. Grove, Reference 4, page 193, Figure 6.27.
31. See, for example, B. G. Streetman, Reference 3, Section 5.4.2; S. M. Sze, Reference 7, Section 2.5.3.
32. S. M. Sze, Reference 7, Section 2.4.4.
33. W. Shockley, Reference 16, pages 328–333.
34. J. L. Moll, Reference 8, pages 121–123.
35. A. S. Grove, Reference 4, Chapter 6.
36. B. G. Streetman, Reference 3, Section 5.6.4.
37. B. G. Streetman, Reference 3, Section 5.5.3.
38. J. Millman and C. C. Halkias, *Integrated Electronics*, McGraw-Hill, New York (1972), pages 71–73.
39. S. Wang, Reference 13, pages 321–331.
40. R. S. Muller and T. I. Kamins, *Device Electronics for Integrated Circuits*, Second Edition, John Wiley, New York (1986), Section 5.4.

Suggested Reading

B. G. Streetman, *Solid State Electronic Devices*, Second Edition, Prentice-Hall, New York (1980). Chapter 5 of this text discusses $p-n$ junctions, and our treatment draws heavily on Streetman's Sections 5.2 and 5.3. This treatment of the $p-n$ junction is, in the author's opinion, the best available at the introductory level.

S. WANG, *Solid State Electronics*, Mc Graw-Hill, New York (1966). Chapter 5 of this advanced level text discusses the transport and recombination of excess carriers in semiconductors, while Chapter 6 discusses the physics of the *p-n* junction.

A. S. GROVE, *Physics and Technology of Semiconductor Devices*, John Wiley, New York (1967). Chapter 6 of this useful and practical book discusses, with many examples, the physics of the *p-n* junction.

J. L. MOLL, *Physics of Semiconductors*, McGraw-Hill, New York (1964). Chapter 6 of this text discusses nonequilibrium carrier densities, and Chapter 7 treats the *p-n* junction with special attention to the approximations made.

A. VAN DER ZIEL, *Solid State Physical Electronics*, Second Edition, Prentice-Hall, New York (1968). Sections 15.1 and 15.2 discuss the *p-n* junction in a terse and mathematical style.

S. M. SZE, *Physics of Semiconductor Devices*, Second Edition, John Wiley, New York (1981). Sections 1.7 and 2.1–2.4 of this large advanced treatise treat the mathematics of the *p-n* junction in detail, but without much discussion of the physics.

N. W. ASHCROFT AND N. D. MERMIN, *Solid State Physics*, Holt, Rinehart, and Winston, New York (1976). This textbook of solid state physics at the advanced level includes, in Chapter 29, a careful discussion of the physics of the *p-n* junction.

C. M. WOLFE, N. HOLONYAK, AND G. E. STILLMAN, *Physical Properties of Semiconductors*, Prentice-Hall, Englewood Cliffs (1989). Our discussion in this chapter has been concerned exclusively with the semiconductor *p-n* homojunction, in which the material is the same on each side of the junction. Of increasing interest is the *p-n* heterojunction, in which the materials forming the junction are different, as in, for example, a GaAs–AlAs heterojunction. This graduate-level text includes a chapter on this topic.

W. SHOCKLEY, *Electrons and Holes in Semiconductors*, Van Nostrand, New York (1950). This is the book that began it all, written by one of the pioneers of semiconductor physics.

3

Semiconductor *p–n* Junction Devices

Introduction

The aim of this chapter is a discussion of the physics of a number of devices based on the semiconductor *p–n* junction. First is the *p–n* junction itself, used as a diode. Second is the important bipolar junction transistor. The emphasis is on a qualitative discussion of current amplification in this device, followed by a brief calculation of the currents involved. The tunnel diode, or degenerate diode, is treated next in a qualitative manner which obtains the negative dynamic resistance characteristic of this device. Finally, the junction field effect transistor is discussed, also qualitatively, and its function as a voltage-controlled amplifier is treated.

Semiconductor *p–n* Junction Diodes

Consideration of the highly asymmetric current–voltage characteristic shown in Figures 2.28 and 2.29 of Chapter 2 and expressed by equation (2.134) suggests the use of the semiconductor *p–n* junction as a diode. For use as a diode,[1,2] either as a rectifier or in other applications, the reverse saturation current of the diode should be small. From equation (2.135)

$$J(\text{generation}) \equiv e\left(\frac{D_n}{L_n}\, n_{p0} + \frac{D_p}{L_p}\, p_{n0}\right) \tag{2.135}$$

we can see the factors[†] that govern the magnitude of the reverse saturation current J(generation). First, we see that small values of the equilibrium minority carrier densities n_{p0} and p_{n0} will favor a small magnitude of J(generation). We recall from equation (2.136) that

$$p_{n0} = n_i^2/N_d, \qquad n_{p0} = n_i^2/N_a \tag{3.1}$$

where the intrinsic constant n_i^2 is the product of the intrinsic electron and hole densities defined by equation (1.31). Therefore

$$n_i^2 \equiv 4(k_B T/2\pi\hbar^2)^3(m_e^* m_h^*) \exp(-E_g/k_B T) \tag{3.2}$$

where E_g is the semiconductor energy gap and m_e^* and m_h^* are, respectively, the electron and hole effective mass values. From equations (3.1) and (3.2) we see that a large semiconductor energy gap E_g will favor small values of the minority carrier densities and hence will favor small values of the reverse saturation current for given values of the impurity densities N_d and N_a and for a given value of the temperature. For this reason, silicon ($E_g = 1.11$ eV at 300 K) p–n junctions have a smaller reverse saturation current than germanium ($E_g = 0.67$ eV at 300 K) p–n junctions and hence are preferable as diodes.

We note also from (2.135) that small values of the ratio D/L for both electrons and holes favor small values of J(generation). Since $L^2 = D\tau$ from equation (2.83), and using the Einstein relation[(3)]

$$D = (k_B T/e)\mu \tag{3.3}$$

between the carrier mobility μ and its diffusion coefficient D, we find that

$$D/L = (k_B T/e)^{1/2}(\mu/\tau)^{1/2} \tag{3.4}$$

for both electrons and holes. Equation (3.4) tells us that small carrier mobilities and large minority carrier lifetimes will favor small values of D/L and hence small values of the reverse saturation current density. As discussed in Chapter 1, the mobility will be inversely proportional to the carrier effective masses, and effective masses are usually larger in semiconductors with larger energy gaps. Thus we see that a large value of E_g

---

[†] As pointed out in Chapter 2, equation (2.135) was derived using an idealized model of a *p–n* junction which neglected, among other factors, the generation of carriers in the space charge region. For this reason, equation (2.135) is not strictly true for silicon diodes, in which such generation is important. See References 22 and 35 of Chapter 2 for a discussion.

generally means lower carrier mobilities and hence lower values of D/L. Further, as also discussed in Chapter 1, an indirect band gap will favor longer carrier lifetimes; in this case, the lifetime of interest is the minority carrier lifetime. Finally, lower impurity densities in the junction will also favor longer minority carrier lifetimes, but this gain would have to be weighed against the fact that lower impurity densities will, as seen from equation (2.135), increase the equilibrium minority carrier densities.

In summary, we see that a larger energy gap will favor a smaller value of the reverse saturation current density at a given temperature. For this reason, silicon is preferred to germanium for making rectifying semiconductor diodes, and is also preferred for high-temperature operation.

Diodes are also used in switching circuits, in which applications they are alternately forward and reverse biased. The speed with which they can assume a new bias condition will determine their performance. One of the main factors determining this speed is the phenomenon of minority carrier storage,[4] discussed briefly in Chapter 2.

Finally, it should be mentioned that the critical voltage V_B for Zener breakdown[5] can be quite reproducible and is nondestructive. Such diodes[6] are called Zener diodes, and the reproducible breakdown voltage V_B serves as a voltage reference in many circuits.

The Bipolar Junction Transistor

We will study the bipolar junction transistor (BJT) because it is an important active amplifying device and its treatment is based on the p–n junction physics we developed in Chapter 2.

Consider first a p–n junction in which the p side is much more heavily doped than the n side, so, if N_a and N_d are the densities of ionized acceptors and donors, respectively, we are considering the situation for which

$$N_a \gg N_d$$

Since the intrinsic constant n_i^2 is a constant at a given temperature, the equilibrium minority hole density p_{n0} and minority electron density n_{p0} are, from equation (3.1), such that

$$p_{n0} \gg n_{p0} \tag{3.5}$$

meaning that there are more minority holes on the n side of the junction than there are minority electrons on the p side. In this case, then, the current density J, given by

$$J = e\left(\frac{D_p}{L_p} p_{no} + \frac{D_n}{L_n} n_{po}\right)(e^{eV_{a'}/k_BT} - 1) \qquad (3.6)$$

becomes, because of the condition (3.5),

$$J \cong e\left(\frac{D_p}{L_p} p_{no}\right)(e^{eV_a/k_BT} - 1) \qquad (3.7)$$

Equation (3.7) tells us that, in this kind of a junction, called a p^+n junction, the bulk of the injection current is carried by holes because, from (3.5), the density of minority electrons is negligibly low. (The symbol "+" in p^+n means that the p side of the junction is much more heavily doped than the n side.) Further, we can see, from equations (2.70) and (2.71),

$$x_p = \frac{W}{1 + N_a/N_d} \qquad (2.70)$$

$$x_n = \frac{W}{1 + N_d/N_a} \qquad (2.71)$$

where $W = x_n + x_p$ is the total width of the space charge region, that $x_n \cong W$ and $x_p \cong 0$ because $N_a \gg N_d$. As expected then, the space charge region of the p^+n junction is almost entirely in the n-type side of the junction.

From these considerations, we see that a forward-biased p^+n junction is a good way to inject minority holes into n-type material because equation (3.7) tells us that, in such a junction, many more holes are injected from p^+ to n than electrons are injected from n to p^+.

Suppose we now mentally construct a p–n–p junction transistor by taking a forward-biased p^+n junction and coupling to it a reverse-biased pn junction. The result is shown in Figure 3.1. The forward-biased p^+n junction, labeled E, is called the emitter junction, while the reverse-biased np junction, labeled C, is called the collector junction. The n-type region between the two junctions is called the base (labeled B) of the transistor. Figure 3.2 shows the band diagram of the pnp transistor with no external

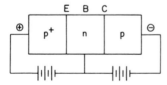

Figure 3.1. Schematic representation of a *pnp* junction transistor. The forward-biased p^+n emitter junction is labeled E, the reverse-biased np collector junction is labeled C, and the base is labeled B.

Figure 3.2. Band diagram of a *pnp* junction
transistor with no applied bias. The shaded
regions are filled valence band states, and
electron energy increases upward. The Fermi level is denoted by E_F, and the junction
labels are as in Figure 3.1.

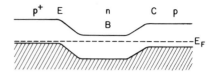

voltages applied to it, in the equilibrium state, where E, B, and C refer,
respectively, to the emitter junction, base region, and collector junction.
Figure 3.3 shows the same band diagram with forward bias applied to the
emitter junction and reverse bias applied to the collector junction. Since
the p^+n emitter junction E is forward biased, the energy barrier, between
p^+ and n, for holes has decreased. Since the np collector junction C is
reverse biased, the energy barrier to the diffusion of holes from p to n has
increased. Figure 3.3 also shows, schematically, the injection of holes into
the base B by the forward-biased p^+n emitter junction. The built-in electric
field $\mathscr{E}$ at the collector junction is also shown.

We now consider the flow of particles in the *pnp* transistor; the directions of these particle flows are shown in Figure 3.4. In this figure, particle
flow (1) consists of holes injected into the base by the emitter. Particle
flow (2) shows the diffusion of these injected holes across the base region.
At the collector junction, the built-in electric field there sweeps these holes
into the collector, producing particle flow (3), as the holes fall down the
energy hill for holes shown in Figure 3.3. Some of the holes diffusing across
the base will recombine with majority electrons in the base; this is indicated
as the hole flow (4). This means electrons, shown as particle flow (5), are
supplied from the external circuit to replace those lost due to recombination
with holes. Finally, back at the p^+n emitter junction, there will be some
injection of electrons from n to p^+ even though the p^+ doping means that
most of the current in the emitter junction will be due to holes. The electrons
injected from n to p^+ are shown as particle flow (6) and are also replaced
from the external circuit. (We are neglecting the small reverse current in
the reverse-biased collector junction. This is composed of thermally generated electrons drifting from p to n and thermally generated holes drifting

Figure 3.3. Band diagram of the *pnp* junction
transistor with forward bias applied to the emitter junction E and reverse bias applied to the
collector junction C. The electric field at the
collector junction is $\mathscr{E}$, and the injection of
holes ($\bigcirc$) into the base by the emitter is indicated schematically by an arrow.

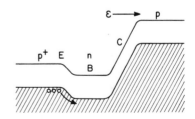

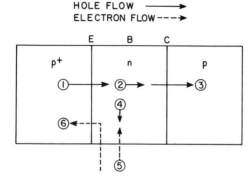

Figure 3.4. Particle flows in the biased *pnp* transistor. Hole flows are indicated by the solid arrows, and electron flows by the dashed arrows.

from *n* to *p*.) The same particle flows are shown on the transistor band diagram in Figure 3.5, in which the symbol ○ indicates a hole and the symbol ● indicates an electron, and where the particle flows are numbered as in Figure 3.4. The electric field $\mathscr{E}$ at the collector junction is also indicated.

We saw from Figure 3.4 that there is a flow of electrons into the base of the *pnp* transistor under bias. This flow comes from the external circuit and makes up the base current I_B in the transistor. [The direction of the current I_B is opposite to the direction of the electron flows (5) and (6) which constitute it.] The other currents are hole currents and are in the same direction as their respective particle flows. The hole flow (1) is the emitter hole current I_{E_p} and hole flow (3) is the collector hole current I_{C_p}. The emitter electron current I_{E_n} is due to electron flow (6) and is opposite in direction to that flow; I_{E_n} is thus in the same direction as the emitter hole current I_{E_p}. As mentioned earlier, we are neglecting the reverse current (both electrons and holes) at the collector junction.

In terms of these currents, we may redraw Figure 3.4 as shown in Figure 3.6. Also shown (at the "contacts" to the device, indicated by the symbol △) are the currents in the external circuit. The current I_E, where

$$I_E = I_{E_p} + I_{E_n} \tag{3.8}$$

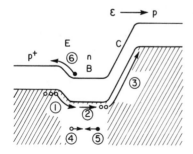

Figure 3.5. Band diagram indicating the particle flows in the biased *pnp* transistor of Figure 3.3. Holes are represented by ○, and electrons by ●; $\mathscr{E}$ is the electric field at the collector junction.

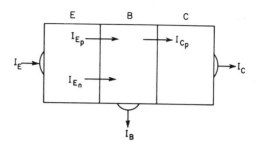

Figure 3.6. Current directions in the *pnp* transistor.

is the emitter current; the current I_C, where

$$I_C = I_{C_p} \qquad (3.9)$$

is the collector current; and I_B is the base current. The currents I_E, I_C, and I_B are the currents in the external circuit of the transistor.

Amplification in the Bipolar Transistor

Given the currents flowing in the *pnp* transistor, it is useful to discuss physically why the device acts as a current amplifier. First, however, some comments about amplification in general[7] are appropriate here.

Generally, an amplifier is a device that increases the amplitude of an AC signal. Since the laws of thermodynamics tells us that we cannot get more energy out of our device than we originally put in, what is really going on in an amplifier is the conversion of energy, put into the device as DC current, into an output AC current which is proportional to the input AC current. Since the device has effectively "multiplied" the input current, we speak of the amplifier as having current gain, which is, of course, the constant of proportionality between the output and input AC signals. There are two types of amplifying devices. In the first, the output current is controlled by the input current. Such a device, of which the bipolar junction transistor is an example, is a current-controlled amplifier. In the second type of amplifying device, the output current is controlled by the input voltage. This type of device is called a voltage-controlled amplifier; an example (which we will discuss later) is the field-effect transistor.

We now examine the physical reasons why the bipolar junction transistor is a current-controlled amplifier, in which a small base current I_B controls a larger collector current I_C. Since the collector current I_C is a flow of holes across the base, and the base current I_B is a flow of electrons into the base, we can see why the electron flow I_B will have an effect on the hole flow I_C. The basic idea[8] is the existence of electrical neutrality in the base region of the transistor. If we increase the flow of electrons into the base, the flow of holes from the emitter (and hence to the collector) will increase in order to maintain space charge neutrality. If the flow of electrons into the base is reduced, the hole flow from the emitter will decrease. We see, then, why the electron flow into the base (the base current I_B) modulates the hole flow from emitter to collector (the collector current I_C) and we have a device in which the output current I_C is controlled by the input current I_B. The device is thus a current-controlled amplifier.

We must next examine why the bipolar junction transistor is such that the constant of proportionality between the output current and the input current can be greater than unity, i.e., why this amplifier can exhibit current gain greater than unity. For this to be true, one electron in the base or input current must "affect" or modulate *more* than one hole in the collector or output current. To see that this can be the case, suppose the lifetime τ_n of an electron in the base is longer than the time τ_t it takes an injected hole to diffuse from the emitter to the collector. In such a situation, the electron lives long enough to provide space charge electrical neutrality while several holes traverse the base. Each electron put into the base is able to electrostatically compensate for τ_n/τ_t holes passing through. Given the requirement of space charge neutrality in the base, we would expect the ratio of the hole current I_C in the base to the electron current I_B in the base to be equal to τ_n/τ_t (assuming for the moment that all of the holes injected into the base are collected). We see then that, if the device is such that τ_n is longer than τ_t, I_C/I_B will be greater than unity, and we will have current gain greater than unity.

Current Gain in the Bipolar Transistor

We now want to calculate the ratio I_C/I_B. First, we note that the collector current I_C will be the fraction of the injected hole current I_{E_p} that crosses the base without recombining. If we call this fraction the base transport factor B, then

$$I_C = BI_{E_p} \tag{3.10}$$

The total emitter current I_E is made up of a hole current I_{E_p} and an electron current I_{E_n}, and was given by

$$I_E = I_{E_p} + I_{E_n} \tag{3.8}$$

so we define the emitter injection efficiency γ by the relation

$$\gamma \equiv I_{E_p}/I_E = I_{E_p}/(I_{E_p} + I_{E_n}) \tag{3.11}$$

If the electron component of the emitter current were zero, then we would have $I_{E_n} = 0$ in (3.11) and the efficiency γ would be equal to unity, meaning that the current injected by the emitter was composed entirely of holes.

The base current I_B is calculated as follows. The base current is, as shown in Figure 3.4, composed of electron flow (6) injected from n to p^+ at the emitter junction and electron flow (5) entering the base to replace electrons lost due to recombination with holes. The first component (6) of the base current is thus equal to I_{E_n}, the electron current through the emitter junction. The second component (5) is numerically equal to the fraction of the injected hole current that does *not* traverse the base. This hole current is equal to $(1 - B)I_{E_p}$ because, from (3.10), BI_{E_p} is the fraction of the injected hole current that *does* cross the base. Adding the two components of the base current, we have the result that

$$I_B = I_{E_n} + (1 - B)I_{E_p} \tag{3.12}$$

We now define the base-to-collector current gain β as

$$\beta \equiv \frac{I_C}{I_B} = \frac{BI_{E_p}}{I_{E_n} + (1 - B)I_{E_p}} \tag{3.13}$$

where we have used (3.10) and (3.12). Substituting the definition (3.11) of γ into equation (3.13) gives

$$\beta = \frac{B\gamma}{1 - B\gamma} \tag{3.14}$$

for the base-to-collector current gain β. Since the base transport factor B is less than unity and the emitter injection efficiency γ is less than unity, $B\gamma$ is less than unity. If $B\gamma$ is close to unity, the current gain β will be large; for $B\gamma$ greater than 0.5, the current gain β will be greater than unity and current amplification will take place. A typical value of β for an ordinary transistor is $\beta = 100$.

We now consider the factors that influence B and γ. The base transport factor B is the fraction of injected holes that cross the base by diffusion and reach the collector. For B to be close to unity, one wants a narrow base width and a large value of the hole diffusion length L_p (where $L_p^2 = D_p \tau_p$, D_p is the minority hole diffusion coefficient, and τ_p is the minority hole lifetime in the base). The width of the base of the transistor is a question of fabrication technology[9,10] and of the doping of the base relative to the emitter and collector regions. If the base is lightly doped relative to the emitter and collector regions, then the space charge layers at the two junctions will extend relatively far into the base region, thereby making narrower the neutral *n* region which the minority holes must traverse. To have a long diffusion length L_p, one must have large values of the diffusion constant D_p and long values of the minority carrier lifetime τ_p. The value of D_p is fairly well determined by the semiconductor from which the transistor is made, but, from the Einstein relation (3.3), large values of the minority hole mobility will favor large values of D_p. While scattering of carriers by phonons is a process fixed by the fundamental characteristics of the material, light doping in the base will decrease carrier scattering by ionized impurities. Such light doping will thus serve to increase carrier mobility and hence will favor large values of D_p. The minority carrier lifetime τ_p is (as discussed in Chapter 1) favored by a band structure with an indirect energy gap; examples are germanium and silicon. Since electron–hole recombination in indirect gap semiconductors is predominantly via an intermediate state (e.g., a defect level), lower doping of the base also favors a longer minority carrier lifetime. In summary, a value of the base transport factor close to unity is favored by a long hole diffusion length and hence is favored by the conditions leading to a longer minority carrier lifetime.

The emitter injection efficiency γ can be shown[11] to approach unity as the ratio of the doping of the emitter to the doping of the base becomes large. The p^+n junction is therefore expected to have an injection efficiency close to unity.

We see then, from (3.14), that large values of the current gain β are favored by values of $B\gamma$ close to unity, and that values of $B\gamma$ close to unity are favored by, among other factors, a long minority carrier lifetime in the base and the use of a p^+n junction as an efficient hole injection at the emitter junction.

Circuit Configurations for Amplification with the Bipolar Transistor

Two circuit configurations for amplification[12] using the bipolar junction transistor are shown in Figures 3.7 and 3.8. In these configurations, the emitter junction is forward biased and the collector junction is reverse biased. This biasing scheme is called the forward active mode. The arrange-

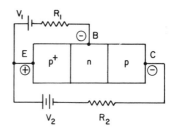

Figure 3.7. Common emitter circuit configuration for amplification with a *pnp* bipolar transistor.

ment in Figure 3.7 is called the common emitter configuration and that in Figure 3.8 is called the common base configuration.

In the common emitter circuit, the base current is determined by battery V_1 (e.g., 10 V) and resistor R_1 (e.g., $10^5\,\Omega$), where we are neglecting the small voltage appearing across the forward-biased emitter junction. Almost all of the voltage between the collector and emitter appears across the relatively high resistance reverse-biased collector junction. The voltage V_2 might be 10 V and resistor R_2 5000 Ω. Note that, given the base current I_B, the collector current I_C is determined by the current gain β through equation (3.13),

$$I_C = \beta I_B \tag{3.13}$$

and not by the parameters V_2 and R_2 of the collector circuit, as long as the latter maintain the reverse bias at the collector junction. In the circuit in Figure 3.7, the voltage V_2 will be divided between the resistance R_2 and the resistance of the reverse-biased collector junction.

Using the typical values for the circuit elements quoted above, and using a current gain $\beta = 100$, one finds (with the approximations mentioned) that $I_B = 0.1$ mA and $I_C = 10$ mA. Suppose an AC current of the form $i_1 = i_0 \sin \omega t$, with $i_0 = 0.05$ mA, is applied as an input to the base, superposed on the DC bias current I_B. Then the total base current $i_B = I_B + i_1$ will have the form shown in Figure 3.9a. The total collector (output) current will have the form shown in Figure 3.9b, which is an AC component $i_2 = (\beta i_0) \sin \omega t$ superposed on the DC collector current I_C, where $I_C = \beta I_B$. This simple example exhibits the current gain by showing that

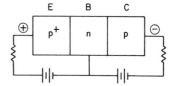

Figure 3.8. Common base circuit configuration for amplification with a *pnp* bipolar transistor.

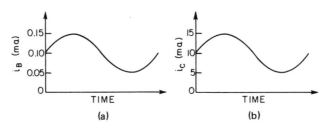

Figure 3.9. (a) Base (input) current i_B as a function of time. (b) Collector (output) current i_C as a function of time.

the output current i_C is equal to the input current i_B multiplied by the current gain β. Since the value of the input current i_B determines the value of the output current i_C, the transistor is, as expected from the discussion earlier, acting as a current-controlled amplifier.

In the common base configuration of Figure 3.8, the emitter-to-collector current amplification factor α is defined by

$$\alpha \equiv \frac{I_C}{I_E} = \frac{BI_{Ep}}{I_{Ep} + I_{En}} \tag{3.15}$$

using (3.10) and (3.8). Since the emitter injection efficiency γ is defined by (3.11), equation (3.15) becomes

$$\alpha = B\gamma \tag{3.16}$$

where B is the base transport factor. Since both B and γ are less than unity, the current amplification represented by α is amplification by a factor less than unity. However, the emitter-to-collector current is the transport of holes from the forward-biased, low-resistance emitter junction to the reverse-biased, high-resistance collector junction. Since this current flows in the high-resistance direction, the current flow represents power gain or amplification even though the current gain α is less than unity.[13] Finally, combining (3.16) with (3.14) gives

$$\beta = \frac{\alpha}{1 - \alpha} \tag{3.17}$$

as the relation between the base-to-collector current gain β and the emitter-to-collector current gain α. Equation (3.17) shows that, as α approaches its upper limit of unity, β becomes large, corresponding to the physical situation in which most of the holes injected by the emitter reach the collector.

Quantitative Discussion of the Bipolar Transistor

We now make a simple analysis of the bipolar transistor. This analysis will be sufficient for the situation in which the transistor is used as an amplifier, but is not the most general possible, in that a number of assumptions and approximations are made. The reader is referred to the literature for a discussion[14] of deviations from our idealized case, as well as a treatment of the important use of the bipolar transistor as a switch.[15,16] However, our discussion will provide, within the approximations made, interesting quantitative results.

We simplify the problem by making the following assumptions. First, holes diffuse from the emitter to collector in our *pnp* transistor. This is equivalent to assuming that no electric field, and hence no drift current of holes, exists in the base. (This assumption is the same as the one we made in Chapter 2 concerning the neutral regions of the *p–n* junction.) Second, the emitter current I_E is composed only of holes, so the emitter injection efficiency γ is taken as equal to unity. Third, we assume that the reverse saturation current in the collector junction can be neglected. Fourth, the transistor has a uniform cross section area, so the problem becomes one-dimensional. Fifth, we neglect generation and recombination of carriers in the space charge regions of the emitter and collector junctions. Sixth, all currents and voltages are taken to be steady state values.

Given these approximations, the calculation will be in three steps. First, we calculate the distribution in space of the excess holes injected into the base. Second, we calculate the emitter current I_E and the collector current I_C as diffusion currents due to the concentration gradients of injected holes at the emitter and collector boundaries of the base. The base current I_B is calculated from I_E and I_C. Third, we calculate the base-to-collector current gain β and the collector-to-emitter current gain α.

We set up the geometry of the problem as shown in Figure 3.10. The space charge regions of the emitter (E) and collector (C) junctions are delineated by vertical dashed lines. The neutral base region extends from $x = 0$ to $x = W_b$, so the base width is equal to W_b. We take $x = 0$ at

Figure 3.10. *pnp* junction transistor geometry. The space charge regions of the emitter junction E and the collector junction C are delineated by vertical dashed lines. The width of the neutral base region is W_b.

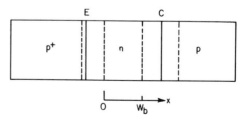

the edge of the emitter junction space charge region in the base and $x = W_b$ at the edge of the collector junction space charge region in the base. We use the symbol Δp_E for the injected excess hole density at the emitter junction at $x = 0$, and Δp_C for the excess hole density at the collector junction at $x = W_b$. If we use V_E for the forward bias voltage applied to the emitter junction and V_C for the reverse bias voltage applied to the collector junction, then equation (2.108) gives us

$$\Delta p_E = p_{no}(e^{eV_E/k_BT} - 1) \qquad (3.18)$$

$$\Delta p_C = p_{no}(e^{eV_C/k_BT} - 1) \qquad (3.19)$$

where p_{no} is the equilibrium minority hole density in the base. Since the emitter junction is forward biased, V_E is positive, and, since the collector is reversed biased, V_C is negative.

The differential equation describing the diffusion of injected holes away from the emitter junction and into the base is, in the steady state, equation (2.84),

$$\frac{d^2}{dx^2}(p - p_0) = \frac{1}{L_p^2}(p - p_0) \qquad (2.84)$$

where $p(x)$ is the nonequilibrium hole density at point x, p_0 is the (spatially constant) equilibrium hole density, and $p - p_0$ is the excess hole density at point x. As before, L_p is the diffusion length for minority holes in the base. The boundary conditions on equation (2.84) are given by the requirements that, at $x = 0$, $p - p_0$ must equal Δp_E, and, at $x = W_b$, $p - p_0$ must equal Δp_C. These conditions are expressed by the equations

$$p(0) = p_0 + \Delta p_E \qquad (3.20)$$

$$p(W_b) = p_0 + \Delta p_C \qquad (3.21)$$

A solution of (2.84) is

$$p(x) = p_0 + C_1 e^{x/L_p} + C_2 e^{-x/L_p} \qquad (3.22)$$

where the boundary conditions determine the constants C_1 and C_2. From (3.20) and (3.21), we obtain

$$C_1 + C_2 = \Delta p_E \qquad (3.23)$$

$$C_1 \exp(W_b/L_p) + C_2 \exp(-W_b/L_p) = \Delta p_C \qquad (3.24)$$

Solving (3.23) and (3.24) for C_1 and C_2 gives

$$C_1 = [\Delta p_C - \Delta p_E e^{-W_b/L_p}](e^{W_b/L_p} - e^{-W_b/L_p})^{-1} \qquad (3.25)$$

$$C_2 = [-\Delta p_C + \Delta p_E e^{W_b/L_p}](e^{W_b/L_p} - e^{-W_b/L_p})^{-1} \qquad (3.26)$$

for the constants in equation (3.22) for $p(x) - p_0$, the excess hole density at point x in the base.

To find the diffusion current $J_p(x)$ of holes flowing from the emitter to the collector we use the result from (2.77) that

$$J_p(x) = -eD_p \frac{d}{dx}[p(x)] \qquad (2.77)$$

This leads, on differentiating $p(x)$ given by (3.22) and substituting the expressions (3.25) and (3.26) for C_1 and C_2, to the result

$$J_p(x) = \frac{-eD_p/L_p}{e^{W_b/L_p} - e^{-W_b/L_p}}$$
$$\times [\Delta p_C e^{x/L_p} - \Delta p_E e^{(x-W_b)/L_p} + \Delta p_C e^{-x/L_p} - \Delta p_E e^{-(x-W_b)/L_p}] \qquad (3.27)$$

for the diffusion current $J_p(x)$ of holes at any point x of the base.

To find the emitter current density J_E, we evaluate (3.27) at the edge $x = 0$ of the emitter junction space charge region (remembering that we assumed $\gamma = 1$ so J_E is composed only of holes), and find

$$J_E = J_p(0) = \frac{eD_p/L_p}{e^{W_b/L_p} - e^{-W_b/L_p}} [-2\Delta p_C + \Delta p_E(e^{W_b/L_p} + e^{-W_b/L_p})] \qquad (3.28)$$

Equation (3.28) can be rewritten using the identities

$$\coth x = (e^x + e^{-x})/(e^x - e^{-x}) \qquad (3.29)$$

and

$$2/(e^x - e^{-x}) = (\sinh x)^{-1} = \operatorname{csch} x \qquad (3.30)$$

so we obtain

$$J_E = e(D_p/L_p)[\Delta p_E \coth(W_b/L_p) - \Delta p_C \operatorname{csch}(W_b/L_p)] \qquad (3.31)$$

To find the collector current density J_C, we evaluate (3.27) at the edge $x = W_b$ of the collector junction space charge region in the base, so

$$J_C = J_p(W_b) = \frac{eD_p/L_p}{e^{W_b/L_p} - e^{-W_b/L_p}} [2\Delta p_E - \Delta p_C(e^{W_b/L_p} - e^{-W_b/L_p})] \qquad (3.32)$$

which, on rearranging to introduce hyperbolic functions, becomes

$$J_C = e(D_p/L_p)[-\Delta p_C \coth(W_b/L_p) + \Delta p_E \operatorname{csch}(W_b/L_p)] \quad (3.33)$$

To find the base current density J_B, we note that the current density J_E of holes into the base must equal the total current density $J_B + J_C$ of holes out of the base. Thus we have $J_B = J_E - J_C$, and

$$J_B = e(D_p/L_p)\{[\coth(W_b/L_p) - \operatorname{csch}(W_b/L_p)](\Delta p_E + \Delta p_C)\} \quad (3.34)$$

Using the identity $\coth x - \operatorname{csch} x = \tanh(x/2)$, equation (3.34) becomes

$$J_B = e(D_p/L_p)[(\Delta p_E + \Delta p_C) \tanh(W_b/2L_p)] \quad (3.35)$$

Equations (3.31), (3.33), and (3.35) give the emitter, collector, and base currents as functions [through Δp_E and Δp_C given by (3.18) and (3.19)] of the voltages V_E and V_C applied to the emitter and collector junctions. These expressions for J_E, J_C, and J_B are therefore correct for any bias voltages applied to the junctions of the transistor, and not just for the "usual" voltages in which the emitter junction is forward biased and the collector junction is reverse biased.

We now specialize equations (3.31), (3.33), and (3.35) to the particular case of "normal" or "active" biasing of the transistor, in which the emitter junctions is forward biased and the collector junction is reverse biased. For values of V_E and $|V_C|$ larger than $k_B T/e$, equations (3.18) and (3.19) become

$$\Delta p_E \cong p_{n0} \exp(eV_E/k_B T) \quad (3.36)$$

and

$$\Delta p_C \cong -p_{n0} \quad (3.37)$$

the negative sign in (3.37) meaning that carriers are extracted. Note that, in this situation, the total hole density $p_{n0} + \Delta p_C$ at the collector junction is approximately zero because of minority hole extraction. From (3.36) and (3.37), we see that, for the case of normal transistor bias, we have

$$\Delta p_E \gg \Delta p_C \quad (3.38)$$

and we can neglect the terms in Δp_C in the expressions (3.31), (3.33), and (3.35) for the currents in the device. We thus obtain

$$J_E = e(D_p/L_p) \Delta p_E \coth(W_b/L_p) \quad (3.39)$$

$$J_C = e(D_p/L_p) \Delta p_E \operatorname{csch}(W_b/L_p) \quad (3.40)$$

$$J_B = e(D_p/L_p) \Delta p_E \tanh(W_b/2L_p) \quad (3.41)$$

where Δp_E is given by (3.36), so (3.39)–(3.41) give the device currents as a function of the forward bias V_E on the emitter junction. We may exhibit that functional dependence explicitly by substituting (3.36) into (3.39) through (3.41), obtaining

$$J_E = e(D_p/L_p)[\coth(W_b/L_p)]p_{n0}\exp(eV_E/k_BT) \qquad (3.42)$$

$$J_C = e(D_p/L_p)[\operatorname{csch}(W_b/L_p)]p_{n0}\exp(eV_E/k_BT) \qquad (3.43)$$

$$J_B = e(D_p/L_p)[\tanh(W_b/2L_p)]p_{n0}\exp(eV_E/k_BT) \qquad (3.44)$$

We see from (3.42) through (3.44) that, in this approximation of neglecting Δp_C compared to Δp_E, the collector voltage V_C does not appear in the expressions for the currents. In particular, the collector current J_C is independent of the voltage across the collector junction in this approximation. Finally, we remind ourselves that the minority hole density p_{n0} in the base is equal to (n_i^2/N_d), where N_d is the doping in the base region.

We now use equations (3.42)–(3.44) to calculate the current gain factors α and β. From (3.15) and (3.13), we obtain

$$\alpha \equiv J_C/J_E = [\cosh(W_b/L_p)]^{-1} \qquad (3.45)$$

$$\beta \equiv J_C/J_B = [\operatorname{csch}(W_b/L_p)]/[\tanh(W_b/2L_p)] \qquad (3.46)$$

showing how the current gain varies with the ratio W_b/L_p of the base width W_b to the minority hole diffusion length L_p. Figure 3.11 is a plot of the collector to base current gain β as a function of L_p/W_b, showing how β increases as the ratio of the hole diffusion length L_p to the base width W_b

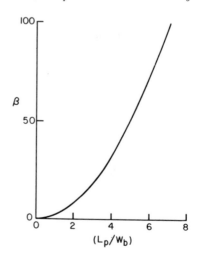

Figure 3.11. Current gain β, given by equation (3.46), plotted as a function of L_p/W_b, where L_p is the hole diffusion length and W_b is the width of the base.

increases. Since $L_p = (D_p\tau_p)^{1/2}$, this curve tells us that the current gain β increases with increasing minority hole lifetime τ_p.

Finally, for large values of L_p/W_b, W_b/L_p is less than unity, so we may expand the hyperbolic functions in (3.46) in series, where, for $x \ll 1$,

$$\tanh x \cong x - (1/3)x^3 + \cdots \tag{3.47}$$

$$(\sinh x)^{-1} \cong 1/x - (1/6)x + \cdots \tag{3.48}$$

Using the first term of the series expansions above in the expression (3.46) for β gives

$$\beta \cong 2L_p{}^2/W_b{}^2 = 2D_p\tau_p/W_b{}^2 \tag{3.49}$$

This result shows that, for $L_p \gg W_b$, the current gain β is proportional to the minority hole lifetime τ_p.

We note that the results (3.45) and (3.46) of our treatment predict that the current gains α and β are independent of the current densities. This is not actually true in real devices.[17,18]

Summary of the Physics of Amplification in the Bipolar Transistor

Several points may be mentioned. First, the action of the bipolar transistor as an amplifier depends directly on the simultaneous coexistence of electrons and holes in the base of the transistor. In a *pnp* transistor, this coexistence allows the base current of electrons to control the collector current of holes. Second, the bipolar transistor is a useful amplifier because a small base current can control a large collector current. This is equivalent to saying that large values of the current gain β are possible. This effect is, in turn, due to the fact that minority carriers in an indirect gap semiconductor like silicon have a lifetime long enough to yield usefully long values of the minority hole diffusion length, thereby allowing minority holes to cross the base of the transistor and be collected. Finally, the bipolar transistor is a minority carrier device, in that it depends on the existence of minority carriers in a semiconductor to produce amplification.

Tunnel Diodes

The tunnel diode is a particular type of semiconductor *p–n* junction which exhibits a negative differential conductivity. By this, it is meant that this device shows a decrease in current with increasing voltage over at

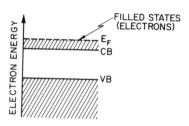

Figure 3.12. Band diagram showing free electrons in the conduction band of a degenerate n-type semiconductor. Filled electron states are shown as shaded; E_F is the Fermi level; CB and VB are the conduction and valence band edges, respectively.

least a part of its current–voltage characteristic. These negative resistance diodes, or tunnel diodes as they are more commonly called, have a number of useful circuit applications[19,20] because of their negative dynamic resistance.

Consider an n-type semiconductor so heavily doped with impurities that it is degenerate. Then, as discussed in Chapter 1 and exhibited in Figures 1.15 and 1.16, the Fermi level is above the conduction band minimum and there is a density of free electrons in the conduction band. For a degenerate p-type semiconductor, there is a density of free holes in the valence band. Band diagrams exhibiting these effects are showing in Figures 3.12 and 3.13. Consider next a p–n junction, at equilibrium, in which both the n-type and p-type sides are degenerate. The band diagram of such a "degenerate junction," at equilibrium, is shown in Figure 3.14. The junction is assumed to be ideally abrupt and, since the impurity density is high in both n-type and p-type regions, the width W of the space charge region will be small for this type of junction. A representative value[19,21] of W of the order of 100 Å for a degenerate silicon p–n junction with donor and acceptor densities of 10^{19} cm^{-3}.

If we consider the possibility of the tunneling of electrons from one side of the junction to the other, the width W of the depletion layer will be the width of the energy barrier to tunneling. Since W will be small for the degenerate junction, electron tunneling is possible, and we examine next the factors affecting tunneling.

What factors will affect the probability of tunneling by an electron? We expect first that the current of tunneling electrons will increase as the

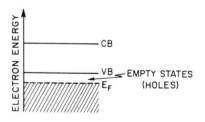

Figure 3.13. Band diagram showing free holes in the valence band of a degenerate p-type semiconductor. Filled electron states are shown as shaded; E_F is the Fermi level; CB and VB are the conduction and valence band edges, respectively.

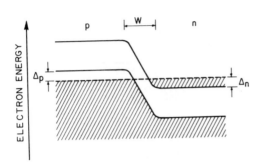

Figure 3.14. Degenerate _p–n_ junction at equilibrium. The junction is assumed to be abrupt and the width W of the space charge region is exaggerated. The Fermi level is the dashed line and filled electron states are shown as shaded. The energies to which the conduction and valence bands are filled with electrons and holes are denoted by Δ_n and Δ_p, respectively.

width W of the barrier decreases since we expect[22] that the transmission coefficient for the tunneling process will behave in this manner. While the calculation of the tunneling probability will depend on the details of the energy barrier,[23] we can see qualitatively (say from[24] Fermi's golden rule number 2) that the probability of tunneling per unit time will be directly proportional to the density of final states available to the electron. For a given tunneling probability per unit time, we expect that the number of electrons tunneling will be proportional to the number of initial states occupied by electrons. In summary, we expect the current of tunneling electrons to increase with increasing density of empty final states, with increasing number of occupied initial states from which tunneling may take place, and with decreasing barrier width.

Given the degenerate junction band diagram shown in Figure 3.14 how will a voltage V_a applied to the junction affect the factors above which influence the magnitude of the tunnel current? One would expect that forward bias would decrease the width of the space charge region somewhat, but not greatly because W is already relatively small owing to the high doping. Forward bias will, however, increase the electron energies on the _n_-type side of the junction, thus "raising" the electron energy states on the _n_-side relative to the _p_-side. This effect is exhibited in Figure 3.15, which shows the band diagram with an applied forward bias. The symbols Δ_n and Δ_p are used, as in Figure 3.14, for the energies to which the conduction and valence bands are, respectively, filled with electrons and holes. Figure 3.15 is drawn for an applied forward bias $V_a = \Delta_n/e$. We can see from Figure 3.15 that, since the tunneling process takes place at constant electron energy, forward bias will increase both the number of possible initial states and the number of available final states.

Beginning with an applied potential $V_a = 0$, as shown in Figure 3.14, we see that no filled state on the _n_ side is at the same energy as an empty state on the p side. Further, no filled state on the p side is at the same energy as an empty state on the n side. At $V_a = 0$, then, no tunnel current of electrons flows. Figure 3.15 shows the application of a nonzero forward bias V_a. Since the electron states on the n side are "raised," there are now filled electron states at the same energy as empty states on the p side. Electrons can now tunnel from n to p, and a tunnel current of electrons flows through the junction.

At some value of the applied potential V_a, a maximum in the joint "overlap" of filled states on the n side with empty states on the p side will take place. Calling this value V_a', we can see intuitively that the value of V_a' will be related to the fillings Δ_n and Δ_p of the bands in the junction. Even though V_a' does not turn out to be a simple function[25] of Δ_n and Δ_p, we do expect that the tunnel current of electrons from n to p will reach a maximum at some value V_a' of the applied forward bias. At values of the forward bias larger than V_a', the joint "overlap" of filled states on the n side and empty states on the p side begins to decrease, until this "overlap" becomes zero at a value V_a'' of forward bias, where

$$eV_a'' = \Delta_n + \Delta_p \tag{3.50}$$

We expect, then, that the tunnel current of electrons from n to p will go to zero at a value of forward bias V_a equal to V_a''. A schematic plot of electron tunnel current density J_t as a function of forward bias V_a is shown in Figure 3.16. We note that, for values of the forward bias between V_a' and V_a'', the dynamic conductance dJ_t/dV_a is negative. This means that, for $V_a' \leq V_a \leq V_a''$, the device current decreases for increasing values of the applied forward bias V_a.

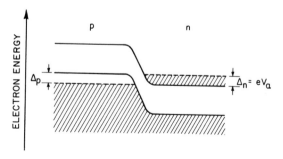

Figure 3.15. Band diagram of a degenerate p–n junction under an applied forward bias V_a. The filled electron states are shown as shaded and the energies to which the conduction and valence bands are filled with electrons and holes are Δ_n and Δ_p, respectively. The figure is drawn for an applied forward bias $V_a = \Delta_n/e$.

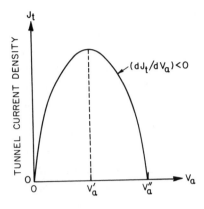

Figure 3.16. Variation (schematic) of the tunnel current density J_t with applied forward bias V_a. The voltage V_a'' at which J_t goes to zero is given by equation (3.50), and the dynamic conductance dJ_t/dV_a is negative for the voltage range $V_a' < V_a \leq V_a''$.

For all values of the applied forward bias, there will also be a forward injection current of electrons from n to p and holes from p to n, over their respective energy barriers. The total current density J will be the sum of the tunnel and injection currents and the behavior of J with applied forward bias is shown schematically in Figure 3.17. The total current–voltage characteristic has a minimum at a value V_V of the applied forward bias. A typical value[26] of the "valley voltage" V_V is 0.5 V for a GaAs tunnel diode. The ratio of the current density J_P (at the peak of the curve) to the current density J_V (at the valley voltage V_V) is typically[26] 15 for a GaAs device.

The discussion of tunneling here has tacitly assumed that the semiconductor has a direct band gap (e.g., GaAs) and that the electron tunneling process takes place between band extrema at the same point in the Brillouin zone. For indirect tunneling, in which the initial and final electron states are at different points in the Brillouin zone (e.g., silicon), the plot[27] of tunnel current as a function of forward bias is qualitatively similar to that for direct tunneling.

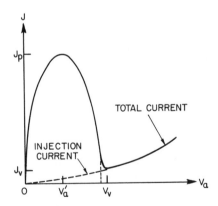

Figure 3.17. Schematic variation of the total current density J, as a function of applied forward bias V_a, for a tunnel diode.

For values of the forward bias larger than V_a'', one would expect only injection current to flow in the tunnel diode. However, the observed currents at such values of forward bias are larger than the expected injection current. This excess is referred to as the "excess current" in the tunnel diode and is, at least partially, due[28] to carrier tunneling through intermediate states in the semiconductor band gap.

Finally, for reverse bias, one can see from Figure 3.14 that there is "overlap" of states permitting tunneling of electrons from p to n for all values of the reverse bias. Since the "overlap" increases monotonically with increasing magnitude of the reverse bias, we expect that the reverse current–voltage characteristic will show a monotonic increase in the current with increasing reverse bias. This characteristic resembles that of a resistor.

It is also appropriate to remark on the fact that the process of tunneling by an electron is fast[29] compared to the drift or diffusion motion of an electron. This makes the tunnel diode a device that can shift rapidly from one point to another on its current–voltage characteristic, so it is very useful in high-speed circuits,[30] including switching circuits and high-frequency oscillators. This utility in high-speed applications, in addition to its negative dynamic resistance,[31] makes the tunnel diode a useful device.

Finally, it is useful to summarize how the physics of the heavily doped p–n junction gives the tunnel diode its useful properties of high speed and negative dynamic resistance. First, it is the rapidity of electron tunneling, compared to electron drift or diffusion, that gives the device its high speed. Second, the high doping in the junction does two things. It creates empty states in the valence band on the p side and filled states in the conduction band on the n side, thereby making possible electron tunneling from n to p under forward bias. (If this were not true, only electron tunneling from p to n under reverse bias would be possible.) The heavy doping also decreases the width of the energy barrier to tunneling by making the width of the space charge region small. Third, the negative resistance portion of the tunnel diode current–voltage characteristic comes about because the probability of tunneling decreases for values of applied forward bias greater than a critical value, thereby decreasing the device tunnel current for increasing magnitude of the forward bias.

The Junction Field Effect Transistor (JFET)

We consider next another solid state device that acts as an amplifier, the junction field-effect transistor[32,33] (or JFET). This device is one in which

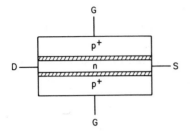

Figure 3.18. Schematic structure of a junction field-effect transistor. The shaded areas are the depletion layers of the reverse-biased p^+n junctions; S is the source, D the drain, and G the gate electrodes of the device.

the device current is controlled by an applied voltage, so it is a voltage-controlled amplifier.

The basic idea of the junction field-effect transistor is to vary the spatial extent of the depletion layer of a reverse-biased $p–n$ junction by varying the magnitude of the applied reverse bias. The spatial extent of the depletion layer, in turn, controls the conductance of a channel in a semiconductor, thereby controlling the current flowing through this channel. Figure 3.18 shows a schematic picture of a JFET structure. In Figure 3.18, the depletion layers of the reverse-biased p^+n junctions are shown as shaded, and the source, drain, and gate electrodes of the device are marked S, D, and G, respectively. The n-type region is called the channel; its conductivity is determined by its doping. If the p^+n junctions are reverse biased, the width of the depletion layer (almost all of which is on the n side because of the high doping on the p side) will be increased. The shaded regions of Figure 3.18 are depleted of electrons and are thus of high resistance, thereby decreasing the effective cross-section area of the n-type channel and increasing the resistance of the channel. It is clear that the current through the channel will decrease with increasing magnitude of the reverse bias applied to the p^+n junctions. In this manner the voltage applied to the gate electrode controls the current through the device, so the junction field-effect transistor is a voltage-controlled amplifier.

Physical Basis of the Current–Voltage Characteristic of the JFET

Consider a JFET connected as shown in Figure 3.19. The potential of the drain relative to the source is $(+V_{DS})$ and V_{GS} is the potential of the gates relative to the source. The n-type channel is of length L, and the shaded regions represent the equilibrium depletion layers in the n channel due to the reverse-biased p^+n junctions. We consider first the situation when $V_{GS} = 0$. The reverse bias applied to the junctions is then of magnitude

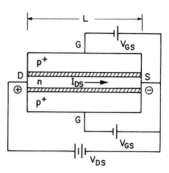

Figure 3.19. Junction field-effect transistor with circuit connections; the gates G are connected to the source S. The shaded areas are the equilibrium depletion layers of the p^+n junctions; the n-type channel is of length L.

$|V_{DS}|$. As shown in Figure 3.19, the channel current I_{DS} flowing from drain to source is composed of electrons flowing from source to drain.

What happens to the current I_{DS} as the magnitude of V_{DS} increases? We assume that the heavily doped p^+ gate regions have such a high conductivity that they are essentially metallic, so that the electrostatic potential is the same throughout the gate regions. Since the gates in Figure 3.19 are connected to the source, the potential throughout the gates is the source potential, which we choose to be zero. Since the n channel is lightly doped, the channel will act as a distributed resistor. For low currents, voltage V_{DS} varies approximately linearly across the length L of the channel. If we call $V(x)$ the potential at point x of the channel (where the source is at $x = L$ and the drain at $x = 0$), $V(L) = 0$ at the source and $V(0) = +V_{DS}$ at the drain. We see that $V(x)$ will have a larger magnitude near the drain than near the source. In this configuration (with no bias applied to the gates G) the voltage $V(x)$ is the magnitude of the reverse bias applied at point x to the p^+n junctions. The conclusion is that a larger magnitude reverse bias is applied to the p^+n junctions near the drain D than is applied near the source S.

The result of this conclusion is that the junction depletion layers extend further into the channel near the drain than they do near the source. For a given value of the voltage V_{DS}, the depletion layers in the JFET look (schematically) as shown in Figure 3.20, where the depletion layers are

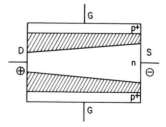

Figure 3.20. Depletion layers (schematic) of JFET for a given value of V_{DS} and no voltage applied to the gates G ($V_{GS} = 0$). The depletion layers are shown as shaded.

the shaded areas. As the magnitude of V_{DS} is increased the depletion layers extend further into the channel, increasing the resistance R of the channel. Since $R = dV_{DS}/dI_{DS}$ and R increases with increasing V_{DS}, a plot of V_{DS} as a function of I_{DS} will have a positive slope which increases with increasing V_{DS}. A plot of I_{DS} as a function of V_{DS} will then have a slope $1/R = dI_{DS}/dV_{DS}$ which decreases with increasing V_{DS}, as shown in Figure 3.21. This figure shows the current–voltage characteristic of a JFET with no voltage applied to the gates.

Figure 3.21 shows the device current I_{DS} saturating with increasing magnitude of the voltage V_{DS}. Is that result physically reasonable? The answer is yes, as can be seen from the shape of the depletion layers in the drawing in Figure 3.20. As V_{DS} is increased, the depletion layers extend further into the channel until, at a particular value of V_{DS}, called V_P, the two depletion layers meet near the drain. This effect is called "pinching-off" the conducting part of the channel, and V_P is called the "pinch-off" voltage. (The physics involved in pinch-off may be more complicated than simple merging of the depletion layers.[34]) When pinch-off takes place, the current I_{DS} does not increase significantly with increasing V_{DS} above the pinch-off voltage V_P, resulting in the approximate saturation of the device current shown in Figure 3.21.

Consider next the application of a negative potential $V_{GS} = -V_1$ between the gates G and the source S. Since this potential makes the gates (p^+ regions) more negative with respect to the source, application of a negative voltage to the gates increases the magnitude of the reverse bias applied to the p^+n junctions (for a given value of V_{DS}). The result is an increase in the penetration of the junction depletion layer into the channel for a given value of the voltage V_{DS}. Pinch-off thus takes place at a smaller values of V_{DS} when a negative bias is applied to the gates of the device. This effect is shown in Figure 3.22, where we see that pinch-off occurs at smaller values of V_{DS} for increasing magnitude ($V_2 > V_1$) of the negative potential V_{GS} applied to the gates. The locus of the points showing the

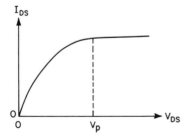

Figure 3.21. Current I_{DS} as a function of voltage V_{DS} for JFET with $V_{GS} = 0$. The slope of the curve decreases with increasing V_{DS} until the curve saturates at a value of $V_{DS} = V_P$, the pinch-off voltage.

Figure 3.22. Current I_{DS} as a function of voltage V_{DS} for JFET for $V_{GS} = 0$, $-V_1$, $-V_2$, where V_2 is greater than V_1. The black dots indicate the pinch-off voltage V_p at each value of V_{GS}; the pinch-off curve is shown as a dashed curve.

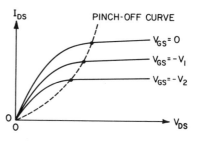

pinch-off voltage V_P on the I_{DS}–V_{DS} curves is called the pinch-off curve of the device. We can see that the channel resistance $R = dV_{DS}/dI_{DS}$ is larger (for a given value of V_{DS}) for larger magnitudes $|V_{GS}|$ of the negative potential applied to the gates. Since the slope dI_{DS}/dV_{DS} of the curves in Figure 3.22 is equal to $1/R$, this fact means that the curves in Figure 3.22 have smaller slopes (at a given value of V_{DS}) for larger magnitudes $|V_{GS}|$ of the gate voltage.

Examining the current–voltage plots of Figures 3.22 for values of V_{DS} above the pinch-off voltage, we see that the JFET is a device in which the current I_{DS} is (for a given value of V_{DS}) controlled by the magnitude of the gate voltage V_{GS}. The JFET is thus a voltage-controlled amplifier in which the gate voltage V_{GS} controls the device current I_{DS} through the effect of the reverse bias on the channel resistance.

The JFET device current I_{DS} is composed only of majority carriers which, in this example of an n channel, are electrons. The JFET is therefore a majority carrier device, as opposed to the bipolar junction transistor, which depends for its operation on the coexistence of majority and minority carriers in the base.

Since the gate voltage V_{GS} is applied across a reverse-biased p–n junction, the input impedance of a JFET is high, typically many megohms. The control of I_{DS} by the gate voltage V_{GS} is represented by the mutual transconductance g_m of the device, where

$$g_m = (dI_{DS}/dV_{GS})_{V_{DS}} \tag{3.51}$$

The amplification factor μ of the JFET is defined by

$$\mu = g_m r_d \tag{3.52}$$

where the drain resistance $r_d = (dV_{DS}/dI_{DS})_{V_{GS}}$ is what we have been calling the channel resistance R. Typical values[35] of these parameters are $g_m =$

10^{-3} A/V and $r_d = 5 \times 10^5 \, \Omega$, giving a value of $\mu = 500$. The use of the JFET as an amplifier is discussed in various electrical engineering texts;[36] however, the JFET is not as important commercially as another type of field-effect transistor (the metal-oxide–semiconductor field effect transistor, or MOSFET), which we will discuss in a later chapter.

Problems

3.1. *Band Diagram of npn Transistor.* (a) Draw band diagrams for an *npn* transistor at equilibrium and under the normal active bias conditions for amplification. (b) Draw a figure showing the currents flowing and use it to discuss the injection and collection of electrons. (c) What are the components of the base current in an *npn* transistor?

3.2. *Transistor at High Temperature.* A silicon *pnp* transistor is operated at 150°C. Discuss qualitatively the changes you would expect in its characteristics relative to those at room temperature.

3.3. *Reverse Current of a Tunnel Diode.* Using a band diagram of a degenerate junction, discuss the current flow in a tunnel diode under reverse bias. Sketch qualitatively the current–voltage characteristic for reverse bias.

3.4. *High-Temperature Rectifier.* Discuss the design of a possible *p–n* junction diode for use as a high-temperature rectifier. Suggest a specific material for a diode with a reverse saturation current density no greater than 10^{-12} A/cm² at 150°C (for reverse voltages less than the breakdown value). Discuss doping and other variables for such a design.

3.5. *Depletion Layer Width in a Tunnel Diode.* Calculate the width of the depletion layer in a degenerate *p–n* junction in silicon (of dielectric constant 11.7). Take the donor and acceptor densities as equal to 2×10^{19} cm⁻³.

References and Comments

1. W. G. Oldham and S. E. Schwarz, *Electrical Engineering*, Holt, Rinehart, and Winston, New York (1984), Section 11-2.
2. J. Millman and C. C. Halkias, *Integrated Electronics*, McGraw-Hill, New York (1972), Chapter 4.
3. See, for example, C. Kittel, *Introduction to Solid State Physics*, Third Edition, John Wiley, New York (1966), page 323.
4. See References 38–40 of Chapter 2.
5. See References 26 and 27 of Chapter 2.

6. J. Millman and C. C. Halkias, Reference 2, pages 73–77.

7. B. C. Streetman, *Solid State Electronic Devices*, First Edition, Prentice-Hall, New York (1972), pages 278–285.

8. B. G. Streetman, *Solid State Electronic Devices*, Second Edition, Prentice-Hall, New York (1980), pages 231–238.

9. B. G. Streetman, Reference 8, pages 239–241.

10. A. S. Grove, *Physics and Technology of Semiconductor Devices*, John Wiley, New York (1967), pages 1–88 and 208–209.

11. S. M. Sze, *Physics of Semiconductor Devices*, First Edition, John Wiley, New York (1969), pages 270–272.

12. B. G. Streetman, Reference 8, pages 231–238; W. G. Oldham and S. E. Schwarz, Reference 1, page 526; J. Millman and C. C. Halkias, Reference 2, pages 126–134.

13. See, for example, C. Kittel, *Introduction to Solid State Physics*, Second Edition, John Wiley, New York (1956), pages 397–398.

14. S. Wang, *Solid State Electronics*, McGraw-Hill, New York (1966); pages 340–349; see also A. S. Grove, Reference 10, pages 222–228.

15. W. G. Oldham and S. E. Schwarz, Reference 1, pages 556–560.

16. B. G. Streetman, Reference 8, page 254.

17. S. M. Sze, *Physics of Semiconductor Devices*, Second Edition, John Wiley, New York (1981), pages 140–147.

18. S. Wang, Reference 14, page 346; J. Millman and C. C. Halkias, Reference 2, page 133.

19. J. Millman and C. C. Halkias, Reference 2, pages 77–79.

20. K. K. N. Chang, *Parametric and Tunnel Diodes*, Prentice-Hall, New York (1964), pages 147–212; H. C. Okean in *Semiconductors and Semimetals*, R. K. Willardson and A. C. Beer (editors), Academic Press, New York (1971), Volume 7, Part B, pages 473–624.

21. A. S. Grove, Reference 10, page 163, Figure 6.9.

22. L. I. Schiff, *Quantum Mechanics*, Third Edition, McGraw-Hill, New York (1968), page 103, equation (17.7).

23. S. M. Sze, Reference 17, pages 516–527; S. Wang, Reference 14, pages 368–372.

24. L. I. Schiff, Reference 22, page 295, equation (35.14).

25. See S. M. Sze, Reference 17, page 526, for a particular case; see also S. Wang, Reference 14, pages 378–379.

26. J. Millman and C. C. Halkias, Reference 2, page 79.

27. S. M. Sze, Reference 17, page 526, Figure 7a.

28. S. M. Sze, Reference 17, pages 527–531, discusses excess current.

29. K. K. Thornber, T. C. McGill, and C. A. Mead, *Journal of Applied Physics*, **38**, 2384–2385 (1967) give a discussion of the tunneling time of an electron.

30. J. Millman and C. C. Halkias, Reference 2, page 78.

31. See, for example, J. J. Brophy, *Basic Electronics for Scientists*, Third Edition, McGraw-Hill, New York (1977), pages 266–270.

32. B. G. Streetman, Reference 8, page 286; A. S. Grove, Reference 10, pages 243–252.

33. R. F. Pierret, *Field Effect Devices*, Addison-Wesley, Reading, Mass. (1983), Chapter 1.

34. See B. G. Streetman, Reference 8, page 288.

35. J. Millman and C. C. Halkias, Reference 2, pages 318–321.

36. J. Millman and C. C. Halkias, Reference 2, Chapter 10; J. J. Brophy, Reference 31, Chapter 6.

Suggested Reading

B. G. Streetman, *Solid State Electronic Devices*, Second Edition, Prentice-Hall, New York (1980). Our discussion of junction devices parallels that of Streetman to some extent. This text is excellent as an introduction to semiconductor devices.

S. M. Sze, *Physics of Semiconductor Devices*, Second Edition, John Wiley, New York (1981). Once you have learned the basic physical ideas behind a semiconductor device, this formidable treatise will furnish the details and the mathematics, in addition to references to the original literature.

A. S. Grove, *Physics and Technology of Semiconductor Devices*, John Wiley, New York (1967). Recommended for both device physics and fabrication technology, especially for discrete silicon devices.

J. Millman and C. C. Halkias, *Integrated Electronics*, McGraw-Hill, New York (1972). This electrical engineering textbook covers circuit applications of all sorts of devices.

W. G. Oldham and S. E. Schwarz, *Electrical Engineering*, Holt, Rinehart, and Winston, New York (1984). This electronics text is also suggested for device applications.

J. J. Brophy, *Basic Electronics for Scientists*, Third Edition, McGraw-Hill, New York (1977). The two textbooks immediately above are primarily for electrical engineers who will be concerned with the design of devices and circuits. This electronics text is aimed at the science student who wishes to understand the operation of electronic instruments.

R. F. Pierret, *Field Effect Devices*, Addison-Wesley, Reading, Massachusetts (1983). This short book, one of a series on solid state devices, includes a discussion of the junction field effect transistor.

4

Physics of Metal–Semiconductor and Metal–Insulator–Semiconductor Junctions

Introduction

This chapter discusses some topics in the physics of metal–semi-conductor junctions and metal–insulator–semiconductor (MIS) junctions. The former topic is treated first in a discussion of the metal–semiconductor junction band diagram at equilibrium and under an applied potential. The results are used to show that the current–voltage characteristic is asymmetric and to treat metal contacts to semiconductors. The latter topic is then discussed using the band diagram of an idealized MIS structure. The key result is the formation of an inversion layer at the insulator-semiconductor interface when an appropriate potential is applied. This result will be used in the next chapter to study two important devices, the induced-channel field-effect transistor and the charge-coupled device.

The Metal–Semiconductor Junction at Equilibrium

We consider a junction between a metal and a semiconductor,[1-4] often called a Schottky junction or a Schottky diode. These are useful solid state devices and this topic will also give us some information on metal contacts to semiconductors.

In the following discussion we neglect the existence of surface states at the semiconductor surface. (In Chapter 6, a brief discussion of surface physics is presented, and it will consider the effect of surface states on the

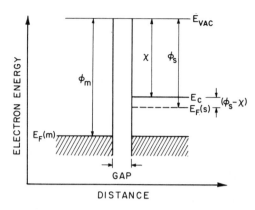

Figure 4.1. Band diagrams of a metal and an n-type semiconductor before contact is made, and with a gap or space between them. The work function of the metal is ϕ_m, the work function of the semiconductor is ϕ_s, χ is the electron affinity of the semiconductor, and ϕ_m is larger than ϕ_s. The semiconductor Fermi level is $E_F(s)$, E_C is the energy of the conduction band edge, E_V is the energy of the valence band edge, and $E_F(m)$ is the Fermi level in the metal. Filled electron states are shaded, and the position of $E_F(s)$ relative to E_C is exaggerated for clarity. (The fact that $E_F(m)$ is at the same energy as the valence band maximum is merely coincidental.)

properties of metal–semiconductor contacts.) We can derive the band diagram[5] of a metal–semiconductor junction at equilibrium by the following "thought experiment." Suppose a piece of metal and piece of n-type semiconductor are separated in space, and not in contact, but are connected by a metal wire so electric charge can flow between them. Figure 4.1 shows the band diagrams of the metal and of the semiconductor before contact is made and with a small gap or space between them. The notation used in Figure 4.1 is that $E_F(m)$ denotes the Fermi level of the metal, $E_F(s)$ is the Fermi level of the semiconductor, and E_{VAC} is the vacuum energy, defined[6] as the energy of an electron at rest (and hence with zero kinetic energy) just outside the crystal surface and not interacting with the crystal. Because of the lack of interaction, the energy E_{VAC} has the same value for the metal and for the semiconductor. The work function ϕ_m of the metal is defined by

$$\phi_m \equiv E_{\text{VAC}} - E_F(m) \tag{4.1}$$

and is the energy necessary to move an electron from the Fermi energy in the metal to the vacuum energy E_{VAC}. Similarly, the work function ϕ_s of the semiconductor is defined by

$$\phi_s \equiv E_{\text{VAC}} - E_F(s) \tag{4.2}$$

and is the energy difference between the Fermi level in the semiconductor and the vacuum energy. Since the position of the semiconductor Fermi level $E_F(s)$ depends on the doping-impurity content, the work function ϕ_s is also a function of the doping. The electron affinity χ of the semiconductor is defined as

$$\chi \equiv E_{\text{VAC}} - E_C \qquad (4.3)$$

where E_C is the energy of the conduction band minimum. The electron affinity is the energy required to move an electron from the bottom of the conduction band to the vacuum energy. If we choose the zero of energy at the conduction band edge, so $E_C = 0$, then equations (4.2) and (4.3) can be combined to give

$$\phi_s = \chi - E_F(s) \qquad (4.4)$$

where $E_F(s)$ is a negative quantity. Both the semiconductor[7] and metal[8] work functions and the semiconductor electron affinity[7] are of the order of a few electron volts.

We now consider the formation of a metal–n-type-semiconductor junction for the case, shown in Figure 4.1, for which the metal work function ϕ_m is larger than the semiconductor work function ϕ_s. We (mentally) allow the flow of charge between the metal and the semiconductor through the wire connecting them, until the chemical potentials (Fermi levels) of the electrons are the same in the metal and in the semiconductor. This process transfers electrons from the semiconductor to the metal because (in this example) the electron Fermi level (chemical potential) is higher in the semiconductor than in the metal. The result is an excess of negative charge on the metal surface and a depletion of negative charge in a semiconductor region of width W near its surface. These excess charges produce an electric field $\mathcal{E}$, directed from semiconductor to metal, and located mainly in the gap between them. This electric field corresponds to a gradient of electron electrostatic potential energy in the gap, with the potential energy being higher on the metal surface. This energy gradient in the gap is indicated in Figure 4.2, and is a reflection of the fact that the vacuum energy E_{VAC} now varies with position.

We continue the "thought experiment" by bringing the metal and semiconductor closer together, until the gap between them is infinitesimally small, as shown in Figure 4.3. Since the gap between them has now almost vanished, the electric field $\mathcal{E}$ is located almost entirely in the semiconductor depletion layer. (The energy barrier in the gap is now of infinitesimal thickness and so may be neglected.) The electric field $\mathcal{E}$ corresponds now to a gradient of electron potential energy across the depletion layer in the semiconductor, as also shown (schematically) in Figure 4.3. The energy E_C' of an electron (at the conduction band minimum) at the semiconductor–metal interface is larger than the energy E_C of an electron (also at the conduction band minimum) in the bulk of the semiconductor, outside of the depletion layer. The transfer of electrons from semiconductor to metal

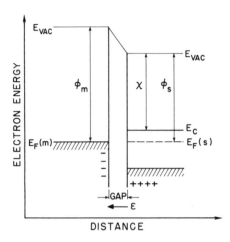

Figure 4.2. Band diagram of a metal-n-type-semiconductor junction with a gap between them. The Fermi levels are shown as equal due to transfer of electrons from semiconductor to metal. The resulting electric field $\mathcal{E}$ is located mainly in the gap between the metal and the semiconductor and is due to the positive charges (indicated by + signs) in the depletion layer of the semiconductor and the negative charges (indicated by − signs) on the metal surface.

continues until the Fermi level in the bulk of the semiconductor is the same as the Fermi level in the metal, as shown in Figure 4.3.

In the limit, as the gap between metal and semiconductor vanishes, the value of the energy difference, *at the surface*, between the vacuum level E_{VAC} and the conduction band minimum E_C' is equal to χ. (This is equivalent to the statement[1] that, as the gap vanishes, the value of the separation of E_C' and $E_F(s)$, at the surface, is equal to the metal work function less the semiconductor electron affinity.) This is shown in Figure 4.3, and the energy barrier ΔE_{ms} between metal and semiconductor is thus $(\phi_m - \chi)$. As seen further from Figure 4.3, $(\phi_m - \chi)$ is also the energy dif-

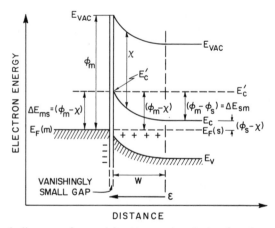

DISTANCE

Figure 4.3. Band diagram of a metal-n-type-semiconductor junction with an infinitesimally small gap between them. The width of the positive space charge region (indicated by + signs) in the semiconductor is W, $\mathcal{E}$ is the electric field, ΔE_{ms} is the energy barrier between metal and semiconductor, and ΔE_{sm} is the energy barrier between semiconductor and metal.

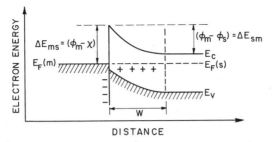

Figure 4.4. Simplified version of the metal–n-type-semiconductor junction band diagram shown in Figure 4.3 at equilibrium, with the gap reduced to zero, for the case in which ϕ_m is larger than ϕ_s.

ference between the Fermi energy $E_F(s)$ in the bulk semiconductor and the energy E_C' of the conduction band minimum at the semiconductor surface.

In the bulk of the semiconductor, we see from Figure 4.3 that the energy difference between the conduction band edge E_C and $E_F(s)$ is equal to $(\phi_s - \chi)$. Therefore, the energy barrier ΔE_{sm} for an electron going from an energy E_C in the bulk semiconductor into the metal is given by

$$\Delta E_{sm} = (\phi_m - \chi) - (\phi_s - \chi) = \phi_m - \phi_s \qquad (4.5)$$

We conclude from these considerations that the energy barrier ΔE_{ms} for an electron going from metal to semiconductor is $(\phi_m - \chi)$, while the barrier ΔE_{sm} for an electron going from the semiconductor bulk into the metal is $(\phi_m - \phi_s)$. Figure 4.4 shows, in the conventional manner, the band diagram based on these results, with the gap or space between metal and semiconductor reduced to zero and the vacuum energy E_{VAC} omitted. The width W of the depletion layer may be found by solving[9] Poisson's equation using the charge density $+ eN_d$ in the space charge region, where N_d is the density of ionized donors in the semiconductor.

Figure 4.5. Band diagrams of a metal and an n-type semiconductor before contact is made, and with a gap or space between them. The metal work function ϕ_m is smaller than the semiconductor work function ϕ_s in this case. The notation is the same as that of Figure 4.1.

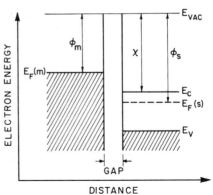

We consider next the metal–n-type-semiconductor junction for the situation in which the semiconductor work function ϕ_s is larger than the metal work function ϕ_m. (The previous discussion treated the opposite case, in which ϕ_m is the larger.) Figure 4.5 shows the band diagram before contact, with a gap between the metal and the semiconductor; the symbols are the same as those used in Figures 4.1–4.4. The metal Fermi energy $E_F(m)$ is now higher than the Fermi level $E_F(s)$ in the semiconductor, so, as the gap between metal and semiconductor is decreased, electrons are transferred to the semiconductor. This is shown in Figure 4.6, in which the Fermi level in the semiconductor is raised, and is equal to the Fermi level in the metal.

At the semiconductor surface, the energy $E_C{}'$ of the conduction band minimum is still an energy χ below the vacuum energy at the surface, as shown in Figure 4.6. Thus we have

$$\chi = E_{\text{VAC}} - E_C{}' \tag{4.6}$$

at the surface. Since equation (4.1) is also still true at the metal surface, we have, combining equations (4.6) and (4.1),

$$\chi - \phi_m = E_F(m) - E_C{}' \tag{4.7}$$

at the joint metal–semiconductor interface, so the conduction band edge energy $E_C{}'$ at the interface is an energy $\chi - \phi_m$ below the Fermi level in the metal.

We now calculate the energy barrier

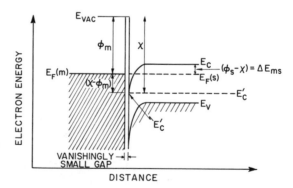

Figure 4.6. Band diagram of a metal–n-type-semiconductor junction, with an infinitesimally small gap between them, for the situation in Figure 4.5. The energy $E_C{}'$ is the energy of the conduction band minimum at the surface and is an energy χ below the vacuum energy at the surface, as shown in the drawing. (The vacuum energy inside the semiconductor is not shown.) The energy barrier between metal and semiconductor is $\Delta E_{ms} = \phi_s - \chi$.

$$\Delta E_{ms} \equiv E_C - E_F(m) \qquad (4.8)$$

for an electron going from the metal to the conduction band edge in the bulk semiconductor, where its energy will be E_C. Since, at equilibrium, $E_F(s)$ equals $E_F(m)$, equation (4.8) becomes

$$\Delta E_{ms} = E_C - E_F(s) = -E_F(s) \qquad (4.9)$$

since we defined E_C (in the bulk) as the energy zero, so $E_C = 0$, and $E_F(s)$ is negative. Substituting equation (4.4) into (4.9) gives

$$\Delta E_{ms} = (\phi_s - \chi) \qquad (4.10)$$

as the energy barrier ΔE_{ms} between the metal and the bulk of the semiconductor. The result (4.10) is shown in Figures 4.6 and 4.7; the latter figure shows the results of Figure 4.5 as the gap between metal and semiconductor goes to zero, again, as is common, with the vacuum energy E_{VAC} omitted. The band diagram of Figure 4.7 is that of a metal-n-type-semiconductor junction, at equilibrium, for the case for which ϕ_s is larger than ϕ_m.

Since electrons are transferred from metal to semiconductor in the situation in which ϕ_m is less than ϕ_s, a negative surface charge develops on the semiconductor. The negative charge is a surface charge because, as shown in Figure 4.7, the semiconductor Fermi level $E_F(s)$ crosses the conduction band edge near the semiconductor–metal junction. This results in an accumulation layer of filled electron states at those electron energies below the Fermi level, and the electron accumulation layer is, as shown in Figure 4.7, confined in space to the region of the semiconductor–metal

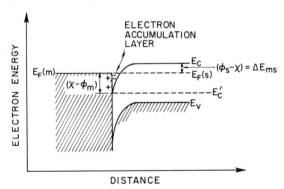

Figure 4.7. Simplified version of the metal–n-type-semiconductor junction band diagram shown in Figure 4.6, at equilibrium, with the gap reduced to zero, for the case in which ϕ_m is smaller than ϕ_s. An electron accumulation layer has formed in the semiconductor near the junction. The spatial extent of the accumulation layer is schematic and has been exaggerated for clarity.

junction surface. (This should be compared with the extended positive space charge region in the semiconductor shown in Figure 4.4 for the case in which ϕ_m is greater than ϕ_s.) This negative surface charge is compensated by a positive surface charge on the metal.

We summarize these results concerning the metal–n-type-semiconductor junction at equilibrium. If the metal work function ϕ_m is larger than the semiconductor work function ϕ_s, then the band diagram of the junction at equilibrium is shown in Figure 4.4. If ϕ_m is smaller than ϕ_s, then the band diagram is shown in Figure 4.7, in which an electron accumulation layer is formed in the semiconductor near the junction. The corresponding results for the metal–p-type-semiconductor junction are left as an exercise.

Effect of an Applied Potential on the Metal–Semiconductor Junction

We next look into the effect of an applied potential on the energy barrier ΔE_{ms} to electron flow from the metal to the semiconductor and the energy barrier ΔE_{sm} to electron flow from the semiconductor to the metal. We consider the metal–n-type-semiconductor junction, so forward bias is a negative potential of magnitude V_a applied to the semiconductor and reverse bias is a positive potential V_a applied to the semiconductor.

In the case in which ϕ_m is larger than ϕ_s, there is a positively charged depletion layer in the semiconductor, as shown in Figure 4.4. We assume that all of the applied potential V_a appears across the depletion layer, so no "band tilting" takes place in the bulk semiconductor. Figure 4.8 shows the effect on the band diagram of Figure 4.4 of applying a forward bias V_a. The energy barrier ΔE_{sm} to the flow of electrons from semiconductor to metal is decreased from its equilibrium value $\phi_m - \phi_s$ to the smaller value $\phi_m - \phi_s - eV_a$. The energy barrier ΔE_{ms} to electron flow from metal to

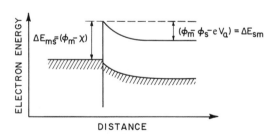

Figure 4.8. Band diagram of the metal–n-type-semiconductor junction in Figure 4.4 (for which ϕ_m is larger than ϕ_s) under an applied forward bias (semiconductor negative) of magnitude V_a.

semiconductor is unchanged from its equilibrium value $\phi_m - \chi$ by forward bias. Figure 4.9 shows the effect of a reverse bias of magnitude V_a on the band diagram in Figure 4.4. The energy barrier ΔE_{ms} is again unchanged, while the energy barrier ΔE_{sm} is increased to $\phi_m - \phi_s + eV_a$.

We consider next the net currents through the junction. These currents are primarily majority carriers[10] in a metal–semiconductor junction, so we neglect any hole currents[11] in our treatment of the metal–n-type-semiconductor junction. The net electron current through the junction is the difference between two components, J_1 and J_2. The current J_1 is the electron current from metal to semiconductor, and J_2 is the electron current from semiconductor to metal. At equilibrium, when the applied potential $V_a = 0$, the net current $J = J_1 + J_2$ must vanish. Choosing the direction from semiconductor to metal as positive, we write

$$J_2 = J_0, \qquad J_1 = -J_0 \qquad (4.11)$$

and at equilibrium, $J = J_1 + J_2 = 0$, as required.

To find the effect of V_a on the electron current J, we note from the lack of effect of V_a on the energy barrier ΔE_{ms} that the electron current J_1 will be essentially unchanged by the applied potential V_a. Since V_a decreases the barrier ΔE_{sm}, we expect that the component J_2 will increase exponentially with increasing magnitude of the forward bias V_a. These results are expressed as

$$J_1 = -J_0$$
$$J_2 = J_0 \exp(eV_a/k_BT)$$

giving the effect of the applied potential V_a on the currents J_1 and J_2. The total current $J = J_1 + J_2$ is then

$$J = J_0[\exp(eV_a/k_BT) - 1] \qquad (4.12)$$

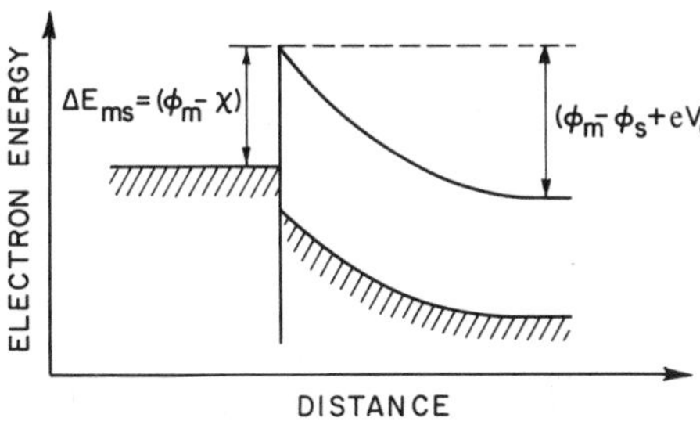

Figure 4.9. Band diagram of the metal–n-type-semiconductor junction in Figure 4.4 (for which ϕ_m is larger than ϕ_s) under an applied reverse bias (semiconductor positive) of magnitude V_a.

giving the total current J as a function of applied bias V_a. We note that for large magnitudes of reverse bias, $J = -J_0$, so J_0 is a reverse saturation current. The overall result is that the total electron current is large for forward bias applied to the metal–semiconductor junction, and is small for reverse bias. The $J - V_a$ characteristic expressed by equation (4.12) is highly asymmetric,[12] just as it is for the semiconductor p–n junction. The calculation of the reverse current J_0 for the metal–semiconductor junction is more complex than it is for the semiconductor p–n junction, and depends on the width of the space charge region relative to the carrier mean free path. For a discussion of the calculation of J_0 for metal–semiconductor junctions, the reader is referred to the literature.[13]

We consider next the effect of an applied potential V_a on a metal–n-type-semiconductor junction in which ϕ_m is smaller than ϕ_s, so the band diagram at equilibrium is that of Figure 4.7. In this case[14] there is no extended space charge layer in the semiconductor (as there was for the situation in which ϕ_m is larger than ϕ_s). We must therefore consider the applied potential V_a as appearing across the entire bulk of the semiconductor when discussing the effect of V_a on the band diagram in Figure 4.7. If V_a is a negative potential applied to the n-type semiconductor (relative to the metal), the result is an increase in the electron potential energy in the semiconductor, i.e., the "band tilting" discussed in Chapter 2. The result of such an applied bias on the band diagram of Figure 4.7 is shown in Figure 4.10. From the figure, we see that there is no energy barrier to electron flow from the semiconductor to the metal when the n-type semiconductor is negative. The band diagram under an applied bias with the n-type semiconductor positive is shown in Figure 4.11, here the electron energy in the semiconductor is decreased. The flow of electrons is from metal to semiconductor when the semiconductor is positive, over the small energy barrier shown in Figure 4.11. The barrier is expected to be small if the quantity $(\phi_s - \chi)$ is small, as it usually is in a doped n-type semiconductor. We conclude that there will be little difference in the electron flow for the two

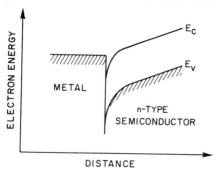

Figure 4.10. Band diagram of a metal–n-type-semiconductor junction (for which ϕ_m is smaller than ϕ_s) under an applied bias in which the semiconductor is negative with respect to the metal.

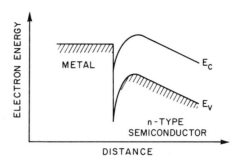

Figure 4.11. Band diagram of a metal–n-type-semiconductor junction (for which ϕ_m is smaller than ϕ_s) under an applied bias such that the semiconductor is positive.

conditions of bias for the metal–n-type-semiconductor junction in which ϕ_m is smaller than ϕ_s. We expect therefore that such a metal–semiconductor junction will form an ohmic (i.e., nonrectifying) contact to the n-type semiconductor.

We may combine our results on the metal–n-type-semiconductor junction as follows. If the metal work function ϕ_m is larger than the semiconductor work function ϕ_s, then the current–voltage characteristic is asymmetric. We therefore expect a metal–n-type-semiconductor contact to be rectifying if ϕ_m is larger than ϕ_s. Conversely, we expect such a contact to be ohmic if ϕ_m is smaller than ϕ_s. These results are an approximate "rule" for the kind of contact, ohmic or rectifying, formed by a metal of work function ϕ_m on an n-type semiconductor of work function ϕ_s. One may show[13] that this "rule" is reversed for metal contacts to a p-type semiconductor. In that case, the contact is ohmic if ϕ_m is larger than ϕ_s, and is rectifying if ϕ_m is smaller than ϕ_s. It should be noted that this "rule" is only approximate and does not always predict the type of contact obtained experimentally. One reason[16] for this is the neglect of the effect of surface states at the semiconductor surface. This point will be treated in Chapter 6 when surface states are discussed.

Physics of the Metal–Insulator–Semiconductor Structure

We now discuss the physics[17–20] of a structure composed of a metal, an insulator, and a semiconductor, as shown schematically in Figure 4.12. This structure is usually referred to as an MIS structure or junction. While the physics of this structure is interesting for its own sake, the MIS junction

Figure 4.12. Metal–insulator–semiconductor (MIS) structure (schematic). The relative thicknesses of the metal and insulator layers are exaggerated.

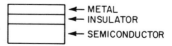

is the basis of important solid state devices, including charge coupled devices (CCD) and the insulated-gate field effect transistor (IGFET), both of which will be discussed later.

We consider first the band diagram, shown in Figure 4.13, of a simple metal–insulator–semiconductor of the type shown in Figure 4.12. We consider the case in which the semiconductor is n type. Figure 4.13 shows the equilibrium band diagram of an "*ideal*" MIS structure, in which the work function ϕ_m of the metal is equal to the work function ϕ_s of the semiconductor. While this "ideal" case is oversimplified,[21,22] it will give us the central physical results necessary for the applications we are interested in. A further idealization is made by neglecting surface states at the interfaces of the structure.

We may now make the following points regarding the equilibrium band diagram in Figure 4.13. The metal and semiconductor Fermi energies are equal at equilibrium in this ideal case. There are assumed to be no free charges in the insulator, so the only mobile charges present are those in the semiconductor and those in the metal. Since there is no charge transfer through the insulator, the Fermi level in the semiconductor remains "flat."[23] However, the mobile charges in the semiconductor may rearrange themselves under the influence of an applied electric field and the semiconductor energy bands may bend up or down (relative to the "flat" Fermi level) to reflect such a rearrangement.

We now consider applying a negative potential $-V_a$ to the metal, relative to the semiconductor. (We may consider the MIS structure as a capacitor, one of whose plates is the semiconductor. Then the MIS structure is connected to a battery, with the metal "plate" of the capacitor nega-

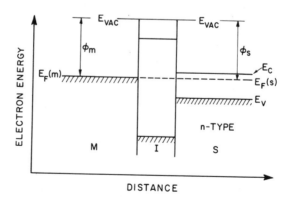

DISTANCE

Figure 4.13. Band diagram at equilibrium of an "ideal" MIS structure, in which the work function ϕ_m of the metal (M) is equal to the work function ϕ_s of the semiconductor (S). The notation is the same as that used in Figure 4.1; the insulator is denoted by I, and its bands and Fermi level are not explicitly labeled.

tive.) Choosing the potential of the metal as zero, the semiconductor is now at a potential $+V_a$ relative to the metal. This means that there is now a difference of potential energy equal to $(-e)(V_a) = -eV_a$ between electrons in the semiconductor and electrons in the metal. The Fermi level in the semiconductor is now an energy eV_a below the Fermi level in the metal. This is shown in Figure 4.14 depicting the situation before any rearrangement of charge in the semiconductor. Figure 4.14 shows the Fermi level in the semiconductor at an energy eV_a below the Fermi level in the metal. The electron energy bands in the insulator are ignored because there are assumed to be no free electrons in the insulator. Figure 4.14 also shows the electric field $\mathscr{E}$, due to the applied potential, which is directed from the semiconductor toward the metal.

Even though no charge flows through the insulator, the mobile charges (i.e., mostly majority electrons) in the semiconductor rearrange themselves under the influence of the electric field $\mathscr{E}$ which exists in the semiconductor and insulator. Electrons are repelled from the insulator–semiconductor interface, producing a region depleted of electrons; part of the potential difference V_a appears across this depletion layer in the semiconductor, and part across the insulator. (We may think of the electrons repelled from the insulator–semiconductor interface as traveling around the external circuit, through the battery supplying the external potential, and being deposited[24,25] on the metal, where they form a surface charge at the metal–insulator interface.) The fact that electrons are moving away from the insulator–semiconductor interface means that the applied potential makes the electron energy higher at that interface than it is in the bulk of the semiconductor. This means that the energy bands in the semiconductor bend upward at the interface with the insulator, as shown in Figure 4.15. Also shown in Figure 4.15 is the depletion layer of width x_d in the semiconductor, and the negative surface charge at the metal–insulator interface.

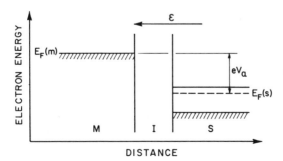

Figure 4.14. Band diagram of the ideal MIS structure of Figure 4.13 with a negative potential $-V_a$ applied to the metal. The diagram is shown before the rearrangement of the mobile charges in the semiconductor under the influence of the electric field $\mathscr{E}$.

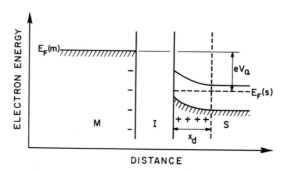

DISTANCE

Figure 4.15. Band diagram of the ideal MIS structure of Figure 4.14 after rearrangement of electrons in the semiconductor. The negative potential $-V_a$ applied to the metal produces a positive space charge region of width x_d in the semiconductor and a corresponding negative surface charge density at the metal–insulator interface.

The space charge density $\varrho(x)$ in the depletion layer is given by

$$\varrho(x) = + eN_d \tag{4.13}$$

where N_d is the density of donors in the n-type semiconductor. Figure 4.16 shows a plot of $\varrho(x)$ as a function of x; $x = 0$ is taken at the insulator–semiconductor interface. The negative surface charge density at the metal–insulator interface is also indicated in Figure 4.16. We could find the magnitude x_d of the depletion layer in the semiconductor by solving Poisson's equation [for the electrostatic potential $V(x)$ in the depletion layer and the insulator] using the charge density (4.13). However, we will merely note that the solution would give V as a function of x, and vice versa, so we expect that $x = x(V)$, meaning that the extent x_d of the depletion layer will depend on the magnitude $|-V_a|$ of the negative potential applied to the metal. If the magnitude $|-V_a|$ is increased, the extent of the depletion layer will increase (just as for a reverse-biased p–n junction), to the larger value x_d'. Increasing the magnitude $|-V_a|$ further will increase the energy difference eV_a between the metal and the semiconductor. We thus expect the bands in the semiconductor depletion layer to bend further upward

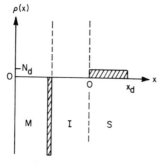

Figure 4.16. Space charge density $\varrho(x)$ as a function of distance x for the ideal MIS structure (n-type semiconductor) with a negative potential applied to the metal. The quantity x_d is the extent of the positive space charge density in the semiconductor and N_d is the ionized donor density. The space charge at the metal–insulator interface represents a surface density of negative charge.

when the magnitude $|-V_a|$ is increased. These effects are shown in Figure 4.17, in which the depletion layer in the semiconductor is now of width x_d'. The upward bending of the semiconductor energy bands may, with increasing magnitude $|-V_a|$, produce a situation in which the valence band edge at the semiconductor surface comes close to, or even crosses, the Fermi level $E_F(s)$.

This crossing of the Fermi level by the valence band edge at the surface means that the semiconductor becomes effectively p-type near the surface, forming an inversion layer of p-type material in the n-type semiconductor. If there are sufficient minority holes available, they will accumulate near the surface, thereby forming the inversion layer (if there are not sufficient minority holes available, the inversion layer will not form) as minority holes diffuse to, and accumulate in, the potential energy well formed by the band bending at the surface.[26] The formation of the inversion layer modifies the plot of space charge density $\varrho(x)$ in Figure 4.16. There would now be an additional component of $\varrho(x)$, extending the width of the inversion layer from $x = 0$, representing the positive space charge density due to accumulated holes in the surface inversion layer.[27]

Further, as the magnitude $|-V_a|$ increases, the band bending at the insulator–semiconductor interface increases. This makes the potential energy well for holes deeper, allowing the well to accomodate more holes. This in turn leads to an increased hole density and an increased conductivity of the inversion layer. Qualitatively then, we expect an increase in $|-V_a|$ to increase the conductivity of the inversion layer.

We may summarize the physical results embodied in the Figure 4.17 as follows. If a negative potential $-V_a$ is applied to a metal–insulator–

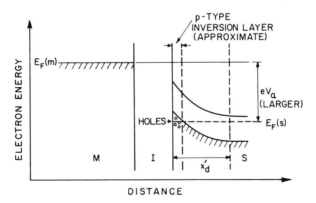

Figure 4.17. Band diagram of the ideal MIS structure of Figure 4.15 with a negative potential $-V_a$ of larger magnitude applied to the metal. The width of the positive space charge layer in the semiconductor has increased to x_d' and a p-type inversion layer has formed at the insulator–semiconductor interface.

n-type-semiconductor structure, a depletion layer of positive space charge will form in the semiconductor. For sufficiently large values of $| - V_a |$, a *p*-type inversion layer will form at the surface of the *n*-type semiconductor if sufficient minority holes are available. Increasing $| - V_a |$ increases the conductivity of the inversion layer.

In the following chapter, we discuss a number of electronic devices based on metal-semiconductor and metal–insulator–semiconductor junctions.

Problems

4.1. *Metal–p-Type-Semiconductor Junction.* (a) Draw band diagrams for the junction between a metal and a *p*-type semiconductor. (b) If ϕ_m is the metal work function and ϕ_s is the semiconductor work function, verify that the junction is a rectifying contact if ϕ_s is larger than ϕ_m, and an ohmic contact if ϕ_s is smaller than ϕ_m. (Neglect surface states in your treatment.)

4.2. *Inversion Layer in MIS Structure.* Using band diagrams, discuss the formation of an *n*-type inversion layer when a positive potential is applied to the metal in an MIS structure in which the semiconductor is *p* type. Use the "ideal" MIS structure in which the work functions of the metal and the semiconductor are equal, and neglect surface states at the *I–S* interface.

4.3. *Nonideal MIS Structure.* Consult the literature (e.g., Reference 4, Sec. 8.1) and discuss the attainment of equilibrium via charge transfer in a nonideal MIS structure, for which the metal and semiconductor work functions are not equal.

4.4. *Charge Distribution in MIS Structure.* Make a plot, analogous to Figure 4.16, of charge density $\varrho(x)$ as a function of distance x for an MIS structure in which the inversion layer shown in Figure 4.17 has formed. On your plot, label the different charge density regions.

References and Comments

1. H. K. Henisch, *Semiconductor Contacts*, Oxford University Press, London (1984), Section 1.3.1, pages 26–29.
2. J. P. McKelvey, *Solid State and Semiconductor Physics*, Harper and Row, New York (1966), Chapter 16.
3. A. van der Ziel, *Solid State Physical Electronics*, Second Edition, Prentice-Hall, New York (1968), Sections 5.4 and 14.1.
4. R. S. Muller and T. I. Kamins, *Device Electronics for Integrated Circuits*, Second Edition, John Wiley, New York (1986), Chapter 3.
5. See, for example, M. S. Tyagi, "Physics of Shottky Barrier Junctions", in *Metal–Semiconductor Shottky Barrier Junctions and their Applications* (B. L. Sharma, editor), Plenum Press, New York (1984), pages 2–11.

6. C. Kittel, *Introduction to Solid State Physics*, Sixth Edition, John Wiley, New York (1986), pages 534–536.

7. See, for example, J. I. Pankove, *Optical Processes in Semiconductors*, Prentice-Hall, New York (1971), page 298, for values of the work function and of the electron affinity for semiconductors.

8. See, for example, A. J. Dekker, *Solid State Physics*, Prentice-Hall, New York (1957), page 223, for representative values of the work function for metals.

9. See, for example, A. van der Ziel, Reference 3, Section 14.1a.

10. S. M. Sze, *Physics of Semiconductor Devices*, Second Edition, John Wiley, New York (1981), Section 5.4.

11. J. P. McKelvey, Reference 2, Section 16.2, pages 482–485.

12. A. Y. C. Yu, "The Metal–Semiconductor Contact: An Old Device with a New Future," *IEEE Spectrum*, **7**, 83–89 (March 1970), gives several examples.

13. S. M. Sze, Reference 10, Section 5.4; A. van der Ziel, Reference 3, pages 266–274; M. S. Tyagi, Reference 5, pages 35–53.

14. A. van der Ziel, Reference 3, pages 100–101.

15. A. van der Ziel, Reference 3, pages 102–103.

16. A. van der Ziel, Reference 3, page 101; J. P. McKelvey, Reference 2, Section 16.3, pages 485–489.

17. A. Goetzberger and S. M. Sze, "Metal–Insulator–Semiconductor (MIS) Physics," in *Applied Solid State Science*, R. Wolfe (editor), Academic Press, New York (1969), Volume 1, pages 154–238.

18. S. M. Sze, Reference 10, Section 7.2.

19. R. S. Muller and T. I. Kamins, Reference 4, Chapter 8.

20. A. S. Grove, *Physics and Technology of Semiconductor Devices*, John Wiley, New York (1967), page 317–327.

21. See A. S. Grove, Reference 20, pages 278–285, for a discussion of the effect of deviations from the "ideal" MIS model. See also S. M. Sze, Reference 10, Section 7.3.3, and A. Goetzberger and S. M. Sze, Reference 17, pages 205–210, for similar discussions.

22. See Reference 17 for a discussion of the important nonideal $Al–SiO_2–Si$ structure.

23. S. M. Sze, Reference 10, Section 7.2; A. S. Grove, Reference 20, pages 265–266.

24. B. G. Streetman, *Solid State Electronic Devices*, Second Edition, Prentice-Hall, New York (1980), Section 8.3.2.

25. A. S. Grove, Reference 20, Section 9.1.

26. W. S. Boyle and G. E. Smith, "Charge-Coupled Devices—A New Approach to MIS Device Structures," *IEEE Spectrum*, **8**, 18–27 (1971), especially pages 20–21 and Figure 5.

27. See, for example, S. M. Sze, Reference 10, page 370, Figure 6(b) for this plot, but drawn for the case of an *n*-type inversion layer in a *p*-type semiconductor.

28. R. S. Muller and T. I. Kamins, Reference 4, page 426.

Suggested Reading

H. K. HENISCH, *Semiconductor Contacts*, Oxford University Press, Oxford (1984). This is an advanced monograph on the physics of metal–semiconductor contacts and surface properties.

A. G. MILNES AND D. L. FEUCHT, *Heterojunctions and Metal–Semiconductor Junctions*, Academic Press, New York (1972). Chapters 6 and 7 of this monograph discuss metal–semiconductor junctions and a number of their applications.

A. VAN DER ZIEL, *Solid State Physical Electronics*, Second Edition, Prentice-Hall, New York (1968). This electrical engineering text discusses metal–semiconductor junctions and contacts in a terse and mathematical manner.

R. S. MULLER AND T. I. KAMINS, *Device Electronics for Integrated Circuits*, Second Edition, John Wiley, New York (1986). Chapters 3 and 8 of this electrical engineering text discuss metal–semiconductor and metal–oxide–semiconductor structures, including the technologically important metal–SiO_2–silicon system.

S. M. SZE, *Physics of Semiconductor Devices*, Second Edition, John Wiley, New York (1981). This treatise compiles a great deal of information on semiconductor devices, as well as many references. However, the explanations of the physics are very brief.

B. L. SHARMA, editor, *Metal–Semiconductor Shottky Barrier Junctions and their Applications*, Plenum Press, New York (1984). This monograph consists of review articles, by different authors, covering many aspects of its subject.

C. M. WOLFE, N. HOLONYAK, AND G. E. STILLMAN, *Physical Properties of Semiconductors*, Prentice-Hall, Englewood Cliffs (1989). This recent text at the graduate level includes a chapter on surface structures, which discusses, among other topics, metal–semiconductor and metal–insulator–semiconductor junctions.

5

Metal–Semiconductor and Metal–Insulator–Semiconductor Devices

Introduction

This short chapter discusses the basic ideas of some of the applications of metal–semiconductor and metal–insulator–semiconductor structures. These include the Schottky metal–semiconductor diode, various insulated-gate field-effect transistors, and charge-coupled devices. The myriad applications of these devices, especially the field-effect transistor, are left to the literature for lack of space, time, and knowledge on the author's part. However, the basic ideas are discussed, some applications are mentioned, and a number of references to this important and changing field are given.

Metal–Semiconductor (Schottky) Diodes

As pointed out in Chapter 4, the current–voltage characteristic of a metal–semiconductor junction is of the form

$$J = J_0[\exp(eV_a/k_BT) - 1] \qquad (5.1)$$

where V_a is the applied voltage and J_0 is a reverse saturation current. The asymmetric current–voltage characteristic is similar to that of a semiconductor p–n junction. For this reason, the Schottky diode can be used as a rectifier.

As pointed out in Chapter 2, minority carrier storage limits the speed with which a semiconductor *p–n* junction can be switched from a condition of forward bias to one of reverse bias. The metal–semiconductor junction, as noted in Chapter 4, is essentially a majority carrier device. For this reason, minority carrier storage is not a problem in the Schottky diode, and an important application[1–4] of such diodes is their use at high (e.g., microwave) frequencies in mixers, etc.

The Insulated-Gate Field-Effect Transistor (IGFET)

In Chapter 3, we discussed the junction field-effect transistor (JFET). A second type of field-effect transistor has a metal gate electrode insulated from the semiconductor and is usually called an insulated-gate field-effect transistor[5–7] or IGFET. Since the semiconductor involved is generally silicon, and the insulator is then silicon dioxide, this type of FET is often called a metal–oxide–semiconductor field-effect transistor, or MOSFET. Finally, this general class of devices is often called metal–insulator–semiconductor, or MIS, devices. The great importance of these devices is that they can be made very small in integrated circuits used in a wide variety of applications, especially digital circuits.

There are several kinds of insulated-gate field-effect transistors. We consider first the diffused-channel MOSFET, which is shown schematically in Figure 5.1. The source *S* and drain *D* are n^+ regions fabricated by diffusion into the *p*-type semiconductor substrate. A channel of *n*-type material between source and drain is also made by diffusion. An insulating layer (usually SiO_2) separates the channel from the metal gate. There are also diffused-channel MOSFET devices with *p*-type channels. We will use the *n*-channel device as our example.

A potential of either polarity may be applied to the gate; the appropriate source and drain polarities for an *n*-channel device are shown in Figure 5.1. We consider first a negative potential applied to the gate and look upon the metal gate and *n*-type semiconductor of the channel as forming a parallel plate capacitor in which the oxide layer is the dielectric

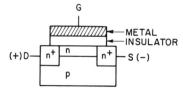

Figure 5.1. Schematic drawing of a diffused *n*-channel MOSFET or IGFET. The drain, source, and gate are labeled *D*, *S*, and *G*, respectively.

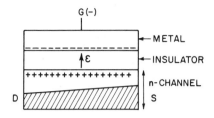

Figure 5.2. Schematic view of the metal gate (G), insulator, and n-type channel MOSFET with a negative potential applied to the gate G; the electric field in the insulator is $\mathscr{E}$. The undepleted part of the channel is shown shaded.

between the plates. This is shown in Figure 5.2. An electric field $\mathscr{E}$ will exist in the insulator, maintained between the metal and the semiconductor of the n channel. The applied negative potential will repel electrons from the n channel, leaving behind ionized donors bearing a positive charge. These positive charges (shown in Figure 5.2) and the induced negative charges at the metal–insulator interface maintain the electric field $\mathscr{E}$ in the insulator. The effect of the applied negative potential on the gate G is to deplete part of the n channel of mobile electrons. In Figure 5.2 the *un*depleted portion of the channel is shown as shaded. Since the source to drain current in the n-channel MOSFET is composed of majority electrons in the n channel, we see that the negative potential applied to the gate depletes the channel of current carriers, thereby reducing the effective cross-section area of the channel in a manner analogous to that taking place in the JFET. The resistance of the n channel is thus increased by the negative potential applied to the gate, and the electron current I_{DS} from source to drain is decreased. Just as in the JFET (discussed in Chapter 3), the depletion region extends further into the channel near the drain, as indicated schematically in Figure 5.2. The overall effect of the negative potential applied to the gate is to deplete the n channel of carriers, so we speak of this n-channel MOSFET operating in the depletion mode.

If a positive potential is applied to the gate, the opposite happens. Electrons enter the channel from the source, increasing the conductivity of the n channel. This is an n-channel MOSFET operating in the enhancement mode, in which the conductivity of the n channel is increased by the application of a positive voltage to the gate.

We see then that the magnitude and sign of the potential V_{GS} applied to the gate (relative to the source) will determine the conductivity of the n channel and hence will determine the current I_{DS} flowing between source and drain for a given value of the voltage V_{DS} between source and drain. To find the current–voltage characteristic of the n-channel MOSFET, we know that the current I_{DS} will be larger (for a given value of the voltage V_{DS}) in the enhancement mode than it will in the depletion mode. We thus expect that the current I_{DS} will vary with the potential V_{GS} (between gate

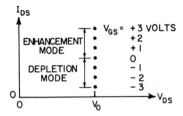

Figure 5.3. Schematic variation of current I_{DS} in an *n*-channel MOSFET for a particular value V_0 of the source-to-drain voltage V_{DS}. The enhancement and depletion modes of operation are indicated.

and source), for a fixed value V_0 of V_{DS}, approximately as shown in Figure 5.3. In that figure, we see that I_{DS} increases with increasingly positive values of the gate voltage V_{GS} for a fixed value of V_{DS}.

We consider next how I_{DS} will vary with the source to drain voltage V_{DS} for a given value of the gate voltage V_{GS}. We expect that (as for the JFET) the relatively lightly doped *n* channel will act as a distributed resistance and the voltage V_{DS} across the channel will, as it did for the JFET, produce a higher potential at a point in the channel near the drain than near the source. In the depletion mode, we expect that the channel resistance (dV_{DS}/dI_{DS}) will increase with increasing V_{DS} for a given value of the gate voltage V_{GS}. We thus expect that the I_{DS}–V_{DS} characteristic of an *n*-channel MOSFET operating in the depletion mode will resemble the JFET characteristic. This is shown in Figure 5.4.[8] In the enhancement mode, the shape of the I_{DS}–V_{DS} curves is quite similar[9] to that of the curves in the depletion mode. The channel resistance increases with increasing magnitude of V_{DS} for any value of the gate voltage.

Comparing Figure 5.4 for the MOSFET with Figure 3.22 for the JFET, we see that the MOSFET is also a voltage-controlled amplifier in that the gate voltage V_{GS} controls the device current I_{DS}, just as in the JFET. We note also that the device current I_{DS} in the MOSFET is composed of majority

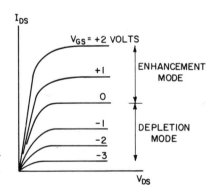

Figure 5.4. Current–voltage (I_{DS}–V_{DS}) characteristics for a diffused *n*-channel MOSFET. (After Streetman.[8])

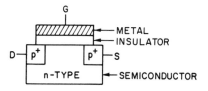

Figure 5.5. Schematic view of an MIS induced-channel MOSFET. The symbols S, D, and G refer to source, drain, and gate, respectively.

carriers (in this case, electrons in the n channel) only. The same is true of the JFET.

While we have discussed the n-channel MOSFET, all of the ideas obtained pertain also to the p-channel MOSFET on making the necessary sign changes in the device currents and voltages.

The diffused-channel IGFET or MOSFET may therefore have either an n or a p channel, and may be operated in either enhancement or depletion modes. The input resistance (to the gate) is very high (typically many megohms) because of the insulator layer under the gate.

The Induced-Channel MOSFET

The induced-channel (or enhancement mode) MOSFET[10–13] is an insulated-gate field-effect transistor which does not, at equilibrium, have a diffused channel between source and drain. It is a member of the general class of metal–insulator–semiconductor (MIS) devices whose band diagram was considered in Chapter 4. We consider the MIS structure shown schematically in Figure 5.5, which is the same as Figure 4.11 with a source S, drain D, and gate G added. The metal gate in Figure 5.5 is separated from the n-type semiconductor by an insulator layer.

The device in Figure 5.5 functions as a field-effect transistor in the following manner. We saw in Chapter 4, Figure 4.16, that the application of a negative potential to the gate G of the metal–insulator–n-type-semiconductor device can produce a p-type inversion layer. This layer, at the insulator–semiconductor interface, is in addition to a depletion layer of

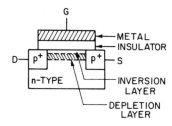

Figure 5.6. Schematic view of the formation of an induced p-type channel in the MOSFET of Figure 5.5 by the application of a negative potential to the gate.

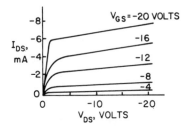

Figure 5.7. Current–voltage (I_{DS}–V_{DS}) characteristics of an induced p-channel MOSFET. (After Millman and Halkias.[12])

positive fixed space charge extending into the n-type semiconductor. These results are shown schematically[†] in Figure 5.6. The p-type inversion layer in the n-type semiconductor acts as a channel between the source and drain of the device, so one speaks of an induced-channel[‡] MOSFET.

As we saw in Chapter 4, the increasing magnitude of the negative potential applied to the gate increases the conductivity of the induced channel. The drain-current–drain-voltage (I_{DS}–V_{DS}) characteristic[12,13,15] of the induced-channel MOSFET will be similar to the characteristics of the other field-effect transistors we have discussed, including the saturation[15a] of the device current with increasing drain voltage. A representative characteristic[12] is shown in Figure 5.7, from which we see that the induced-channel MOSFET also functions as a voltage-controlled amplifier like the other field-effect transistors discussed.

Summary of the Physics of Field-Effect Transistors

All types of FET are voltage controlled amplifiers because an applied electric field controls the flow of majority carriers in a channel which is either diffused or induced. All types of FET have a high input impedence. In the JFET the control voltage is applied to a reverse biased p^+n junction. In the IGFET, the control voltage is applied between gate and source separated by a high-resistance insulator. All types of FET operate by controlling the conductivity of a channel (diffused or induced) with an applied electric field, hence the name field-effect transistor. The IGFET is particularly ap-

---

[†] The shapes of the inversion layer and the depletion layer are shown as uniform for simplicity in Figure 5.6. This will be true only for small drain voltages; the actual shapes for various drain voltages are given by Muller and Kamins.[14]

[‡] The name "enhancement-mode" MOSFET is more commonly used in electrical engineering, but the term "induced-channel" will be retained here because of its physical descriptiveness.

propriate for fabrication in small size, making the principal use of the MOSFET in integrated circuits. For this reason, the IGFET is more important than the JFET.

Applications of the MOSFET

While MOSFET transistors are used in various amplification configurations,[16,17] the most common application of the MOSFET is in digital electronics as an electronically controlled switch.[18] Figure 5.8 shows[19] a partial set of current–voltage (I_{DS}–V_{DS}) curves, for two values of the gate voltage V_{GS}, for a p-channel MOSFET typical of those used in a digital integrated circuit. We can see from Figure 5.8 that the resistance dV_{DS}/dI_{DS} can be changed from its lower value for $V_{GS} = -9$ V to its higher value for $V_{GS} = -4$ V (for a fixed value of V_{DS}). The device can therefore be switched from an "on" state of lower resistance dV_{DS}/dI_{DS} to an "off" state of higher resistance by changing the magnitude of the control voltage V_{GS}. The device is therefore bistable and can be used in a variety of digital circuits.[20]

The primary use of the MOSFET is in integrated circuits for digital applications. An integrated circuit (as opposed to an assembly of the discrete individual devices discussed in this book) is a single crystal chip of silicon on which are fabricated both active and passive components and their interconnections. These will not be discussed here for two reasons. First, this vast technology is in a more or less constant state of change. Second, there are many expositions[10] of the device physics relevant to integrated circuit technology. (Even though it is more than a decade old, the September 1977 issue of *Scientific American* on microelectronics can still be recommended to students of solid state physics. Especially pertinent is the article[21] on microelectronic circuit elements. This article discusses, among other things, a comparison of MOS devices of various types. The use of the MOSFET in semiconductor computer memories[22-24] is treated in another article in the same issue.)

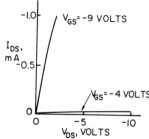

Figure 5.8. Current I_{DS} versus voltage V_{DS} for two values of the gate voltage V_{GS} for a p-channel MOSFET. (After Oldham and Schwarz.[19])

Charge-Coupled Devices

We consider in this section a type of device, based on the MIS structure, whose function could not be duplicated[21] using discrete components, but which is instead inherently based on the concept of the integrated circuit. This is the charge-coupled device (abbreviated CCD), which is also called a charge transfer device.

We return to the result, illustrated in Figure 4.16, that the application of a negative potential $-V_a$ to the gate of a metal–insulator–n-type-semiconductor structure can produce a p-type inversion layer at the insulator–semiconductor interface. Figure 4.16 shows the upward bending of the electron energy bands and the accumulation of holes producing the inversion layer. Since Figure 4.16 is a plot of electron energy as a function of distance, we may (mentally) convert it to a plot of hole energy as a function of distance by multiplying by -1. If we do so to the valence band contour of Figure 4.16, we obtain the curve shown in Figure 5.9. This figure shows a schematic plot of hole energy as a function of distance in the n-type semiconductor of an MIS structure with a negative potential $-V_a$ applied to the metal; the zero of distance is at the insulator–semiconductor interface. Figure 5.9 shows a potential energy well, of depth D, for holes. Further, we saw from the discussion of Figure 4.16 that the larger the magnitude of V_a, the greater the band bending in the semiconductor and the greater the depth D of the potential energy well. Our conclusion is that a negative bias $-V_a$ applied to the gate of the MIS structure produces an energy well for holes at the I–S interface, and, the larger the magnitude of V_a, the deeper the energy well.

Given the existence of this energy well for holes in the biased MIS structure, we see that minority holes in the n-type semiconductor will accumulate there until a condition is reached in which the diffusion of holes away from the surface is just balanced by the drift of holes toward the surface. This steady state is referred to as the condition of saturation.

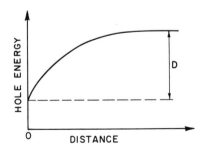

Figure 5.9. Energy well for holes, of depth D, in a metal–insulator–n-type-semiconductor structure with a negative potential applied to the metal gate.

Figure 5.10. Schematic view of a charge-coupled device. The metal gates are shaded. The applied potentials are such that $V_2 > V_1 = V_3 = V_4$. Holes are shown in the deeper energy well under gate 2. The dashed line indicates (approximately) the potential energy contour for holes as a function of distance.

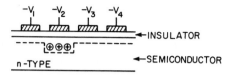

(If the minority hole density corresponding to saturation is exceeded, the excess holes flow back into the bulk semiconductor where they vanish by recombination with majority electrons.) The accumulation and storage of minority carriers in an energy well at the semiconductor surface in a biased MIS device is the basis of the charge-coupled device.[25-27]

Given the possibility of accumulation and storage[28] of carriers in potential energy wells, we may consider charge-coupled devices based on the transfer of minority carriers between different potential wells at different parts of the device. The MIS structure in Figure 5.10 has a series of closely spaced metal gates on a common insulating layer on an n-type semiconductor substrate. The potential $-V_a$ applied to the four electrodes has the values $-V_1, -V_2, -V_3,$ and $-V_4$, where V_2 is larger than V_1 and $V_1 = V_3 = V_4$. The dashed line in Figure 5.10 represents the variation of the hole potential energy with distance along the structure; the larger the magnitude of the applied potential, the deeper the energy well for holes. The deeper well is under the gate 2 and is the one in which any available (or introduced) minority holes will accumulate. The regions under gates 1, 3, and 4 are shown free of minority holes, a situation which can occur because silicon can be made with a low density of surface and bulk generation centers. While not really a steady state situation, these wells can remain empty for times of the order of seconds[29] before thermally generated minority carriers accumulate.

In Figure 5.10, we have a situation in which holes have been introduced (in some manner) into the potential energy well under gate 2 and are stored there. Suppose next that V_3 is increased so V_3 is larger than V_2. Then the potential energy contour will change to that shown in Figure 5.11, in which the energy well under gate 3 is now deeper than that under gate 2. Holes will be transferred into the deeper well, as indicated schematically in Figure

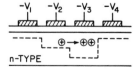

Figure 5.11. Schematic view of transfer of holes between energy wells in a charge-coupled device. The applied potentials are such that $V_3 > V_2 > V_1 = V_4$, so the well under gate 3 is deeper than the well under gate 2.

5.11. The key conclusion is that, in this type of MIS structure, minority carriers (holes in this case) can be transferred between potential energy wells by appropriate manipulation of the bias voltages applied to the metal electrodes. Since the presence or absence of charge carriers in a particular well constitutes information, this device allows the transfer of information along the semiconductor surface. This is the basic idea of the charge transfer or charge-coupled device.

Various methods[27,30] of introducing the minority carriers are in use. These include injection from a *p–n* junction and optical generation by incident photons. Among the methods[27,30] of detecting the minority carriers is, for example, injecting the holes into the *n* region of a reverse-biased diode, thereby producing a current pulse in the external circuit. The literature[25-27] also gives information on "clocking" sequences of voltages applied to the device electrodes in order to transfer the minority carriers along the series of potential energy wells.

The relative simplicity and small size of the charge-coupled device have made it attractive for a number of applications.[27] In particular, the presence or absence of charge in a given potential well provides two stable states, so the charge-coupled device is used as the basis of semiconductor computer memories.[31-33] Another application is the use of charge-coupled devices in imaging,[34,35] i.e., the conversion of a pattern of light and dark areas into an electrical signal. In this application, minority carriers are introduced into the device by optical absorption. These carriers then accumulate in the potential wells of the device, and the resulting charges are transferred and detected. The result is an output whose amplitude varies with the light intensity at different points of the original light pattern.

Problems

5.1. *Area Imaging with a* CCD. Using the literature, discuss area imaging with a charge-coupled device. In particular, since the fabrication of a CCD avoids the diffusion steps necessary in making a bipolar or FET device, consider the use of "exotic" semiconductors (i.e., anything other than silicon) for a CCD imaging device. Choose a particular "exotic" semiconductor for imaging outside the visible, and suggest an application.

5.2. *Speed of* MOS *Devices.* Read the article by Meindl,[21] considering especially the fact that electron mobilities are generally larger than hole mobilities. This means that MOS devices with *n*-type channels tend to be faster than *p*-channel defices. From this point of view, discuss the possibility of using a semiconductor (e.g., GaAs) with an electron mobility larger than that of silicon for a MOSFET. Consider as many semiconductor parameters as you can in your discussion of the feasibility and desirability of such a device.

References and Comments

1. R. S. Muller and T. I. Kamins, *Device Electronics for Integrated Circuits*, Second Edition, John Wiley, New York (1986), pages 157–160.
2. J. Millman and C. C. Halkias, *Integrated Electronics*, McGraw-Hill, New York (1972), pages 228–230.
3. D. Cooper, B. Bixby, and L. Carver, "Power Shottky Diodes," *Electronics*, February 5, 1976, pages 85–89.
4. A. Y. C. Yu, "The Metal–Semiconductor Contact: An Old Device with a New Future," *IEEE Spectrum*, 7, 83–89 (March 1970).
5. B. G. Streetman, *Solid State Electronic Devices*, First Edition, Prentice-Hall, New York (1972), pages 293–301.
5a. B. G. Streetman, *Solid State Electronic Devices*, Second Edition, Prentice-Hall, New York (1980), Section 8.3.
6. S. M. Sze, *Physics of Semiconductor Devices*, Second Edition, John Wiley, New York (1981), Chapter 8.
7. A. S. Grove, *Physics and Technology of Semiconductor Devices*, John Wiley, New York (1967), Chapter 11.
8. B. G. Streetman, Reference 5, page 295, Figure 8-15(a).
9. J. Millman and C. C. Halkias, Reference 2, page 326, Figure 10-13(a).
10. R. S. Muller and T. I. Kamins, Reference 1, Chapters 8, 9, 10.
11. B. G. Streetman, Reference 5, pages 293–301; Reference 5a, Section 8.3.
12. J. Millman and C. C. Halkias, Reference 2, pages 322–324.
13. W. G. Oldham and S. E. Schwarz, *Electrical Engineering*, Holt, Rinehart, Winston, New York (1984), pages 482–491.
14. R. S. Muller and T. I. Kamins, Reference 1, Figure 9.3, page 425, and Figure 9.4, page 428.
15. R. S. Muller and T. I. Kamins, Reference 1, Section 9.1.
15a. R. S. Muller and T. I. Kamins, Reference 1, page 429.
16. J. Brophy, *Basic Electronics for Scientists*, Third Edition, McGraw-Hill, New York (1977), pages 143–150 and 156–169.
17. J. Millman and C. C. Halkias, Reference 2, Chapter 10.
18. W. G. Oldham and S. E. Schwarz, *An Introduction to Electronics*, Holt, Rinehart, Winston, New York (1972), pages 559–562.
19. W. G. Oldham and S. E. Schwarz, Reference 18, Figure 13.9, pages 557–558.
20. W. G. Oldham and S. E. Schwarz, Reference 18, pages 559–584.
21. J. D. Meindl, "Microelectronic Circuit Elements," in *Scientific American*, **237**, 70–81 (September 1977).
22. D. A. Hoges, "Microelectronic Memories," in *Scientific American*, **237**, 130–145 (September 1977).
23. R. S. Muller and T. I. Kamins, Reference 1, pages 449–454.
24. S. Middelhoek, P. K. George, and P. Dekker, *Physics of Computer Memory Devices*, Academic Press, New York (1976), pages 234–268.
25. W. S. Boyle and G. E. Smith, "Charge-Coupled Devices—A New Approach to MIS Device Structures," *IEEE Spectrum*, **8**, 18–27 (July 1971).
26. R. S. Muller and T. I. Kamins, Reference 1, pages 409–417.
27. C. H. Sequin and M. F. Tompsett, *Charge Transfer Devices*, Academic Press, New York (1975).

28. See References 25 and 26 for a discussion of the problems of, and factors affecting, the times of storage and transfer of minority carriers.
29. W. S. Boyle and G. E. Smith, Reference 25, page 21.
30. W. S. Boyle and G. E. Smith, Reference 25, Figures 12 and 13. Note that the captions of these figures are reversed; Figure 12 shows detection methods and Figure 13 shows input schemes.
31. S. Middelhoek, P. K. George, and P. Dekker, Reference 24, pages 269–281.
32. C. H. Sequin and M. F. Tompsett, Reference 27, pages 236–260.
33. L. Altman, "Charge Coupled Devices Move in on Memories and Analog Signal Processing," *Electronics*, **47**, 91–98 (August 8, 1974).
34. C. H. Sequin and M. F. Tompsett, Reference 27, Chapter 5.
35. M. F. Tompsett, W. J. Bertram, D. A. Sealer, and C. H. Sequin, "Charge Coupling Improves its Image," *Electronics*, **46**, 162–169 (January 18, 1973).

Suggested Reading

Scientific American, September 1977: Special Issue on Microelectronics. Even though now somewhat outdated, this issue of the magazine contains a number of useful articles on the physics and applications of MOSFET transistors.

R. S. MULLER AND T. I. KAMINS, *Device Electronics for Integrated Circuits*, Second Edition, John Wiley, New York (1986). This electrical engineering text stresses the physics of various devices from the point of view of their use in integrated circuits. Chapters 8 and 9 discuss the silicon IGFET.

C. H. SEQUIN AND M. F. TOMPSETT, *Charge Transfer Devices*, Academic Press, New York (1975). This monograph covers the physics and applications of charge-coupled devices in detail.

J. MILLMAN AND C. C. HALKIAS, *Integrated Electronics*, McGraw-Hill, New York (1972). This electrical engineering text gives examples of the uses of many devices.

R. F. PIERRET, *Field Effect Devices*, Addison-Wesley, Reading Mass. (1983). This short book (one of a series on solid state devices) discusses MIS structures and their device applications.

6

Other Semiconductor Devices

Introduction

In this chapter, we discuss some additional topics on semiconductor physics and their applications. After a brief introduction to surface states on semiconductors, some aspects of the band structure at the surface are considered. These concepts are then applied to discussions of metal–semiconductor contacts and photoemission from semiconductors, including the idea of devices with a negative electron affinity. The next topic is the physics of the transferred electron, or Gunn, effect, discussed in terms of band structure using GaAs as the example. One mode (domain formation) of exploitation of the Gunn effect in producing microwave electrical oscillations is considered. Finally, some aspects of the electronic structure of amorphous semiconductors are introduced as the basis for a brief discussion of memory and switching devices.

Semiconductor Surface States

Usually, when we consider the electronic states in a crystal, we neglect the existence of a surface on the crystal. An example would be the Kronig–Penney model[1] of the periodic potential in a perfect crystal. However, surfaces are very important when considering semiconductor devices. In particular, the emission of electrons from semiconductors by incident photons (the photoelectric effect) or electrons (secondary electron emission) is a process that directly involves the semiconductor surface. We now con-

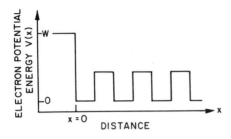

Figure 6.1. Terminated Kronig–Penney model of a one-dimensional crystal, with an energy barrier of height W located at the surface $x = 0$.

sider some basic ideas of the physics of semiconductor surfaces,[2-7] with emphasis on those aspects[8] important in electron emission processes. (In the simple picture we will employ, we neglect many important questions, such as the detailed structure[8a] of the surface of a semiconductor.)

Consider a variant of the Kronig–Penney model in which[7] there is a surface barrier, i.e., a potential energy step of height W at the surface. Figure 6.1 shows both the surface barrier and the variation of the electron potential energy $V(x)$ as a function of distance x. The surface of the one-dimensional crystal is at $x = 0$. Calculation[9] of the energy band structure using the potential energy function of Figure 6.1 yields results different from those of the usual Kronig–Penney model of an infinite one-dimensional crystal. The familiar Kronig–Penney result is that the electronic structure of the crystal is composed of allowed bands of energies with forbidden gaps between them. When the crystal surface is added to the problem by including the surface potential barrier shown in Figure 6.1, the result is the introduction of energy levels in the forbidden gap. These states are due to the presence of the surface itself, and are called[†] surface states. The wave function for such a surface state is localized in space near the surface. There is generally one such surface state per surface atom, resulting in a surface state density[3] of the order of 10^{15} cm^{-2}. The surface states associated with the semiconductor surface itself are usually called *fast* surface states. (The term "fast" comes from the fact[10] that the transition times between such states and the semiconductor bulk are short, of the order of 10^{-6} sec or less.) In high enough densities, the surface state wave functions can overlap to form bands. The surface properties of a particular crystal will therefore depend on the density and energy distribution of the surface states.

A second kind of state, called a *slow* surface state, is found on semiconductor surfaces. These slow states are associated with the oxide layer, or

---

† We ignore the different types[7] (Tamm and Shockley) of surface states since, for our purposes, all we will need is the existence of surface states.

other adsorbed species, generally found on real (as opposed to ideal) semiconductor surfaces. For these states, the transition times with the bulk semiconductor are long (of the order of seconds). The density of slow surface states is believed[5,11] to be larger than that of the fast states, and they are believed also to carry electric charge in a manner similar to donor and acceptor states in a bulk semiconductor. An example is an acceptor surface state due to an oxygen atom on a silicon surface. The oxygen atom can accept an electron from the silicon, creating a hole in the valence band of the bulk silicon crystal. We will consider the effects of both donor and acceptor states on a semiconductor surface, without inquiring as to their detailed identity and nature.[4,12]

Band Structure at the Semiconductor Surface

We discuss the interaction of surface states with the electron energy levels of a bulk semiconductor to investigate how the band structure at the surface[8,13] is modified relative to the bulk band structure. Consider an n-type semiconductor, at 0 K, with a number of acceptor surface states. These acceptor surface states are neutral when empty and are negatively charged when filled. We will assume, for the present, that the number of acceptor surface states is such that, after the rearrangement of charge to be described, all of the acceptor states are occupied by electrons. We assume also that the donor levels in the bulk n-type semiconductor are, initially, all occupied by electrons at the low temperature we are considering. The initial situation is shown in Figure 6.2.

Electrons from bulk donor states flow into the empty acceptor surface states, producing a negative surface charge and leaving behind uncom-

Figure 6.2. Band diagram of an n-type semiconductor with acceptor surface states. The temperature is 0 K, E_{VAC} is the vacuum level, E_F is the Fermi energy, CB and VB refer, respectively, to the conduction and valence bands, and the symbol $(-\bullet-)$ indicates an occupied donor level. The situation pictured is that before any transfer of electrons between bulk states and surface states.

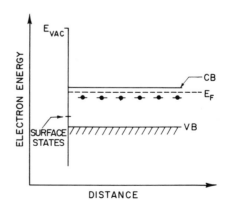

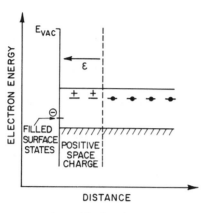

Figure 6.3. Band diagram (schematic) of the situation in Figure 6.2 after transfer of electrons from some donor states to acceptor surface states. Positively charged empty donors are indicated by $\pm$ and $\mathscr{E}$ is the electric field in the positive space charge region. The Fermi level is not shown.

pensated positively charged ionized donors. The result is the creation of a region of fixed positive space charge in the bulk of the semiconductor. (If the temperature is above 0 K, so some of the donors are ionized, the result is still the same. In this case, electrons from conduction band states near the surface fill the acceptor levels. A region of positive space charge in the bulk, and a negative surface charge, still result.) There is, at equilibrium, an electric field $\mathscr{E}$ in the space charge region, as shown in the band diagram in Figure 6.3. The electric field $\mathscr{E}$ repels further electrons from the surface, and equilibrium is attained.

The electric field $\mathscr{E}$ corresponds to a gradient of electrostatic potential $V(x)$ in the space charge region. Since $\mathscr{E}$ is directed toward the surface, $V(x)$ must increase in the direction away from the surface, as shown schematically in Figure 6.4. The electrostatic potential V is thus higher in the bulk of the semiconductor than it is at the surface, by an amount denoted by Φ_s. The quantity Φ_s is called the surface potential. Since the electrostatic potential is higher by Φ_s in the bulk semiconductor, the electron energy in the bulk is lower by an amount $e\Phi_s$ than it is at the surface. Figure 6.5 gives the band diagram, which shows the positive space charge region extending a distance d from the surface into the bulk semiconductor. The energy bands in the n-type semiconductor are thus "bent" upward at the surface due to the existence of surface acceptor states which have been

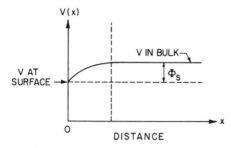

Figure 6.4. Schematic plot of electrostatic potential $V(x)$ as a function of distance x (from the surface taken as $x = 0$). The surface potential Φ_s is the difference in electrostatic potential between bulk and surface. The space charge region is delineated by a vertical dashed line.

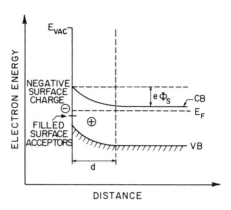

Figure 6.5. Band diagram of the situation in Figure 6.3 after the attainment of equilibrium. The positive space charge region extends a distance d into the semiconductor from the surface. The energy bands are "bent" upward at the surface by an amount $e\Phi_s$, where Φ_s is the surface potential. All surface acceptor states are filled with electrons.

filled with electrons. We note that we have been considering the case in which all of the surface acceptors are full. In that event, as shown in Figure 6.5, the Fermi level at the surface is above the energy of the surface acceptor states since the Fermi function is equal to unity at energies less than the Fermi energy E_F. We summarize the results above by saying that the filled surface acceptors on the n-type semiconductor produce a positive space charge region, a negative surface charge, and upward bending of the energy bands at the surface by an amount $e\Phi_s$, where Φ_s is the surface potential.

In a similar manner, we may examine the surface band structure[8,13] of a p-type semiconductor with donor surface states. Again we assume that, at equilibrium, all of the surface donors are ionized, having given up their electrons to acceptors in the bulk. The result is a positive surface charge and a region of negative space charge in the bulk, the latter being due to fixed ionized acceptors. Because of the space charge, the electrostatic potential is lower in the bulk than at the surface by Φ_s, the surface potential. This in turn means that the electron energy in the bulk is higher by an amount $e\Phi_s$ than it is at the surface. The net result is that the energy bands are "bent" downward at the surface, relative to the bulk, as shown in Figure 6.6. In this figure, we note that the surface donor energy is above the Fermi level at the surface because, in this case, all of the surface donors are empty.

In the cases considered above, both n and p type, the Fermi level position at the surface is determined by the Fermi level position in the bulk because we have assumed a situation in which all of the surface states are ionized. The amount $e\Phi_s$ of band bending (to be discussed later) may be thought of as merely "raising" or "lowering" the conduction and valence band edges at the surface relative to the fixed or "flat" Fermi level.

We consider next the case[14] of a semiconductor in which the number

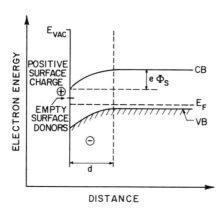

Figure 6.6. Band diagram, at equilibrium, of a p-type semiconductor with donor surface states, all of which are ionized. The negative space charge region extends a distance d into the semiconductor from the surface. The energy bands are "bent" downward at the surface by an amount $e\Phi_s$, where Φ_s is the surface potential.

of surface states is very large, so that *not* all of them are ionized. For concreteness, we consider donor surface states on a p-type semiconductor, all of which lie at roughly the same energy Δ below the conduction band edge. Figure 6.7 shows this situation before the attainment of equilibrium. Electrons flow from the surface donors into the p-type semiconductor bulk *only* until the Fermi level at the surface coincides with the surface donor energy. This means that some of the surface donors will remain un-ionized and still occupied by electrons. This fact *fixes* the position of the Fermi level at the surface at the donor energy, i.e., at an energy Δ below the conduction band edge, as is shown in Figure 6.8. On comparing Figure 6.8 with Figure 6.7, we see that, as equilibrium is attained by the flow of electrons into the bulk, the overall effect is to "raise" the Fermi level in the bulk semiconductor until it coincides with the surface donor energy. The Fermi level position at the surface is determined only by the surface donor energy Δ, and is independent of the doping in the p-type bulk. One says in this

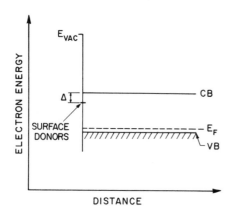

Figure 6.7. Band diagram, before the attainment of equilibrium, of a p-type semiconductor with surface donor states an energy Δ below the conduction band edge.

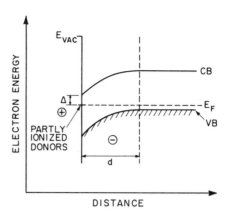

Figure 6.8. Band diagram, at equilibrium, of a *p*-type semiconductor with a sufficiently large number of surface donors, so that the donors are not completely ionized. The surface Fermi level is pinned at the surface donor energy. The negative space charge region extends a distance *d* into the bulk, and there is a positive surface charge.

case that the surface Fermi level is "stabilized" or "pinned"[15,16] at the energy Δ, relative to the conduction band edge.

Calculation of the Amount of Band Bending

We may obtain some simple quantitative results on the amount $(e\Phi_s)$ of band bending, using the situation shown in Figure 6.5. This is the case of an *n*-type semiconductor with acceptor surface states, all of which are ionized, leading to a region of positive fixed space charge extending a distance *d* into the bulk. We will call the bulk donor density N_d, and the density of surface acceptors is n_s. Then the space charge density ϱ in the bulk is given by

$$\varrho = +eN_d \tag{6.1}$$

and Poisson's equation for the electrostatic potential $V(x)$ is

$$\frac{d^2V}{dx^2} = \frac{-4\pi\varrho}{\varepsilon} = \frac{-4\pi eN_d}{\varepsilon} \tag{6.2}$$

where ε is the dielectric constant of the semiconductor and $0 \le x \le d$. The boundary conditions on the problem are as follows. First, there are n_s filled (i.e., ionized) surface acceptors per unit area, so the negative surface charge density σ is

$$\sigma = -en_s \tag{6.3}$$

If we regard the space charge region in the bulk semiconductor as a capacitor filled with a dielectric of dielectric constant ε, then the electric field $\mathscr{E}$ at the semiconductor surface is of magnitude

$$\mathscr{E} = \frac{4\pi\sigma}{\varepsilon} = \frac{-4\pi e n_s}{\varepsilon} \tag{6.4}$$

Since $\mathscr{E} = -dV/dx$, equation (6.4) gives the first boundary condition on the electrostatic potential as

$$\frac{dV}{dx} = \frac{4\pi e n_s}{\varepsilon} \tag{6.5}$$

at the surface (located at $x = 0$) of the semiconductor.

The second boundary condition stems from the assumption that the electric field $\mathscr{E}$ is confined to the space charge region, so $\mathscr{E} = 0$ at $x = d$, the boundary of the space charge layer in the bulk. Thus we have the second condition that, at $x = d$,

$$\frac{dV}{dx} = 0 \tag{6.6}$$

Integrating Poisson's equation (6.2) gives

$$\frac{dV}{dx} = \frac{-4\pi e N_d x}{\varepsilon} + C \tag{6.7}$$

which, combined with the boundary condition (6.6) at $x = d$, leads to the value $4\pi e N_d\, d/\varepsilon$ for the constant C. Equation (6.7) becomes

$$\frac{dV}{dx} = \frac{4\pi e N_d}{\varepsilon}(d - x) \tag{6.8}$$

Applying the condition expressed by equation (6.5) at $x = 0$ gives the result that

$$d = n_s/N_d \tag{6.9}$$

for the width d of the positive space charge region.

We integrate equation (6.8) subject to the following boundary conditions on the electrostatic potential. First, we choose $V = 0$ at the surface $x = 0$. Second, this choice of the zero of potential means that V must equal the surface potential Φ_s in the bulk of the semiconductor, as seen from Figure 6.4. We have then that

$$V(x) = \Phi_s$$

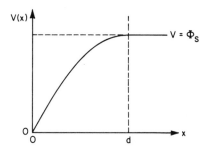

Figure 6.9. Electrostatic potential $V(x)$ as a function of distance x, as given by equation (6.10) for the space charge region $0 \le x \le d$; Φ_s is the surface potential. (Arbitrary units are used for V and x.)

for $x = d$. Integrating (6.8) subject to the second of these conditions gives

$$V(x) = (-4\pi e N_d/2\varepsilon)(d - x)^2 + \Phi_s \qquad (6.10)$$

for the spatial variation of the electrostatic potential. The condition that $V = 0$ at $x = 0$ leads to the expression

$$\Phi_s = 2\pi e N_d\, d^2/\varepsilon \qquad (6.11)$$

for the surface potential. Combining (6.11) with (6.9) gives the alternate form

$$\Phi_s = 2\pi e n_s^2/\varepsilon N_d \qquad (6.12)$$

If we plot $V(x)$ as a function of x using equation (6.10), the curve shown in Figure 6.9 is obtained. A representative value[17] for Φ_s is approximately 0.9 V in silicon (300 K) containing 10^{16} acceptors per cm³. We note also that we could also have chosen $V = 0$ at the point $x = d$, a choice that leads to $V = -\Phi_s$ at the surface $x = 0$. This choice would perhaps be in keeping with the name "surface potential" for Φ_s.

Finally, we multiply (6.11) and (6.12) by e to obtain $e\Phi_s$, the amount of band bending, as

$$e\Phi_s = 2\pi e^2 N_d\, d^2/\varepsilon = 2\pi e^2 n_s^2/\varepsilon N_d \qquad (6.13)$$

We will return to equation (6.13) when we discuss photoemission from semiconductors.

Effect of Surface States on Metal–Semiconductor Contacts

In Chapter 4, we discussed metal contacts to semiconductors in terms of the work functions ϕ_m and ϕ_s of the metal and semiconductor, respectively. It was shown that a metal contact to an n-type semiconductor would

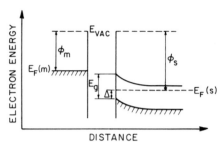

Figure 6.10. Band diagram of a metal (work function ϕ_m) and an n-type semiconductor (work function ϕ_s), where ϕ_m is less than ϕ_s, before equilibrium is established. The surface Fermi level of the semiconductor is pinned at an energy Δ above the surface valence band edge. The semiconductor energy gap is E_g, and the metal and semiconductor Fermi levels are denoted by $E_F(m)$ and $E_F(s)$, respectively. [The position of $E_F(s)$ is exaggerated for clarity.]

be rectifying if ϕ_m were larger than ϕ_s and ohmic if ϕ_m were smaller than ϕ_s; the opposite was true for contacts to a p-type semiconductor. The predictions of these "rules" do not generally work out in practice because the treatment leading to these conclusions neglected the existence of surface states on the semiconductor surface.

Experimentally, it has been found that the height of the energy barrier[18] at the surface of a metal contact to a given semiconductor is approximately independent of the work function ϕ_m of the metal. This is true for a number of different semiconductors, and the explanation[19-22] lies in the "pinning" of the Fermi level at the semiconductor surface. Suppose, in an n-type semiconductor, the Fermi level is pinned at the surface at an energy Δ above the valence band edge at the surface. This situation is (by analogy with Figure 6.8) shown in Figure 6.10, which also shows a metal whose work function ϕ_m is smaller than the work function ϕ_s of the semiconductor, whose energy gap is E_g. Since ϕ_m is smaller than ϕ_s, for an n-type semiconductor one would expect, based on the simpler ideas above, that the energy barrier at the surface would be small (see Figure 4.6), and the contact would be ohmic. However, because the Fermi level in the semiconductor is pinned, then, as equilibrium is established, the metal

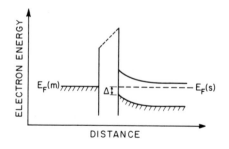

Figure 6.11. Band diagram of Figure 6.10, with a small gap between metal and semiconductor, as equilibrium is attained, and the Fermi levels are equalized.

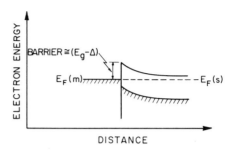

Figure 6.12. Band diagram of the metal–*n*-type-semiconductor contact (when the gap between them has vanished) at equilibrium. The height of the energy barrier at the surface is approximately $E_g - \varDelta$. (The symbols are those used in Figure 6.10.)

Fermi level $E_F(m)$ must equalize with the fixed position of the semiconductor Fermi level $E_F(s)$ at the surface. This is shown in Figure 6.11, in which a gap still exists between the metal and semiconductor as they come together. The equilibrium band diagram is shown in Figure 6.12, and its key feature, when compared with Figure 4.6, is the existence of an energy barrier, whose height is approximately $E_g - \varDelta$, between metal and semiconductor. We would expect the energy band structure in Figure 6.12 to be rectifying, even though it is composed of a metal and *n*-type semiconductor for which ϕ_m is less than ϕ_s. As seen from these band diagrams, we also expect the height $E_g - \varDelta$ of the surface energy barrier to be the same for any value of the metal work function ϕ_m. The barrier height is determined[20] only by the doping and surface properties (i.e., $\varDelta$) of the semiconductor and so is independent of the work function of the metal involved. Analogous results may be obtained for metal contacts to *p*-type semiconductors.

Photoemission from Semiconductors

We now use the results we have obtained on semiconductor surface physics to discuss the photoemission[8,23] of electrons from semiconductors, with some mention of relatively recent developments.

We review in Figure 6.13 the terminology which was introduced in Chapter 4; in the energy band diagram in Figure 6.13, the influence of the surface is, for the moment, neglected. From Chapter 4, we recall the definitions

$$\chi \equiv E_{\text{VAC}} - E_C \qquad (6.14)$$

$$\phi_s \equiv E_{\text{VAC}} - E_F \qquad (6.15)$$

of the electron affinity χ and the work function ϕ_s of a semiconductor.

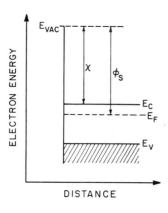

Figure 6.13. Band diagram showing the electron affinity χ and work function ϕ_s of a semiconductor; E_{VAC} is the vacuum level, E_F is the Fermi level, E_C is the energy of the conduction band cdge, and E_V is the valence band edge. (Occupied electron states are shaded.)

In equations (6.14) and (6.15), E_{VAC} is the vacuum energy, E_C is the energy of the conduction band edge, and E_F is the energy of the Fermi level.

We may use Figure 6.13 to discuss the threshold energy E_t for photo-emission. The minimum photon energy $E_t = \hbar\omega_t$ that will excite an electron from the semiconductor into the vacuum is, from Figure 6.13,

$$E_t = \hbar\omega_t = E_g + \chi \tag{6.16}$$

where $E_g \equiv E_C - E_V$ is the energy gap of the semiconductor. For the moment, we consider only direct transitions by electrons at the valence band edge (at energy E_V), so no phonon energies are involved in the expression for the threshold energy. We note also that the threshold, or minimum energy, transition of an electron into vacuum will have as its initial state the highest occupied electron energy level. If, as shown in Figures 6.14 and 6.15, the semiconductor is degenerate n or p type, the expression

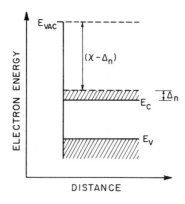

Figure 6.14. Band diagram illustrating photo-emission of electrons from a degenerate n-type semiconductor. The energy threshold $E_t = \chi - \Delta_n$, where Δ_n is the energy, above the band edge energy E_C, to which the conduction band is filled. (Occupied electron states are shaded.)

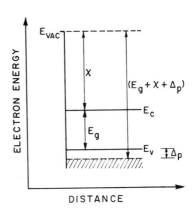

Figure 6.15. Band diagram illustrating photo-emission of electrons from a degenerate p-type semiconductor. The energy threshold $E_t = E_g + \chi + \Delta_p$, where E_g is the energy gap and Δ_p is the depth, below the band edge energy E_V, of the highest occupied electron level. (Occupied electron states are shaded.)

(6.16) for the threshold energy is modified. Equation (6.17),

$$E_t = \chi - \Delta_n \tag{6.17}$$

gives the threshold energy E_t for photoemission of an electron when the conduction band is filled with electrons to an energy Δ_n above the band edge energy E_C. Equation (6.18),

$$E_t = E_g + \chi + \Delta_p \tag{6.18}$$

gives the threshold energy E_t for photoemission of an electron from the valence band when the highest filled electron state is an energy Δ_p below the valence band edge energy E_V. The result is that the threshold energy for photoemission of electrons from a semiconductor varies with the impurity content if the doping is high enough to produce degeneracy.

Effect of the Surface on Photoemission

We now want to discuss the effect of the surface on the photoemission of electrons from a semiconductor. We consider, in Figure 6.16, a p-type semiconductor with donor surface states, in which, as shown in Figure 6.6, the bands are "bent downward" at the surface by an energy $e\Phi_s$, where Φ_s is the surface potential. The width of the space charge layer at the surface is denoted by d, and distance x into the crystal is measured from $x = 0$ at the surface. From the band diagram in Figure 6.16, we see that the electron affinity χ varies with distance x into the crystal, so $\chi = \chi(x)$. If we denote the electron affinity at the surface by χ_s and the electron affinity

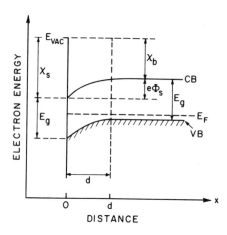

Figure 6.16. Band diagram for a p-type semiconductor showing an amount $e\Phi_s$ of band bending. The electron affinity at the surface and in the bulk are, respectively, χ_s and χ_b, and d is the width of the space charge layer.

in the bulk $(x > d)$ of the crystal by χ_b, then

$$\chi_s = \chi_b + e\Phi_s \tag{6.19}$$

relates χ_s and χ_b. Equation (6.19) says that the electron affinity at the surface is larger than the electron affinity in the bulk by an energy equal to the band bending.

The threshold energy E_t for photoemission of an electron from the valence band edge into the vacuum is, from equation (6.16), given by

$$E_t = E_g + \chi \tag{6.16}$$

Since, from (6.19), χ is larger at the surface than in the bulk, the threshold energy for electron emission is smaller in the bulk than at the surface by an amount $e\Phi_s$. This is seen most clearly on rewriting (6.19) as

$$\chi_b = \chi_s - e\Phi_s \tag{6.20}$$

which states that, if the surface electron affinity χ_s is unchanged,[23] χ_b is decreased. Then the threshold energy E_t for photoemission of electrons from the bulk is given by

$$E_t(\text{bulk}) = E_g + \chi_b = E_g + \chi_s - e\Phi_s \tag{6.21}$$

Equation (6.21) tells us that, all other things being equal, it is desirable to have a large value of $e\Phi_s$ because the bulk electron affinity and the bulk threshold energy for emission are both reduced.

However, all other things are not equal in this case because an electron photoexcited in the bulk by a photon of energy greater than $E_t(\text{bulk})$

must still physically cross the region of the crystal between $x = d$ and $x = 0$ to reach the surface in order to escape into the vacuum. The photo-excited electron can lose energy by phonon emission during this crossing, and, if d is large, may arrive at the surface with an energy less than E_{VAC}; in such a case the electron will not be emitted into the vacuum. It is therefore desirable that d be small so that the escaping photoelectron makes few energy-losing collisions on its way to the surface. If we write the analog of equation (6.13) for a p-type semiconductor, we obtain

$$e\Phi_s = 2\pi e^2 N_a \, d^2/\varepsilon \qquad (6.22)$$

where N_a is the acceptor impurity concentration, so we have

$$d^2 = \varepsilon(e\Phi_s)/2\pi e^2 N_a \qquad (6.23)$$

Equation (6.23) tells us that, for a given amount $e\Phi_s$ of band bending (determined by the physical state of the semiconductor surface[24, 25]) d will be smaller for larger values of the acceptor impurity density N_a. A representative value[24] of d is about 100 Å for acceptor densities in the range 10^{18}–10^{19} cm^{-3}.

As a specific example, consider p-type GaP, whose band diagram is shown in Figure 6.17. The acceptor density is sufficiently high that the Fermi level E_F is very close to the valence band edge; the energy gap $E_g \cong 2.2$ eV. The amount of band bending shown is arbitrary; ϕ is the work funct on $E_{VAC} - E_F$ at the surface, and χ_b is the electron affinity in the bulk. Since the acceptor density is high, we may assume that E_F is sufficiently close to the valence band that we may write

$$\phi \cong E_g + \chi_b \qquad (6.24)$$

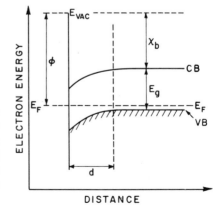

Figure 6.17. Band diagram of a heavily doped p-type semiconductor, with energy gap E_g, surface work function ϕ, and bulk electron affinity χ_b. Since the Fermi level E_F is close to the valence band in this semiconductor, it is approximately true that $\phi \cong E_g + \chi_b$.

Taking a value of $\phi \cong 5$ eV as typical[26] of the III–V semiconductors, we obtain a value of $\chi_b \cong 2.8$ eV as the energy barrier to an electron to be emitted into the vacuum.

Negative Electron Affinity in Semiconductors

It is clear from Figure 6.17 and equation (6.24) that, if the surface work function ϕ were decreased, the electron affinity χ_b in the bulk would be decreased, resulting in a smaller energy barrier to photoexcited bulk electrons emitted into the vacuum. Given this result, the following possibility now presents itself. Suppose a metal with a very small work function ϕ_m could be applied as a thin layer (i.e., thinner than the electron mean free path). Such a metal layer might be expected to reduce the work function ϕ in equation (6.24) to the smaller value ϕ_m, as shown in Figure 6.18. The thickness of the metal is t, and that of the space charge layer in the semiconductor is d. For an electron at the conduction band edge in the bulk of the semiconductor in Figure 6.18, the energy barrier to emission in the vacuum is χ_b, where

$$\phi_m \cong E_g + \chi_b \qquad (6.25)$$

$$\chi_b \cong \phi_m - E_g \qquad (6.26)$$

Again, the height of the Fermi level above the valence band has been neglected in the heavily doped p-type semiconductor. From equation (6.26), we see that the smaller the quantity $\phi_m - E_g$, the smaller the bulk electron affinity χ_b, and the lower the energy barrier between a photoexcited electron and emission into the vacuum. The foregoing discussion presupposes that

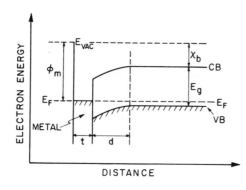

Figure 6.18. Band diagram of a heavily doped p-type semiconductor with a metal layer, of work function ϕ_m, on the surface. The relation $\phi_m \cong E_g + \chi_b$ is approximately true because the Fermi level is close to the valence band.

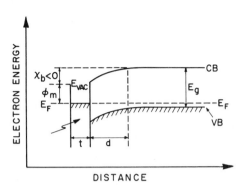

Figure 6.19. Band diagram of a heavily doped p-type semiconductor with a metal of work function ϕ_m on the surface. Since ϕ_m is smaller than the band gap E_g of the semiconductor, χ_b is negative, resulting in a condition of negative electron affinity.

the electron can travel the distance $d + t$ and still arrive at the surface with an energy greater than E_{VAC}, sufficient to escape into the vacuum.

It can also be seen from Figure 6.18 that the threshold energy E_t for photoemission of an electron from the valence band edge into vacuum is

$$E_t = E_g + \chi_b \cong \phi_m \qquad (6.27)$$

so the smaller the value of ϕ_m, the lower the value of the threshold energy E_t.

Examination of equation (6.26) shows the interesting result that, if the energy gap E_g is *larger* than the metal work function ϕ_m, then χ_b is *negative*. The physical meaning of a negative value of χ_b is that, as seen from Figure 6.19, the conduction band edge in the bulk is at a higher energy than the vacuum level at the surface. This means that the energy barrier between a conduction band electron in the bulk and the vacuum level is negative. This situation is referred to as negative electron affinity, and has two main effects. First, the negative energy barrier to emission results in an increase in the number of electrons emitted per absorbed photon. Second, as seen from Figure 6.19, the threshold energy E_t for electron emission is now the band gap energy E_g, since an electron excited only to the bottom of the conduction band now has enough energy to be emitted into vacuum.

Negative electron affinity has been achieved in several semiconductors,[27,28] generally using cesium (or a mixture of cesium and oxygen) as the metal of low work function ϕ_m on the surface of the semiconductor. The procedure for preparing the cesiated surface is generally proprietary, and the Cs–O layer is usually an empirically optimized surface less than an electron mean free path (about 25 Å) in thickness. On p-type GaP, the cesium layer on the surface has a work function[29] $\phi_m \cong 1.2$ eV, resulting, from equation (6.26), in a value of the electron affinity $\chi_b \cong (1.2 - 2.2)$ eV

$= -1.0$ eV. The threshold photon energy for photoemission in GaP with negative electron affinity is[30] at about 2 eV, close to the expected threshold at the band gap energy.

The attainment of negative electron affinity has resulted in improved photoemissive devices, such as photomultiplier tubes, of much higher efficiency. For example, the quantum efficiency[30] (the number of electrons emitted per absorbed photon) is as high as 0.40 for cesiated GaP; this figure is about an order of magnitude higher than that of earlier conventional photoemitters, such as the semiconducting compound K_3Sb. Negative electron affinity has been obtained in other semiconductors, and these have been utilized in various photoemissive devices.[31,32]

We now briefly consider secondary electron emission.[33] When a high-energy electron is incident on a solid semiconductor (or metal), it may excite more than one electron into the vacuum. The electrons so produced are termed secondary electrons. If these secondary electrons are in turn incident on a semiconductor, further "secondaries" can be produced, resulting in a cascade or multiplication of the original electron if there are many stages of secondary electron production. This is the physical idea behind the photomultiplier tube illustrated in Figure 6.20. Photomultiplier tubes produce a cascade of electrons for each photoelectron, and are thus sensitive photon detectors. Recent photomultipliers use photoemissive surfaces (photocathodes) treated to achieve negative electron affinity and thus have high quantum efficiency. If an appropriate semiconductor with a low surface work function or, indeed, with negative electron affinity, is used as the first multiplier stage or dynode, a more efficient secondary emission[32,34] process takes place, resulting in a higher electron multiplication ratio at a lower incident photoelectron energy. Such recent photomultiplier tubes exhibit higher gain and lower noise.

Physics of the Transferred Electron (Gunn) Effect

This section will discuss some aspects of the physics of the transferred electron, or Gunn, effect in semiconductors. This effect is essentially the

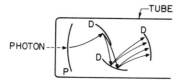

Figure 6.20. Schematic view of a photomultiplier tube with the photoemissive surface P and several secondary emission multiplier stages (dynodes) marked D. The arrows indicate electron paths, showing multiplication.

observation of negative differential conductivity in a bulk semiconductor (in contrast to the negative differential conductivity of a degenerate *p–n* junction). For this reason, one speaks of bulk negative differential conductivity (often abbreviated BNDC). Experimentally, it is found that the electron drift velocity v decreases with increasing applied electric field $\mathscr{E}$ in some *n*-type semiconductors. The curve in Figure 6.21 shows experimental results[35] for *n*-type GaAs at room temperature. The electron velocity increases with increasing electric field intensity, until the velocity reaches a maximum at an electric field of approximately 3200 V cm^{-1}. At higher values of the electric field, the electron velocity *decreases* with increasing electric field. Recalling from equation (1.12) that the magnitude J of the electron current density is given by

$$J = nev \tag{6.28}$$

where n is the electron density, we can write

$$\sigma = dJ/d\mathscr{E} = ne(dv/d\mathscr{E}) \tag{6.29}$$

where σ is the differential electrical conductivity. From Equation (6.29), we see that a negative value of the slope $dv/d\mathscr{E}$, as seen in Figure 6.21 for electric fields larger than 3200 V cm^{-1}, leads to a negative value of the differential conductivity σ.

For this reason, the decrease in electron velocity, with increasing electric field, in a bulk semiconductor is called bulk negative differential conductivity. The value $\mathscr{E}_C$ of the electric field above which $dv/d\mathscr{E}$ is negative is sometimes called the critical value; in Figure 6.21, $\mathscr{E}_C$ is about 3200 V cm^{-1}.

We discuss next the transferred-electron mechanism believed respon-

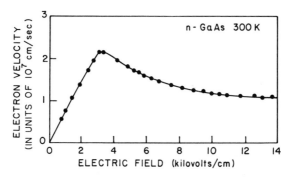

Figure 6.21. Electron velocity as a function of electric field in *n*-type GaAs at room temperature. (After Ruch and Kino.[35])

sible for bulk negative differential conductivity in *n*-type GaAs. We then consider how differential conductivity leads to the Gunn effect, in which the application of a DC electric field of several thousand volts to a bulk sample of GaAs produces electrical oscillations in the microwave frequency range.

To discuss the transferred-electron mechanism leading to negative differential conductivity, we examine the band structure[36-38] of GaAs at room temperature shown in Figure 6.22. The lowest conduction band minimum, at Γ, is 1.42 eV above the valence band edge, which is also at Γ; the energy gap is thus direct at Γ. There are also subsidiary conduction band minima at X, and at L, which are, respectively, 0.48 eV and 0.28 eV above the conduction band minimum at Γ. We will be interested in electrons (in *n*-type GaAs) in the minima at Γ and at L. The effective masses of electrons in the two minima are different. The values,[39] at room temperature, are $m^*(\Gamma) = 0.063m_0$ for electrons in the conduction band at Γ, and $m^*(L) = 0.55m_0$ for electrons in the minimum at L, where m_0 is the free electron mass. Normally, in the absence of an applied electric field, the conduction electrons in GaAs are essentially all in the minimum at Γ, and the electron mobility will be determined by the value of the electron effective mass $m^*(\Gamma)$. However, when an electric field larger than $\mathscr{E}_C$ is applied to the semiconductor, electrons in the conduction band minimum at Γ gain enough energy from the electric field to be transferred (i.e., scattered) into the four higher energy minima at L.

We note next that the combined density of states in the four equivalent minima at L is larger than the density of states in the Γ minimum. This is due to two factors. First, we expect the density of states to increase with increasing electron mass,[40] and $m^*(L)$ is larger than $m^*(\Gamma)$. Second, there are four equivalent minima at L. Qualitatively speaking, we expect that it is relatively unlikely that electrons will be scattered from L (with a higher

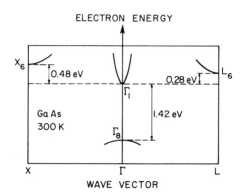

Figure 6.22. Band structure[37] of GaAs at room temperature.[39] The valence band structure is not shown.

density of states) to Γ (with a lower density of states). We conclude that the L minima will remain populated with electrons while the applied electric field remains above the critical value.

The transfer of electrons from Γ to L by the electric field means that the electron mobility in the semiconductor will be (at least partially) due to electrons at L. Since electrons at L have a higher effective mass than electrons at Γ, the transfer of electrons from Γ to L results in a decrease in the electron mobility. This is the reason that the electron drift velocity (or, equivalently, the mobility) decreases with increasing electric field in GaAs, as shown in Figure 6.21. The transferred-electron effect thus results, for large enough electric fields, in a bulk negative differential conductivity in GaAs.

It is clear that the existence of the transferred-electron mechanism in GaAs depends directly on its band structure. The band structure requirements[41] in an n-type semiconductor can be seen to be as follows. The electron effective mass in the lowest conduction band minimum should be low relative to that in the higher conduction band minimum. This will lead to electrons having a lower mobility, and to the density of states being larger, in the upper minimum. The energy separation ΔE between the upper and lower minima should be larger than about $4k_BT$ so that the upper minimum is not occupied at low values of the applied electric field. Further, ΔE should be less than half the energy gap so that the critical field $\mathscr{E}_C$ will be small enough that impact ionization (and its consequent increase in conductivity) will be negligible. Finally, the band gap should be larger than about 1 eV, at room temperature, so that conductivity due to intrinsic carriers will not be a problem. From the band structure parameters for GaAs given above, it can be seen that these criteria are satisfied. Other semiconductors in which the transferred-electron effect and bulk negative differential conductivity have been observed are[41,42] InP, CdTe, and ZnSe. These band structure requirements form a necessary, but not sufficient, condition for the existence of the transferred-electron effect. However, other considerations[41] may prevent the exploitation of the effect.

Given the existence of bulk negative differential conductivity in a semiconductor, we next need to discuss how the application of a sufficiently large DC electric field results in electrical oscillations. This effect, originally observed by Gunn, is only one of a number of ways[42-45] in which negative differential conductivity in a semiconductor may be exploited to produce oscillations at microwave frequencies. The mode to be discussed is not particularly efficient or important technologically, but its physics is quite interesting and worth discussing.

Consider a semiconductor, say GaAs, exhibiting bulk negative differential conductivtiy, and which is under an applied DC electric field. We will show that such a situation leads[44] to the production of space charge instabilities in the semiconductor. To see this, we will discuss first an electrically neutral sample of a normal semiconductor, in which the differential conductivity is positive, under an applied DC electric field. From Gauss's law, we have

$$\text{div } \mathscr{E} = 4\pi\varrho/\varepsilon \tag{6.30}$$

where $\mathscr{E}$ is the electric field in the semiconductor, ϱ is the space charge density, and ε the dielectric constant. Suppose the electric field $\mathscr{E}$ is nonuniform at some point of the semiconductor, possibly due[44] to imperfect doping, or to a statistical fluctuation. Then equation (6.30) says that, if $\mathscr{E}$ is nonuniform, ϱ will be nonzero because div $\mathscr{E}$ is nonzero. Gauss's law thus tells us that a space charge density ϱ will be associated with any nonuniformity of the electric field in the semiconductor. We write the equation of continuity, assuming a constant value of σ in the semiconductor,

$$\partial\varrho/\partial t + \text{div } \mathbf{J} = 0 \tag{6.31}$$

where $\mathbf{J}$ is the current density, and there are no sources or sinks, in the form

$$\partial\varrho/\partial t + \sigma \text{ div } \mathscr{E} = 0 \tag{6.32}$$

by using Ohm's law $\mathbf{J} = \sigma\mathscr{E}$. Combining (6.32) and (6.30) gives

$$\partial\varrho/\partial t + 4\pi(\sigma/\varepsilon)\varrho = 0 \tag{6.33}$$

as the equation governing the time dependence of the space charge density $\varrho(t)$. The solution of (6.33) is

$$\varrho(t) = \varrho_0 \exp(-t/\tau_d) \tag{6.34}$$

where $\tau_d \equiv \varepsilon/4\pi\sigma$ is the dielectric relaxation time,[46] and ϱ_0 is the initial value of the space charge density at time $t = 0$. If, as is usual in a semiconductor, the conductivity σ is positive, the space charge density ϱ will decrease with time as $\exp[-(4\pi\sigma/\varepsilon)t]$. A typical value of τ_d for a usual semiconductor[46] (germanium with a conductivity of $1\,\Omega$ cm) is about 10^{-12} sec. We see then that a localized space charge density ϱ, due say to one of the nonuniformities in the electric field mentioned above, will die out rapidly in a semiconductor with a positive value of the conductivity.

We may see what is happening physically during this process in the following way. Suppose we have a sample of a semiconductor for which σ is positive. Then, from equation (6.29), $dv/d\mathscr{E}$ is also positive, where v is the electron drift velocity, and $\mathscr{E}$ is the magnitude of the electric field. Consider the situation, shown in Figure 6.23, in which the semiconductor sample is under a DC bias such that $\mathscr{E}$ is constant in the semiconductor. Suppose that a random fluctuation produces a local deviation from charge neutrality, so a net space charge ϱ in the form of a separation of positive and negative charge is produced. The dipole is shown schematically in Figure 6.23a, and the charge density ϱ as a function of distance is shown, also schematically, in Figure 6.23b. The electric field $\mathscr{E}$ acts on the electrons, which have a drift velocity v opposite to the direction of $\mathscr{E}$. Because of the space charge in the dipole region, the electric field inside the dipole region will be increased by $\delta\mathscr{E}$, which is also shown in Figure 6.23a. Since σ and $dv/d\mathscr{E}$ are also positive, the electron drift velocity inside the dipole region will be increased by an amount δv, where

$$\delta v = (dv/d\mathscr{E})\,\delta\mathscr{E} \qquad (6.35)$$

The result of the increased electron drift velocity is that electrons inside the dipole region move faster than electrons outside that region, so that the former electrons "overtake" the latter. The effect of this "overtaking" is that the space charge region, or dipole, gradually dies out, restoring space charge neutrality. This is the physical meaning of equation (6.34) with a positive conductivity and hence a positive value of the dielectric relaxation time τ_d.

Now consider the same situation in a semiconductor, like GaAs, in which σ, and hence τ_d, are negative. Formally, this means that the exponent $-t/\tau_d$ is positive in equation (6.34), so that the charge density ϱ *increases* with time. The result is, physically, that a space charge density,

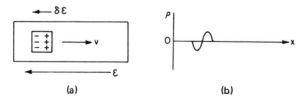

(a)   (b)

Figure 6.23. (a) Schematic view of a dipole region or domain in a semiconductor with a negative value of $dv/d\mathscr{E}$, where v is the electron drift velocity and $\mathscr{E}$ is the electric field in the semiconductor; (b) schematic variation of the space charge density ϱ as a function of distance x.

produced by, say, a fluctuation, grows with time instead of dying out. We may examine the physical situation shown in Figure 6.23a for the case of negative conductivity and hence a negative value of $dv/d\mathscr{E}$. When the dipole forms, the electric field inside the dipole region increases by $\delta\mathscr{E}$, as before. Now, however, δv is negative because $dv/d\mathscr{E}$ is negative, so electrons inside the dipole region have a drift velocity that is smaller than those outside by an amount again given by equation (6.35). As the electrons move in the direction opposite to $\mathscr{E}$, those inside the dipole region "lag behind" those outside. The result of this "lag" is that the negative and positive space charge regions of the dipole increase, and the space charge density increases instead of dying out. The increased value of ϱ in turn increases the increment $\delta\mathscr{E}$ in the electric field, so the total electric field $\mathscr{E} + \delta\mathscr{E}$ inside the dipole region increases still further. The increased electric field in the dipole decreases the electron drift velocity, thereby accelerating the growth of the dipole region. The dipole region, with its space charge density, is called a domain. The domain moves down the length of the semiconductor sample under the influence of the applied electric field.

The growth of the domain is a process which might appear capable of continuing indefinitely. What stops the growth of the domain? As the electric field $\mathscr{E}_i$ inside the domain increases, the field $\mathscr{E}_o$ outside the domain decreases in magnitude. This continues until, as shown schematically in Figure 6.24, $\mathscr{E}_0$ falls to a value less than the critical threshold field $\mathscr{E}_C$. At some pair of values of $\mathscr{E}_i$ and $\mathscr{E}_0$ (where $\mathscr{E}_0 < \mathscr{E}_C$), the electron drift velocity inside the domain will be equal to the drift velocity outside the domain. This is shown in Figure 6.24; the common value of the electron drift velocity inside and outside the domain is denoted by v', and is approximately 10^7 cm sec^{-1} in GaAs. All of the electrons in the sample now have the same drift velocity, and the stable domain moves down the sample with the velocity v' and without further growth.

When the domain reaches the anode at the end of the sample, it pro-

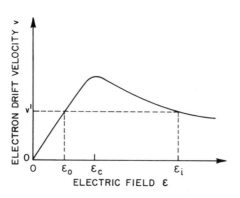

Figure 6.24. Schematic graph of electron drift velocity v as a function of electric field $\mathscr{E}$. The electric fields $\mathscr{E}_i$ and $\mathscr{E}_o$ are, respectively, inside and outside the domain and are shown for the electron velocity $v = v'$, at which the domain is stable. The critical threshold electric field is $\mathscr{E}_C$.

duces a pulse of current in the external circuit. These current pulses are spikes[47] appearing every L/v' seconds, where L is the length of the sample. The transit time τ_t is defined as L/v', so the current pulses have a frequency v'/L. Since v' is of the order of 10^7 cm sec^{-1}, then (for $L = 10^{-1}$ cm), τ_t is of the order of 10^{-8} sec, and the frequency is of the order of 100 MHz. In this way, high-frequency oscillations are produced.

There is a condition which must be fulfilled by the transit time $\tau_t \equiv L/v'$ and the dielectric relaxation time $\tau_d \equiv \varepsilon/4\pi\sigma$ so that the domains will grow and reach a stable size. The transit time must be long compared to the magnitude of the (negative) dielectric relaxation time. Then the domain growth process, described by equation (6.34) with τ_d negative, will have time to grow by many factors of e before the domain reaches the end of the sample. The necessary condition is therefore

$$\tau_t > |\tau_d| \tag{6.36}$$

which can be rewritten as

$$L/v' > \varepsilon/4\pi\sigma = \varepsilon/4\pi ne\mu^* \tag{6.37}$$

on using the relation $\sigma = ne\mu^*$. In equation (6.37), n is the electron density, and μ^* is an average magnitude of the negative electron mobility given by

$$\mu^* = |(dv/d\mathscr{E})|_{\text{av}} \tag{6.38}$$

Using (6.37), the condition (6.36) can be written as

$$nL > \varepsilon v'/4\pi e\mu^* \tag{6.39}$$

For GaAs, the right-hand side of equation (6.39) has a value[48] of about 10^{12} cm^{-2}. When the requirement that nL be larger than 10^{12} cm^{-2} is not met, the formation of domains is not to be expected,[48] and the device is referred to as subcritical.

The mode described above, in which negative differential conductivity leads to the formation and drift of space charge domains, is only one method of exploiting the transferred-electron, or Gunn, effect. Other modes, which are actually more important and efficient, are described in the literature.[42-45]

Physics of Amorphous Semiconductors

An amorphous solid is one that is not crystalline, in that the amorphous solid lacks the long-range order of the periodic crystal lattice. The amorphous solid exhibits short-range order, extending over distances no greater than a few atomic radii. A familiar example of an amorphous solid is quartz (SiO_2), which exists in both amorphous and crystalline forms. There are two main classes of amorphous semiconductors of interest for applications. The first class is the tetrahedrally-bonded covalent semiconductors germanium and silicon, generally written a-Ge and a-Si, respectively. (The "a" stands for "amorphous"; in this context, crystalline silicon is denoted by c-Si.) The second class is the chalcogenide glasses, which are solid mixtures, in amorphous form, of sulfur, selenium, and/or tellurium. These glasses are also covalent amorphous semiconductors, as are the elements S, Se, and Te in amorphous form.

We wish now to present some of the basic ideas[49-55] of the electronic structure of disordered materials in general, and amorphous semiconductors in particular. We begin by considering a basic model of the electronic structure of a disordered solid, and then apply the results to an idealized picture of an amorphous semiconductor.

The basic model[56] of a disordered solid begins with a perfect crystal and examines the nature of the electronic states in a single isolated band. For a perfect crystal, the states in the band are extended, in that an electron may be found with equal probability in any unit cell. An electron in an extended state may therefore move through the crystal and contribute to the conductivity. To discuss a disordered solid, we consider the introduction (by unspecified means) of randomness into the previously perfect crystal. The disordered material now exhibits fluctuations in atomic configuration; these fluctuations are, of course, not present in the perfect crystal. The effect of the fluctuations is, when strong enough, to change the nature of the wave functions in some parts of the band. Figure 6.25 shows[56] the density of electron states $N(E)$ as a function of electron energy E in the single band of the disordered solid. For electron energies greater than a value E_C', and less than a second value E_C, the character of the wave function is no longer extended, but is now localized in space. In this context, a localized wave function is one whose amplitude is nonzero only in a finite region.

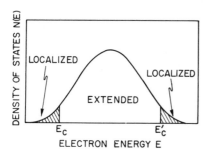

Figure 6.25. Density of states $N(E)$ as a function of electron energy E in an isolated band in a disordered solid. There are tails of localized states (shown shaded) at electron energies less than E_C and greater than E_C'. The band states are extended for energies between E_C and E_C'.

For values of the energy between E_C and E_C', the band states are still extended. Now, however, due to the fluctuations in atomic configuration in the disordered solid, there are "tails" of localized states at the upper and lower edges of the band. These tails are shown as shaded in Figure 6.25.

We now apply the results of this basic picture to a model of an amorphous semiconductor. The model is that of an ideal covalent glass, defined[57] as a covalent network of atoms in which the valence of each atom is satisfied. The ideal glass thus contains no impurities or other defects (e.g., "dangling bonds"); real covalent amorphous solids will contain impurities and defects. It is an experimental fact that the DC conductivity σ of amorphous semiconductors has the form

$$\sigma = \sigma_0 \exp(-\Delta E/k_B T) \qquad (6.40)$$

where ΔE is an activation energy comparable[58] to those in crystals. The form of equation (6.40), along with optical absorption spectra similar to those of intrinsic crystalline semiconductors, suggest a band model for an amorphous semiconductor which includes valence and conduction bands separated by an energy gap. Such a band model for an ideal covalent amorphous semiconductor is shown in Figure 6.26, which, again, is a plot of density of states $N(E)$ as a function of electron energy E. The valence band states are localized for energies greater than the value E_V; the conduction band states are localized for energies less than E_C. [The energy separation $(E_C - E_V)$ is usually referred to as the energy gap of the amorphous semiconductor.] These localized states, shown shaded in Figure 6.26, are those due to the fluctuations inherent in the disordered solid and may be called "intrinsic" localized states. The intrinsic localized states are[59] restricted to narrow (less than 0.1 eV) energy intervals near the band edges.

However, a real amorphous semiconductor will contain impurities and defects such as chain ends, vacancies, dangling bonds, etc. These impurities

Figure 6.26. Band model of an ideal amorphous semiconductor showing density of states $N(E)$ as a function of electron energy E. There are tails of "intrinsic" localized states restricted to narrow (less than 0.1 eV) energy ranges near the band edges. The valence band (VB) states are localized for $E > E_V$; the conduction band (CB) states are localized for $E < E_C$. (After Tauc.[59])

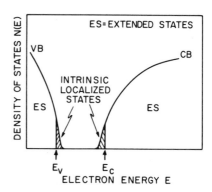

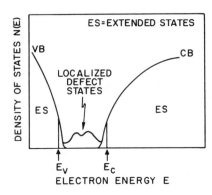

Figure 6.27. Band model of an amorphous semiconductor containing impurities and defects. A nonzero density of localized electron states is present in the gap because of the defects and impurities. The electron states between the energies E_V and E_C are thus localized in nature. The conduction band (CB) and valence band (VB) are labeled. (After Tauc.[59])

and defects give rise to a nonzero density of localized electronic states in the gap between the valence and conduction bands shown in Figure 6.26. The result, shown in Figure 6.27, is that, except for the intrinsic localized states very close to the band edges, the states in the gap are due to impurities or defects. The states between the energies E_V and E_C are thus all localized, while those at energies below E_V and above E_C are extended in nature. In the extended states, the carrier mobility μ is expected to be relatively high; the conduction process in these states is analogous to conduction in crystalline semiconductors. In the localized states between E_V and E_C, on the other hand, the mobility is due[60] to thermally assisted tunneling between localized states. The mobility of carriers, with energies between E_V and E_C, in localized states is thus much lower than that of carriers in the extended states. The carrier mobility $\mu(E)$, plotted as a function of energy E in Figure 6.28, therefore shows sharp decreases at the energies E_V and E_C; these decreases are called mobility edges. For nonzero temperatures, $\mu(E)$ is small but not zero between E_V and E_C; the energy region between E_V and E_C is called the mobility gap. The activation energy ΔE in equation (6.40) thus relates to excitation of carriers from extended states below E_V, across the mobility gap, into extended states above E_C.

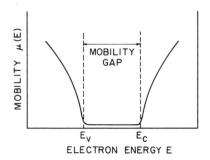

Figure 6.28. Schematic variation of mobility $\mu(E)$ as a function of electron energy E in a covalent amorphous semiconductor at a nonzero temperature.

The activation energy ΔE in an amorphous semiconductor is thus connected with the mobility gap, rather than with the energy gap in the density of states, as it is in a crystalline semiconductor. The mobility gap plays the same role in amorphous semiconductors that the energy gap in the density of states does in crystalline semiconductors. The band model described above is called the Mott–CFO model after its originators, Mott, Cohen, Fritzsche, and Ovshinsky.

From the point of view of the applications to be described, the key result of the Mott–CFO band model is that the conductivity of an amorphous semiconductor will be low. This is because of the existence of the mobility gap shown in Figure 6.29. The current carriers will be found at energies roughly between E_V and E_C, i.e., in states of low mobility. Representative values[61] of the Hall mobility μ_H in chalcogenide glasses are of the order of 0.1 cm^2 V^{-1} sec^{-1}. A theoretical estimate[61] of the conductivity mobility μ is $\mu \approx 100\,\mu_H$, leading to a value of μ of the order of 10 cm^2 V^{-1} sec^{-1}. This value is low compared to mobilities observed in crystalline semiconductors. The conductivity σ varies as shown in equation (6.40), with σ_0 having a value[62] of the order of 1000 Ω^{-1} cm^{-1} for many amorphous chalcogenide semiconductors. The activation energy[62] ΔE is generally between 0.5 and 1.0 eV. The resulting values of the conductivity range from $10^{-3}\,\Omega^{-1}$ cm^{-1} to $10^{-13}\,\Omega^{-1}$ cm^{-1} (at 300 K), depending on the composition and method of preparation of the amorphous semiconductor. These values indicate that the DC conductivity of an amorphous chalcogenide semiconductor at room temperature will be quite low compared to the values we are used to for crystalline semiconductors.

While there is considerable interest in amorphous silicon itself, the more important material from the point of view of applications is hydrogenated amorphous silicon,[63–65] usually written a-Si:H or as $Si_{(1-x)}H_x$, where x ranges from 0.05-0.50.[66] Amorphous silicon (a-Si) itself contains many dangling bonds, which, as defects, give rise to states in the energy gap. The existence of these states in the gap of pure a-Si obscures optical and electrical processes, making pure amorphous silicon unsuitable for applications. Hydrogen atoms are introduced into a-Si to produce what is essentially an amorphous alloy of silicon and hydrogen. In such an alloy, the hydrogen atoms attach themselves to the dangling bonds, thereby removing the states in the energy gap which were due to the presence of these defects. This process is usually referred to as passivation of amorphous silicon. The result of the passivation process,[63] for the alloy $Si_{0.9}H_{0.1}$, is an amorphous semiconductor in which most of the defect states have been removed from the energy gap. The alloys a-Si:H have[63] an activation energy [ΔE in equation (6.40)] equal to about half the energy gap

$(E_C - E_V)$, which ranges from 1.5 eV to 1.9 eV in typical alloys. Since a-Si:H has an energy gap relatively free of states due to defects, it can be doped[67] with the usual shallow donors (such as phosphorus and boron) to produce n- and p-type materials which are extrinsic semiconductors.

Amorphous Semiconductor Devices

There are several types of devices based on amorphous semiconductors. These include memory and switching devices[68–70] based on chalcogenide glasses, solar cells[63] based on a-Si:H, and thin-film transistors.[71] Memory and switching devices will be discussed in this section, and solar cells will be covered in the discussion of photovoltaic devices in Chapter 7.

We consider first the amorphous chalcogenide semiconductor memory device. This device is based on the fact[72,73] that there exists in some of these materials a reversible structure change between the amorphous state of high resistance and a crystalline state of low resistance. It is believed that heating[72,74] of the amorphous state can induce crystallization. Further, the small-grain crystalline region so produced can be returned to the amorphous state on heating followed by rapid cooling. A schematic current–voltage plot of such a memory device is shown in Figure 6.29. It can be seen that the resistance (dV/dI) at zero voltage depends on the previous history of the device. Switching from the amorphous high-resistance OFF state to the low-resistance crystalline ON state takes place when the applied voltage exceeds the threshold value V_t; this is indicated by the dotted lines in Figure 6.29. The conducting state is retained even after removal of the applied voltage. The high-resistance amorphous OFF state can be reestablished by heating (with a short pulse of current), which redissolves the crystalline material into a disordered phase. The memory device thus has two stable states.

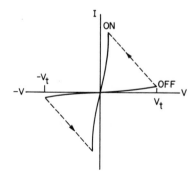

Figure 6.29. Schematic plot of current I versus voltage V for an amorphous semiconductor memory device. The threshold voltage for switching from the high-resistance amorphous OFF state to the low-resistance crystalline ON state is V_t.

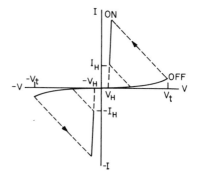

Figure 6.30. Schematic plot of current I versus voltage V for an amorphous semiconductor threshold switching device. The voltage V_t is the threshold for the transition from the OFF state to the ON state; the current I_H is the holding current below which the device reverts to the OFF state. (After Adler *et al.* [75])

The second type of amorphous semiconductor application is a switching device.[52,68,69,75] A schematic current–voltage plot for such a device is shown in Figure 6.30. There is a high-resistance OFF state and a low-resistance ON state. If the threshold voltage V_t is exceeded, the device switches from the OFF state to the ON state; if the current is decreased below the holding current I_H, the device switches to its original OFF state. The switching time[76] is less than a nanosecond, and the resistance ratio[68] between the conducting and nonconducting states is of the order of 10^6. It appears[68,75,77] that the switching mechanism is electronic, not thermal, in nature. Since the details are complex, and not completely settled, the reader is referred to the literature[75,77] for details.

Problems

6.1. *Effect of Surface States on Metal–p-Type-Semiconductor Contacts.* Using drawings analogous to Figures 6.10–6.12, discuss the contact between a metal (of work function ϕ_m) and a p-type semiconductor (of work function ϕ_s), in which the Fermi level is pinned at the surface. Let Δ be the energy of the Fermi level above the valence band edge at the surface, and let E_g be the energy gap of the semiconductor. Consider the case for which ϕ_m is smaller than ϕ_s, and show that one may obtain an ohmic contact in this case, even though the earlier "rule" predicts a rectifying contact.

6.2. *Photoemission in the Infrared.* What semiconductors might be usable as infrared photoemitters? Assume that the Fermi level at the surface of the semiconductor is pinned at an energy $\Delta = E_g/3$ above the valence band edge, where E_g is the band gap. Draw a band diagram, using a value of 1.2 eV for the work function of a Cs + O layer on the semiconductor surface. What is the minimum value of the electron affinity χ? What is the photon energy threshold for electron emission? (For information on actual infrared photoemitters, see R. L. Bell, Reference 13, pages 76–82.)

6.3. *Calculation of Band Bending in GaP.* Suppose the width of the space charge layer of a certain sample of cesiated *p*-type GaP is found to be 300 Å. If the acceptor density is $10^{18}\,\mathrm{cm^{-3}}$ and the dielectric constant is 8.4, calculate the energy through which the bands are bent. Express your result as a fraction f of the energy gap 2.2 eV of GaP, and compare your value of f with that given in the literature.[22]

References and Comments

1. See, for example, C. Kittel, *Introduction to Solid State Physics*, Sixth Edition, John Wiley, New York (1986), pages 164–166.
2. J. P. McKelvey, *Solid State and Semiconductor Physics*, Harper and Row, New York (1966), pages 485–489.
3. A. S. Grove, *Physics and Technology of Semiconductor Devices*, John Wiley, New York (1967), pages 144–145, 282–285, 334–337.
4. A. Many, Y. Goldstein, and N. B. Grover, *Semiconductor Surfaces*, North-Holland, Amsterdam (1965), Chapters 5 and 9.
5. S. Wang, *Solid State Electronics*, McGraw-Hill, New York (1966), pages 294–300.
6. F. Flores, "The Electronic Structure of Surfaces and Interfaces," in *Crystalline Semiconducting Materials and Devices*, P. N. Butcher *et al.*, editors, Plenum Press, New York (1986), Chapter 10.
7. G. Burns, *Solid State Physics*, Academic Press, Orlando, Florida (1985), Chapter 17.
8. J. Van Laar and J. J. Scheer, "Photoemission of Semiconductors," *Philips Technical Review*, **29**, 54–66 (1968), especially pages 56–58.
8a. C. Kittel, Reference 1, Chapter 19; A. L. Robinson, "Consensus on Silicon Surface Structure Near," *Science*, **232**, 451–453 (April 25, 1986); A. L. Robinson, "A Spatially Resolved Surface Spectroscopy," *Science*, **229**, 1074–1076 (September 13, 1985).
9. A Many *et al.*, Reference 4, pages 166–174.
10. A. Many *et al.*, Reference 4, page 348.
11. A. Many *et al.*, Reference 4, pages 357–358.
12. For a review, see S. G. Davison and J. D. Levine, in *Solid State Physics*, H. Ehrenreich, F. Seitz, and D. Turnbull (editors), Academic Press, New York (1970), Volume 25, pages 1–149, especially pages 88–148.
13. H. K. Henisch, *Semiconductor Contacts*, Oxford University Press (1984), pages 29–38.
14. J. Van Laar and J. J. Scheer, Reference 8, pages 56–57.
15. R. L. Bell, *Negative Electron Affinity Devices*, Oxford University Press (1973), pages 17–19.
16. S. G. Davison and J. D. Levine, Reference 12, page 135, gives information on Fermi level pinning in a number of semiconductors.
17. A. S. Grove, Reference 3, pages 267–268, Figure 9.4.
18. See S. M. Sze, *Physics of Semiconductor Devices*, Second Edition, John Wiley, New York (1981), page 291, Table 3, which gives measured barrier heights for a number of metal–semiconductor contacts.
19. J. Bardeen, "Surface States and Rectification at a Metal–Semiconductor Contact," *Physical Review*, **71**, 717–727 (1947).
20. A. M. Cowley and S. M. Sze, "Surface States and Barrier Height of Metal–Semiconductor Systems," *Journal of Applied Physics*, **36**, 3212–3220 (1965).

21. R. S. Muller and T. I. Kamins, *Device Electronics for Integrated Circuits*, Second Edition, John Wiley, New York (1986), pages 154–157.

22. A. Van der Ziel, *Solid State Physical Electronics*, Second Edition, Prentice-Hall, New York (1968), page 101.

23. J. I. Pankove, *Optical Processes in Semiconductors*, Prentice-Hall, New York (1971), Chapter 13, pages 294–295.

24. R. U. Martinelli and D. G. Fisher, "The Application of Semiconductors with Negative Electron Affinity Surfaces to Electron Emission Devices," *Proceedings IEEE*, **62**, 1339–1360 (1974), page 1342.

25. R. L. Bell, Reference 15, Section 6.4, page 74.

26. J. I. Pankove, Reference 23, Table 13-2, page 298.

27. R. U. Martinelli and D. G. Fisher, Reference 24, Table 1, page 1340.

28. R. L. Bell, Reference 15, Chapter 6.

29. J. I. Pankove, Reference 23, page 296.

30. J. I. Pankove, Reference 23, Figure 13-11(b), page 297.

31. R. U. Martinelli and D. G. Fisher, Reference 24, pages 1345–1352.

32. R. E. Simon, "A Solid State Boost for Electron Emission Devices," *IEEE Spectrum*, **9**, 74–78 (December 1972).

33. A. Van der Ziel, Reference 22, Chapter 10.

34. R. U. Martinelli and D. G. Fisher, Reference 24, pages 1352–1356.

35. J. G. Ruch and G. S. Kino, "Transport Properties of GaAs," *Physical Review*, **174**, 921–931 (1968), Figure 5, page 926.

36. D. E. Aspnes, "Lower Conduction Band Structure of GaAs," in *Gallium Arsenide and Related Compounds* (*St. Louis*), 1976, L. F. Eastman (editor), Proceedings of the Sixth International Symposium, Conference Series Number 33B, The Institute of Physics, London (1977), pages 110–119.

37. J. S. Blakemore, "Semiconducting and Other Properties of Gallium Arsenide," *Journal of Applied Physics*, **53**, 123–181 (1982), Figure 45.

38. The room-temperature energy separations shown in Figure 6.22 were calculated by the author from results given in Reference 37, Table 1.

39. D. E. Aspnes, Reference 37, Table 2, page 116.

40. C. Kittel, Reference 1, page 133, equation (20).

41. M. E. Jones and R. T. Bate, "Electronic and Optical Phenomena in Semiconductors," *Annual Review of Materials Science*, **1**, 347–382, (1971), pages 351–356.

42. J. A. Copeland and S. Knight, in *Semiconductors and Semimetals*, R. K. Willardson and A. C. Beer (editors), Academic Press, New York (1971), Volume 7, Part A, pages 3–72.

43. See, for example, B. G. Streetman, *Solid State Electronic Devices*, Second Edition, Prentice Hall, New York (1980), pages 429–434, for a summary.

44. H. Kroemer, "Negative Conductance in Semiconductors," *IEEE Spectrum*, **5**, 47–56 (January 1968).

45. H. Kroemer, "The Gunn Effect—Bulk Instabilities," in *Topics in Solid State and Quantum Electronics*, W. D. Hershberger (editor), John Wiley, New York (1972), pages 20–98.

46. S. Wang, Reference 5, pages 274–275.

47. For examples, see H. Kroemer, Reference 45, page 21; J. A. Copeland and S. Knight, Reference 42, page 33.

48. H. Kroemer, Reference 44, page 53.

49. N. F. Mott and E. A. Davis, *Electronic Processes in Non-Crystalline Materials*, Se-

cond Edition, Oxford University Press, New York (1978).

50. M. H. Cohen, "Theory of Amorphous Semiconductors," *Physics Today*, **24**, 26–32 (May 1971).

51. E. N. Economou, M. H. Cohen, K. F. Freed, and E. S. Kirkpatrick, "Electronic Structure of Disordered Materials," in *Amorphous and Liquid Semiconductors*, J. Tauc (editor), Plenum Press, New York (1974), pages 101–158.

52. J. Tauc, "Amorphous Semiconductors," *Physics Today*, **29**, 23–31 (October 1976).

53. M. H. Brodsky, editor, *Amorphous Semiconductors*, Second Edition, Springer-Verlag, Berlin (1985).

54. H. Fritzsche, "Noncrystalline Semiconductors," *Physics Today*, **37**, No. 10, 34–41 (October 1984).

55. N. Mott, "Electrons in Noncrystalline Materials: The Last Twenty Five Years," *Contemporary Physics*, **26**, 203–215 (1985).

56. E. N. Economou *et al.*, Reference 51, pages 106–108.

57. E. N. Economou *et al.*, Reference 51, page 108.

58. M. H. Cohen, Reference 50, page 26.

59. J. Tauc, Reference 52, page 24.

60. H. Fritzsche, "Electronic Properties of Amorphous Semiconductors," in *Amorphous and Liquid Semiconductors*, J. Tauc (editor), Plenum Press, New York (1974), pages 221–312; page 229.

61. H. Fritzsche, Reference 60, pages 258–262.

62. H. Fritzsche, Reference 60, pages 235–236.

63. R. Zallen, *Physics of Amorphous Solids*, John Wiley, New York (1983), pages 289–292.

64. M. H. Brodsky, Reference 53, pages 3–4.

65. A. Madan *et al.*, editors, *Amorphous Silicon Semiconductors – Pure and Hydrogenated*, Materials Research Society, Pittsburgh (1987). This volume is the proceedings of a recent conference.

66. D. E. Carlson and C. R. Wronski, "Amorphous Silicon Solar Cells," in Reference 53, pages 287–329; page 291.

67. P. G. Le Comber and W. E. Spear, "Doped Amorphous Semiconductors," in Reference 53, pages 251–285.

68. D. Adler, "Amorphous Semiconductor Devices," *Scientific American*, **236**, 36–48 (May 1977).

69. H. Fritzsche, "Switching and Memory in Amorphous Semiconductors," in *Amorphous and Liquid Semiconductors*, J. Tauc (editor), Plenum Press, New York (1974), pages 313–359.

70. D. Adler, "Physics and Chemistry of Covalent Amorphous Semiconductors," in *Physical Properties of Amorphous Materials*, D. Adler, editor, Plenum Press, New York (1985), pages 93–95.

71. D. Adler, editor, *Amorphous Semiconductors for Microelectronics*, SPIE, Bellingham, Washington (1986).

72. H. Fritzsche, Reference 69, page 348.

73. D. Adler, Reference 68, pages 45–46.

74. H. Ehrenreich *et al.*, *Fundamentals of Amorphous Semiconductors*, National Academy of Sciences, Washington, D.C. (1972), Chapter 6.

75. D. Adler, H. K. Henisch, and N. Mott, "The Mechanism of Threshold Switching in Amorphous Alloys," *Reviews of Modern Physics*, **50**, 209-220 (1978).

76. D. Adler, Reference 68, page 44.

77. D. Adler *et al.*, "Threshold Switching in Chalcogenide Glass Films," *Journal of Applied Physics*, **51**, 3289–3309 (1980).

Suggested Reading

A. MANY, Y. GOLDSTEIN, AND N. B. GROVER, *Semiconductor Surfaces*, North-Holland, Amsterdam (1965). This treatise discusses its subject at the advanced level. Chapters 5, on surfaces states, and 9, on the electronic structure of the surface, are particularly pertinent to our device-related discussion.

C. KITTEL, *Introduction to Solid State Physics*, Sixth Edition, John Wiley, New York (1986). Chapter 19 of this text is an introduction to surface physics.

J. VAN LAAR AND J. J. SCHEER, "Photoemission of Semiconductors," *Philips Technical Review*, **29**, 54–66 (1968). This introductory article discusses photoemission and surface effects, including negative electron affinity, in semiconductors.

H. KROEMER, "Negative Conductance in Semiconductors," *IEEE Spectrum*, **5**, 47–56 (January 1968). A good tutorial review article which stresses basic physics. Some of the details of the GaAs band structure are now superseded by new data, but the application to the Gunn effect is well presented.

J. TAUC (editor), *Amorphous and Liquid Semiconductors*, Plenum Press, New York (1974). A set of articles by several authors discussing, at the advanced level, the physics and applications of amorphous semiconductors.

D. ADLER, "Amorphous Semiconductor Devices," *Scientific American*, **236**, 36–48 (May 1977). An article which emphasizes the physics of amorphous semiconductors and devices.

M. H. BRODSKY (editor), *Amorphous Semiconductors*, Second Edition, Springer-Verlag, Berlin (1985). This relatively recent group of review articles covers both basic physics and applications.

R. ZALLEN, *The Physics of Amorphous Solids*, John Wiley, New York (1983). The last chapter of this monograph deals with electrical and optical properties including a clear discussion of hydrogenated amorphous silicon.

R. K. WILLARDSON AND A. C. BEER (editors), *Semiconductors and Semimetals*, Volume 21, Parts A–D, "Hydrogenated Amorphous Silicon," J. I. Pankove, volume editor, Academic Press, Orlando, Florida (1984). This four-part treatise covers all aspects of the physics and applications of a-Si:H.

7

Detectors and Generators
of Electromagnetic Radiation

Introduction

This chapter discusses a number of solid state (mostly semiconductor) detectors and generators of electromagnetic radiation. The absorption of photons by both intrinsic and extrinsic semiconductors is considered first and leads to a discussion of photoconductive and photovoltaic devices. Several important applications of intrinsic photoconductivity (e.g., photography) are treated briefly. Spontaneous photon emission in semiconductors is considered as the basis for p–n junction luminescence in light-emitting diodes. A general discussion of photon amplification by stimulated emission is provided as background for a description of three- and four-level lasers (e.g., ruby). Finally, stimulated emission in semiconductor junction lasers concludes the chapter.

Intrinsic Photon Absorption in Semiconductors

We begin with a brief discussion of processes[1,2] in which the absorption of a photon by a semiconductor results in the production of free charge carriers, either electrons or holes, or both.

Consider first the absorption of a photon by an intrinsic semiconductor[3] with the band structure, at 0 K, shown in Figure 7.1, so the valence band is completely full and the conduction band is completely empty. The band structure in Figure 7.1 is one in which the minimum energy gap E_g is direct. A photon of circular frequency ω_g, where

$$\hbar\omega_g = E_g \tag{7.1}$$

will have just enough energy to excite an electron from the top of the valence band to the bottom of the conduction band. A photon of energy less than E_g will not have sufficient energy to so excite an electron. Excitation of an electron across the energy gap is often called intrinsic excitation and this absorption process in a semiconductor is called intrinsic absorption. The intrinsic absorption process, indicated schematically in Figure 7.1, produces a free electron in the conduction band, and a free hole in the valence band. Intrinsic absorption of a photon thus produces an electron–hole pair in the semiconductor.

Since a photon whose energy is $\hbar\omega_g$ is the lowest-energy photon that can be absorbed by the intrinsic excitation process, there is a threshold for optical absorption by the semiconductor at the photon energy $\hbar\omega_g$. Figure 7.2 shows the optical absorption coefficient α, defined by the relation

$$\alpha \equiv (\omega/c)\varkappa$$

where $\varkappa$ is the imaginary part of the complex refractive index,[4] plotted as a function of photon energy $\hbar\omega$. In Figure 7.2, we note that, above the threshold, the absorption coefficient increases rapidly with photon energy. The functional form $\alpha(\hbar\omega)$ of the dependence of α on $\hbar\omega$ will depend[5] on the details of the band structure of the semiconductors and on the type of transition. The transition of the electron from the valence band to conduction band shown in Figure 7.1 is called a direct transition. This is because

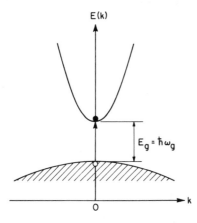

Figure 7.1. Absorption of a photon of energy $\hbar\omega_g = E_g$ by a semiconductor with a minimum direct energy gap equal to E_g located at the zone center. The electron energy $E(k)$ is plotted as a function of wave vector k; filled electron states are shown as shaded. An electron (●) in the conduction band and a hole (○) in the valence band are produced by the absorption of the photon.

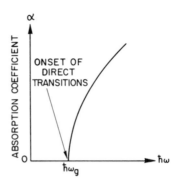

Figure 7.2. Schematic variation of absorption coefficient α as a function of photon energy $\hbar\omega$ for direct transitions. The threshold is located at $\hbar\omega_g = E_g$, where E_g is the minimum direct gap of the semiconductor.

(assuming the photon wave vector to be negligible[†]) the wave vector of the electron does not change in the threshold transition shown, so the transition occurs vertically on the band diagram. Such a direct transition will be the case for a semiconductor (e.g., InSb) in which the minimum energy gap is direct.

We discuss next the more complicated case of intrinsic photon absorption in a semiconductor with an indirect band gap in which the valence band maximum and the conduction band minimum are at different points of the Brillouin zone. Consider such a semiconductor band structure, shown schematically in Figure 7.3. The diagonal arrow in Figure 7.3 shows the excitation of an electron from the valence band maximum at the Γ point of the zone to the conduction band minimum at wave vector k_0 near the zone edge point X, producing an electron–hole pair. Such a transition, involving excitation of an electron across an indirect energy gap, is called an indirect transition and is a transition in which the electron wave vector does change. We can calculate this change because conservation of wave vector requires, for the electron, that

$$\hbar k_i + \hbar k' = \hbar k_f \qquad (7.2)$$

where in equation (7.2) k' is the increase in electron wave vector, k_i and k_f are the initial and final electron wave vectors, and we are again neglecting the wave vector of the photon involved. In Figure 7.3, $k_i = 0$, so $k' = k_f = k_0$.

Where does the "additional" wave vector $k' = k_f - k_i$ come from? It is provided by either the absorption of a phonon of wave vector k', or the emission of a phonon of wave vector $-k'$. Consider first the absorption

[†] See Problem 1 of this chapter for an illustrative calculation and example.

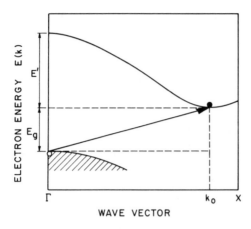

Figure 7.3. Absorption of a photon by a semiconductor with an indirect energy gap, producing a free electron (●) in the conduction band and a free hole (○) in the valence band. The indirect transition is between the valence band maximum at the zone center (Γ) and the conduction band minimum at wave vector k_0 near the zone edge (X).

of a phonon of wave vector k' and energy $\hbar\Omega$. Then the simultaneous conservation of energy and of wave vector requires that

$$E_i + \hbar\Omega + \hbar\omega = E_f \tag{7.3}$$

$$k_i + k' = k_f \tag{7.4}$$

where E_i and E_f are the initial and final electron energies, $\hbar\omega$ is the photon energy, and the photon wave vector has been neglected in equation (7.4). If we rewrite the energy relation (7.3) as

$$E_f - E_i = \hbar\omega + \hbar\Omega \tag{7.5}$$

and note that the threshold electron energy difference $E_f - E_i$ is equal to the minimum energy gap E_g, we see that the threshold photon energy $\hbar\omega_t$ is given by

$$\hbar\omega_t = E_g - \hbar\Omega \tag{7.6}$$

Equation (7.6) tells us that an indirect transition involving the absorption of a phonon of energy $\hbar\Omega$ has its absorption threshold at a photon energy equal to $E_g - \hbar\Omega$, so the threshold in this case is at a photon energy smaller than that of the energy gap. Figure 7.4 shows, schematically, a plot of absorption coefficient α as a function of photon energy $\hbar\omega$ for indirect transitions involving absorption of a phonon of energy $\hbar\Omega$. For a representative semiconductor, $E_g = 1$ eV and $\hbar\Omega$ is about 0.05–0.10 eV. Figure 7.4 is drawn for a hypothetical semiconductor with an indirect band structure (like Figure 7.3) for which $E_g = 0.7$ eV, $\hbar\Omega = 0.1$ eV, and the energy E'

$= 0.15$ eV. Figure 7.4 shows the onset of indirect transitions at a photon energy equal to $E_g - \hbar\Omega$. The figure also shows the onset of direct (vertical) transitions at a photon energy equal to $E_g + E'$; this is seen experimentally as a "knee" in the absorption curve of the semiconductor.[6]

If the indirect transition involves the emission of a phonon of energy $\hbar\Omega$ and wave vector $-k'$, then the threshold photon energy $\hbar\omega_t$ is given by

$$\hbar\omega_t = E_g + \hbar\Omega \qquad (7.7)$$

so the threshold for indirect transitions involving phonon emission is at a photon energy larger than the energy gap.

Finally, it should be remarked that indirect transitions involving the simultaneous participation of more than one phonon have been observed in several semiconductors.[7]

From the point of view of applications, the important results of these considerations are as follows. First, intrinsic excitation in a semiconductor occurs at a photon energy approximately equal to the minimum energy gap. Second, intrinsic excitation produces one free electron–hole pair for each photon absorbed. Third, the direct transition is a two-body process (photon, electron) while the indirect transition is a three-body process (photon, electron, phonon). We therefore expect a lower probability, all other factors being equal, for the indirect transition compared to the direct transition. This is reflected in lower absorption coefficients for indirect transition absorption. Fourth, the photogenerated intrinsic free carriers (electron–hole pairs) are utilized in a variety of applications which we will discuss later.

Figure 7.4. Optical absorption coefficient α as a function of photon energy $\hbar\omega$ for a hypothetical semiconductor with an indirect band gap as in Figure 7.3. The onset of indirect transition at $\hbar\omega = E_g - \hbar\Omega$ due to absorption of phonons of energy $\hbar\Omega$ is shown, as is the onset of direct transitions at $\hbar\omega = E_g + E'$. (Note that the $\hbar\omega$ axis does not go to zero.)

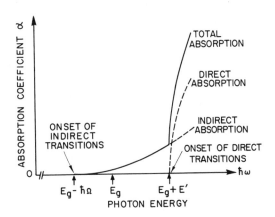

Photon Absorption by Bound States of Impurities in Semiconductors

Impurity conductivity in semiconductors, in which a donor or acceptor impurity is thermally ionized to yield a free electron or hole, was discussed in Chapter 1. At room temperature then, we would expect almost all of the usual donors and acceptors in silicon to be thermally ionized. However, at a low temperature near 0 K, the electron or hole will be bound to the donor or acceptor, as shown schematically in Figures 7.5a and 7.5b. The bound electron or hole may absorb a photon of energy $\hbar\omega$ if $\hbar\omega$ is larger than or equal to E_d for the n-type semiconductor or larger than or equal to E_a for the p-type semiconductor. In the former, the bound electron is excited into the conduction band and, in the latter, the bound hole is excited into the valence band. (Strictly speaking, the process in the p-type case is really the excitation of an electron from a filled state at the top of the valence band into the unoccupied orbital of the acceptor atom, thus creating a hole in the valence band. However, it is usual to speak of the ionization of holes.) In either case a single type of free carrier is created; the processes are indicated symbolically in Figure 7.6. The process of absorption of photons by bound impurities[8] in semiconductors is often called extrinsic absorption or impurity absorption.

The processes described above are excitation of a carrier (electron or hole) from a discrete bound state (donor or acceptor) to a quasicontinuous band of levels (conduction band or valence band). The result is a continuous absorption spectrum with a threshold at the relevant ionization energy. (There may also be observed sharp absorption lines at photon energies less than the ionization energy.[9] These lines are due to transitions between the ground state of the impurity atom and excited, but still bound, states. Such transitions do not produce free carriers because they do not ionize the impurity atom.) Since the density of impurity atoms in a semiconductor (typically in the range 10^{14}–10^{16} cm^{-3} before degeneracy occurs) is quite

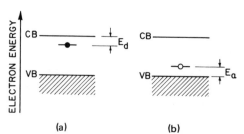

(a)   (b)

Figure 7.5. Band diagrams showing, schematically (a) an electron (–●–) bound to a donor in an n-type semiconductor, and (b) a hole (–○–) bound to an acceptor in a p-type semiconductor, both at low temperatures. The ionization energy of the donor is E_d, and that of the acceptor is E_a. The shaded areas represent filled electron states, and the ionization energies are exaggerated for clarity.

Figure 7.6. Band diagrams showing, schematically, (a) an electron excited from a bound donor state to the conduction band in an n-type semiconductor, and (b) a hole excited from a bound acceptor state to the valence band in a p-type semiconductor.

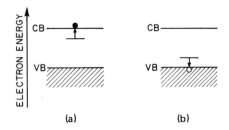

small compared to the density of intrinsic electrons (10^{22} cm^{-3}), the optical absorption coefficient α for extrinsic absorption is quite small (perhaps 10–100 cm^{-1}) compared to the values for intrinsic direct or indirect absorption (typically 10^3–10^4 cm^{-1}).

Finally, it should be noted that the discussion above dealt tacitly with shallow discrete donors and acceptors, principally in germanium or silicon. There are also deeper impurity levels known in these group IV semiconductors; examples are gold and mercury in germanium. Such deeper levels[10] are usually multiple levels and have larger ionization energies than the shallow levels mentioned above.

From the point of view of applications, extrinsic or impurity absorption of photons produces a single type of free carrier if the photon energy is greater than the ionization energy of the bound impurity state. These photogenerated extrinsic free carriers are useful in a variety of applications to be discussed later.

Threshold Energies for Photon Absorption

One of the important attributes of a detector of electromagnetic radiation is the region of the spectrum in which it absorbs. We will consider some examples of intrinsic and extrinsic absorption in semiconductors which are useful detector materials.

A detector for the visible region of the spectrum is cadmium sulfide, CdS. This semiconductor has a hexagonal crystal structure in its common modification, so its band structure[11] and optical absorption[12] are more complicated than those of the usual cubic semiconductors like germanium. A simplified version[13] of the band structure of CdS is shown in Figure 7.7. The energy gap E_g is direct at the center of the Brillouin zone, and E_g has the values[14] 2.58 eV at 0 K and approximately 2.4 eV at 300 K. This means that the energy threshold for photon absorption by CdS at room

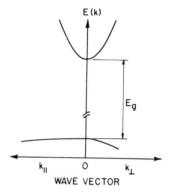

Figure 7.7. Simplified version[13] of the band structure of hexagonal CdS showing electron energy $E(k)$ as a function of wave vectors $k_\|$ parallel to and $k_\perp$ perpendicular to the c axis of the Brillouin zone. The minimum energy gap is direct at the center of the zone.

temperature is at about 2.4 eV. Since a 1-eV photon has a wavelength of 1.240 μm, this absorption threshold in CdS is at a wavelength of approximately 5200 Å, which is in the green region of the visible spectrum. Wavelengths shorter than about 5200 Å are strongly absorbed by CdS, while longer wavelengths (yellow, red) are transmitted. A crystal of CdS thus appears yellowish-orange by transmitted light. Figure 7.8 shows the absorption spectrum[15] of CdS at 300 K, exhibiting the strong absorption at photon energies larger than about 2.4 eV.

There are a number of semiconductors, with energy gaps smaller than about 1 eV, whose absorption thresholds are in the infrared[16,17] region of the spectrum. Several of these are listed in Table 7.1, which shows the energy gap[14] E_g at 0 K, the threshold wavelength λ_t for absorption, and the type of gap.

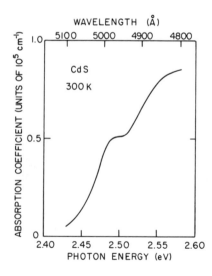

Figure 7.8. Absorption spectrum of CdS at 300 K (for radiation polarized with its electric field vector parallel to the c axis. After Gutsche.[15])

Table 7.1. Semiconductors that Absorb Infrared Photons

Semiconductor	E_g at 0 K (eV)	λ_t (μm)	Type of gap
Si	1.17	1.06	indirect
Ge	0.74	1.68	indirect
InSb	0.24	5.17	direct at Γ
PbSe	0.16	8.27	direct at L

The three lead salt semiconductors PbS, PbSe, and PbTe, all have rather small energy gaps and all are useful as detectors[†] of infrared radiation. Values[18,19] of E_g at 4 K for these semiconductors are as follows: PbS, 0.286 eV; PbSe, 0.165 eV; PbTe, 0.190 eV. The band structures of each of the three are quite similar, and are shown schematically, in the neighborhood of the minimum energy gap, in Figure 7.9. The gap is direct and is located at the L point of the zone, at which the (111) direction intersects the zone edge. The energy gap in the lead salts is thus located at the edge of the Brillouin zone, rather than at the center or Γ point. (The semiconductor SnTe, for which $E_g = 0.3$ eV at 0 K, has a very similar band structure.) The energy gaps of the lead salts provide thresholds for absorption at wavelengths in the infrared as long as approximately 8 μm.

Solid solutions of SnTe in PbTe are semiconductors[19,20] whose energy gap E_g displays unusual behavior as a function of the composition of the solid solution. Figure 7.10 shows a plot of the energy gap E_g (at 12 K) of PbTe–SnTe solid solutions as a function of atom per cent SnTe in the crystal. The curve in Figure 7.10 tells us that, if one prepares a series of crystals of PbTe–SnTe solid solutions containing increasing atomic percentages of SnTe, the variation of the energy gap is as follows. The gap decreases from the value for pure PbTe, and becomes zero for a composition of approximately 34 at. % SnTe. For compositions with SnTe concentrations larger than 34 at. %, the energy gap increases until it has the value for pure SnTe. This unusual behavior of the energy gap is useful because it allows one to choose the value of the gap by specifying the tin–lead ratio in the

---

[†] It should be pointed out that the usual polycrystalline lead salt infrared detectors are not exhibiting simple photoconductivity. The physics of these devices is quite complex and the interested reader is referred to the book by Moss *et al.*, Reference 1, pages 380–384.

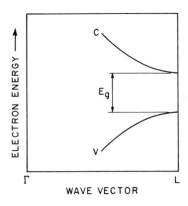

Figure 7.9. Band structure (schematic) of the lead salt semiconductors in the neighborhood of the direct gap E_g at the L point of the Brillouin zone. The conduction and valence bands are labeled C and V, respectively.

PbTe–SnTe solid solution. In particular, it has proved possible to prepare[21] PbTe–SnTe crystals with an energy gap as small as 0.04 eV at 12 K. These semiconductors with very small energy gaps have threshold wavelengths for absorption as long as 30 μm and so are used for detectors in the long-wave infrared region of the spectrum. The same type of variation of energy gap with composition has been observed in semiconducting solid solutions of SnTe in PbSe[22] and in solid solutions[19,23] of CdTe (a semiconductor with the zinc blende crystal structure and a value of $E_g = 1.6$ eV at 0 K) and HgTe (a semimetal with the zinc blende crystal structure).

The threshold for intrinsic absorption of photons is determined by the energy gap E_g of the semiconductor. Since the lower limit on realizable values of E_g is a few hundredths of an eV, one turns to photon absorption by bound impurity states to obtain photon absorption at still longer wavelengths. Shallow impurities (e.g., gallium, antimony) in germanium have ionization energies of about 0.01 eV, so their thresholds[24] for extrinsic absorption are in the far infrared, in the neighborhood of 120 μm. Deeper levels[24,25]

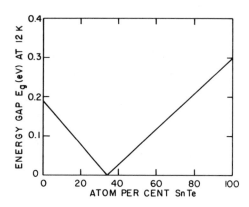

Figure 7.10. Energy gap E_g (at 12 K) of solid solutions of SnTe in PbTe plotted as a function of atomic percent SnTe. (Adapted from I. Melngailis and T. C. Harman.[20])

in germanium are obtained by using impurities such as mercury and gold. Mercury produces an acceptor level at 0.087 eV above the valence band and thus mercury-doped germanium absorbs photons at 14 μm and shorter wavelengths. Impurity levels in doped silicon[26] are also useful, since they have absorption thresholds in the infrared. Shallow impurities (e.g., arsenic, phosphorus) in silicon have ionization energies in the neighborhood of 0.05 eV, so their thresholds are in the 25 μm region of the spectrum.

Photoconductivity in Semiconductors

We have discussed intrinsic and extrinsic absorption of photons in semiconductors and mentioned several examples whose absorption thresholds were determined by the energy gap (intrinsic) or impurity ionization (extrinsic) energy. Intrinsic absorption of a photon produces a photogenerated electron–hole pair, while extrinsic absorption produces free electrons or free holes, respectively, in n- or p-type impurity semiconductors. We turn next to consideration of the use of photogenerated free carriers in various effects used to detect electromagnetic radiation.

The first effect we consider is photoconductivity, which is the change in the conductivity of a solid when irradiated with light. The incident photons produce free charge carriers, electrons or holes, or both, thereby increasing the conductivity of the solid. The conductivity σ of a solid is given by

$$\sigma = q\mu n$$

where q is the electric charge of the carrier, μ is the carrier mobility, n is the density of carriers, and, for simplicity, we assume only one kind of carrier is present. If incident photons produce an additional density Δn of photogenerated carriers, then the increase $\Delta\sigma$ in the conductivity of the solid will be

$$\Delta\sigma = q\mu(\Delta n) \qquad (7.8)$$

If, as shown in Figure 7.11, an electric field is placed across the sample, the current I observed with light on the sample will be larger than the current measured with the sample in the dark. Many semiconductors exhibit photoconductivity,[27, 28] and the phenomenon is the basis of many applications.

We will consider[29] a photoconductor of cross-section area A and length d, as shown in Figure 7.11. A uniform electric field V/d is placed across the sample, which is irradiated uniformly with light such that L photons

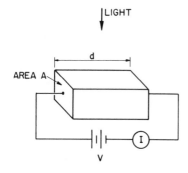

Figure 7.11. Experimental arrangement for observing photoconductivity. The sample is of cross section area A and an electric field V/d is applied across the sample.

are absorbed per second per unit volume of the sample. For simplicity, we consider that these photons produce one kind of photoexcited carrier (say electrons), which then traverse the sample under the influence of the applied electric field. We define the photoconductive gain[30] G of the photoconductor as the ratio

$$G = \frac{\text{number of charges crossing the sample per second}}{\text{number of photons absorbed per second}} \qquad (7.9)$$

The increase $\Delta\sigma$ in the conductivity is given by equation (7.8), so the increase ΔJ in the current density, also called the photocurrent density, is

$$\Delta J = (\Delta\sigma)(V/d) = q\mu(\Delta n)(V/d) \qquad (7.10)$$

The photocurrent $\Delta I = A(\Delta J)$, where A is the cross section-area of the sample. The increase Δn in the carrier density is given by

$$\Delta n = L\tau_e \qquad (7.11)$$

where τ_e is the electron lifetime. Since τ_e is the electron lifetime, L electrons will last τ_e sec. From these results we have that the number of carriers crossing the sample in unit time is

$$(\Delta I/q) = \mu A(\Delta n)(V/d) = \mu AL\tau_e(V/d) \qquad (7.12)$$

and the number of photons absorbed per unit time is LAd because the volume of the sample is Ad. Substituting into the definition (7.9) of the photoconductive gain gives

$$G = \mu\tau_e V/d^2 \qquad (7.13)$$

Since the drift velocity of the electrons is $\mu(V/d)$, the transit time τ_t of the electrons across the sample is

$$\tau_t = \frac{d}{\mu(V/d)} = \frac{d^2}{\mu V} \tag{7.14}$$

and the photoconductive gain G may be expressed as

$$G = \tau_e/\tau_t \tag{7.15}$$

the ratio of the electron lifetime to the transit time across the sample.

We see from equation (7.15) that, if the transit time τ_t is short and the lifetime τ_e is long, many electrons will cross the sample during one lifetime, giving a photoconductive gain greater than unity (since the carrier is replenished at an electrode). A large value of G means a larger current through the photoconductor per absorbed photon, so a large G means a greater photosensitivity for the photoconductor. We can see from equations (7.14) and (7.15) that a high mobility μ and a long carrier lifetime τ_e for a semiconductor favor a high value of the photoconductive gain, which may reach a value[31] of 10^3 in a material in which the carrier lifetime is long.

The sensitivity of a photoconductor is not usually specified in terms of photoconductive gain. Generally, one specifies the incident photon power which will produce a photogenerated signal just equal to the noise of the device under specified conditions.[32,33] This incident power is called the noise equivalent power, abbreviated NEP; the smaller the NEP of a photoconductor, the higher its sensitivity. Another figure of merit that is used is the detectivity D, defined by the relation $D = (\text{NEP})^{-1}$; the larger the detectivity, the higher the sensitivity of the photoconductive device. The responsivity R of a device is the ratio of the signal voltage to the incident power producing the signal.

Figure 7.12 shows an example[34] of photoconductivity in single-crystal solid solutions of SnTe in PbTe. The curves show the responsivity (in volts per watt) as a function of the wavelength of the incident radiation at two temperatures and for two values of the atom percent x of SnTe in the solid solution. The curve at 4.2 K has a threshold at approximately 14 μm in agreement with the energy gap of approximately 0.09 eV of this solid solution with $x = 0.17$. The curves at 77 K show that the threshold for the solid solution with $x = 0.20$ is at a longer wavelength (smaller photon energy) than the threshold for the solid solution with $x = 0.17$. This is as expected since (from Figure 7.10) the solid solution with the larger concentration of SnTe will have the smaller energy gap.

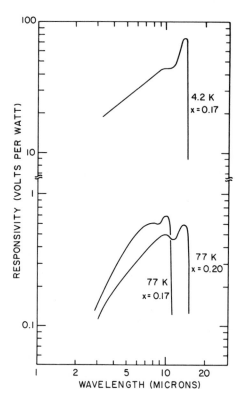

Figure 7.12. Responsivity (in volts per watt) as a function of incident photon wavelength λ (in microns) for solid solutions of SnTe in PbTe; x is the atomic percent of SnTe in the solid solution. (After Melngailis and Harman, Reference 34.)

Photodiodes

Intrinsic photoconductivity in a semiconductor can be exploited also in a nonhomogeneous structure like a semiconductor p–n junction. Consider the photodiode,[35] a reverse-biased p–n junction in which the junction region is illuminated with photons of energy $\hbar\omega$ larger than the band gap of the semiconductor. As shown in Figure 7.13, the absorbed photons generate electron–hole pairs. Those generated within a diffusion length of the depletion layer of the junction (where the photogenerated electrons are on

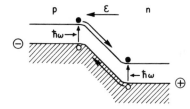

Figure 7.13. Reverse-biased p–n junction as a photodiode. Electrons ($\bullet$) and holes ($\circ$) are generated on both sides of the junction by photons of energy $\hbar\omega$ and diffuse to the junction. The electric field sweeps the minority carriers down the energy barrier, thereby increasing the reverse saturation current.

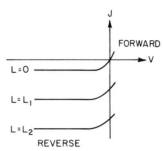

Figure 7.14. Current density J as a function of voltage V shown schematically for an illuminated reverse-biased p–n junction. The illumination intensity L is such that L_2 is larger than L_1.

the p side and photogenerated holes are on the n side) are collected and increase the respective minority carrier densities. These minority carriers are then swept down the energy barrier by the electric field at the junction, thereby increasing the reverse saturation current density. The reverse current through the junction is therefore larger when illuminated than in the dark. This is seen from the current–voltage plot of an illuminated reverse-biased p–n junction, shown schematically in Figure 7.14 for three values $L = 0$, $L = L_1$, and $L = L_2$ of the illumination intensity, where L_2 is greater than L_1. From Figure 7.14, we see that the reverse current through the junction is essentially independent of voltage, but increases with increasing illumination intensity L. For a sensitive detector, one wants a wide depletion layer in which the photogenerated carriers are generated. One way to do this is to dope at least one side of the junction lightly. Another way is to use a p–i–n (where i means intrinsic) structure, in which the intrinsic region has a high resistivity compared to the n and p regions. In that case, the applied reverse bias voltage appears almost entirely across the intrinsic layer, so the width of the depletion layer is about equal to the width of the intrinsic region. In this way, the size of the depletion layer can be built into the detector, keeping in mind that a wider depletion layer means longer transit times across the layer for photogenerated carriers, and hence a device with a longer response time.[36]

The photodiode utilizes intrinsic absorption of photons by a semiconductor, and so will have a threshold for photon absorption determined by the band gap of the semiconductor. Among the materials[37] now used for p–i–n photodiodes are alloys of III–V and II–VI semiconductors (such as AlGaSb and HgCdTe) with band gaps which can be varied by changing the composition of the alloy. The reverse-biased p–n junction can also be used to detect energetic particles (e.g., electrons, protons). Such particles, with energies many times the band gap energy, are absorbed over a large distance in the semiconductor, creating many electron–hole pairs. These pairs are collected in the large depletion layer built into these devices and increase the reverse saturation current density.

Photovoltaic Devices

Consider a p–n junction with no external bias applied, as shown in Figure 7.15. The junction is in equilibrium, the energy barrier between n and p sides is eV_0 (where V_0 is the diffusion potential), and the built-in electric field is $\mathscr{E}$. The junction is then illuminated by photons whose energy is greater than the band gap, creating electron–hole pairs. The minority carriers, as also shown in Figure 7.15, are swept down the energy barrier by the built-in electric field. The net result is the separation of the photo-generated electrons and holes, as shown in Figure 7.16, by the p–n junction. These separated photogenerated carriers set up an electric field $\mathscr{E}'$ which is opposite in direction to the built-in electric field $\mathscr{E}$. The net electric field across the junction is thus reduced to the value $\mathscr{E} - \mathscr{E}'$. This in turn means that the difference in electrostatic potential between the p and n sides of the junction is reduced from its equilibrium value V_0 to a smaller value $V_0 - V_f$, and the energy barrier is reduced to $e(V_0 - V_f)$, as shown in Figure 7.16. The effect is the same as if a forward-bias voltage V_f were applied to the junction. The overall result is the development of an open-circuit voltage V_f, in the forward direction (the n side negative) across the illuminated junction. The appearance of this voltage across an illuminated p–n junction is called the photovoltaic effect.[38]

We can see also from Figure 7.16 that the maximum value of the open-circuit voltage V_f is the diffusion potential V_0, obtained when the illumination is intense enough that the photogenerated electric field $\mathscr{E}'$ cancels completely the built-in field $\mathscr{E}$. Then the energy barrier between the p and n sides goes to zero, or, alternatively, the voltage V_f appearing across the illuminated junction is equal to V_0. Further, the maximum value of V_0 in a given semiconductor is E_g/e, where E_g is the band gap. This is because, from equation (2.4) of Chapter 2,

$$eV_0 = E_F(n) - E_F(p) \tag{2.4}$$

where $E_F(n)$ and $E_F(p)$ are, respectively, the Fermi energies on the n and p sides of the junction in equilibrium. Since $E_F(n) - E_F(p)$ will be largest when the Fermi energies are close to the band edges, the maximum value of

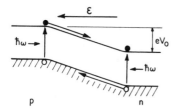

Figure 7.15. Semiconductor p–n junction, without external bias, which is initially at equilibrium, with energy barrier eV_0 (where V_0 is the diffusion potential) and built-in electric field $\mathscr{E}$. Photons of energy $\hbar\omega$ larger than the semiconductor band gap create electrons (●) and holes (○). The minority carriers are swept down the energy barrier.

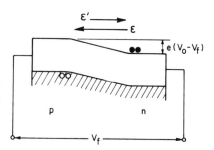

Figure 7.16. Semiconductor p–n junction, without external bias, under illumination. Photogenerated electrons ($\bullet$) and holes ($\bigcirc$) produce an electric field $\mathscr{E}'$ opposite to the built-in electric field $\mathscr{E}$. An open-circuit forward (n side negative) voltage V_f appears across the illuminated junction.

$E_F(n) - E_F(p)$ is the energy gap E_g, so the maximum value of eV_0 is the energy gap. Hence, the larger the energy gap of the semiconductor, the larger the maximum open-circuit voltage in the photovoltaic effect.

The photovoltaic effect in a p–n junction can be used to detect photons in various regions of the spectrum by observing the photogenerated voltage. The old-fashioned exposure meter of photography utilized the photovoltaic effect in the semiconductor hexagonal (grey) selenium, which has an energy gap of about 1.8 eV at 300 K. Junctions fabricated in semiconducting solid solution of SnTe in PbTe have been used as photovoltaic detectors[33] of infrared radiation with wavelengths as long as 30 μm.

If the p–n junction exhibiting the photovoltaic effect is connected across a load resistor, a current will flow in the external circuit when the junction is illuminated. The illuminated junction can therefore deliver power to an external load and the photovoltaic device operating in this manner can convert the energy of the incident photons into electrical energy. Since the junction is now delivering current into an external circuit, the photogenerated voltage appearing across the load will be smaller than the open-circuit photovoltage V_f. The choice of a semiconductor for a specific application of the photovoltaic effect will depend on the distribution in energy of the incident photons. For a particular distribution, there will be an optimum value of the energy gap E_g of the semiconductor. If E_g is too small, some of the energy of the incident photons will be wasted in the production of high-energy hot electron–hole pairs which lose their energy by emitting phonons. If E_g is too large, many photons will pass through the semiconductor without being absorbed.

If the photons whose energy is to be converted into electrical energy come from the sun, then a photovoltaic junction operating as described above is called a solar cell.[40,41] For terrestrial conversion of solar photons,[40] the optimum value[42] of $E_g = 1.4$ eV, suggesting GaAs. However, since many other factors (including economics) influence the choice, silicon ($E_g = 1.1$ eV) and other materials are under intensive development[43] for solar cells for terrestrial use. An interesting recent development is the use of hydrogenated amorphous silicon (a-Si:H), discussed in Chapter 6, for solar cells.[44-47] The introduction of hydrogen into the amorphous silicon network removes dangling-bond states from the energy gap, thereby permitting the observation of intrinisc photoconductivity. Construction of a p–n junction by doping leads to the production of the photovoltaic effect when photons with energies larger than the band gap are absorbed. The structure generally used for amorphous silicon solar cells is a p^+–i–n^+ ("i" stands for intrinsic) junction, but metal–a-Si:H Schottky barrier junctions are also used.

Other Applications of Intrinsic Photoconductivity

In this section, we discuss three familiar processes which depend on intrinsic photoconductivity in different semiconductors. These processes are (1) electrophotography, one form of which is the Xerox process; (2) the light-sensitive unit in a television camera tube; (3) photography.

We begin with a consideration of electrophotography,[48,49,49a] a technique whose perhaps most familiar form is the Xerox process[50,51] of copying. The photoconductor involved is amorphous selenium, a semiconductor with an energy gap[52] of approximately 2.1 eV at room temperature, which is vacuum deposited as a thin film on a metal substrate. The selenium has a very high resistivity, of the order[53] of $10^{16}\,\Omega$ cm, in the dark. The first step in the process is the buildup of a positive charge on the selenium by a high voltage corona discharge, as shown in Figure 7.17. This step is referred to as sensitization. The second step, shown in Figure 7.18, is exposure of portions of the selenium layer to light, generating electron–hole pairs, thereby increasing the conductivity of the illuminated regions. (In the amorphous selenium, the electrons are trapped, and current flow is due to holes alone.) The illuminated regions discharge and the dark regions do not; the resistivity of the semiconductor is so high that charge does not leak off

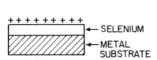

Figure 7.17. First step of the Xerox process, in which a positive charge is built up on the surface of an amorphous selenium film by a corona discharge. (Relative dimensions are not to scale.)

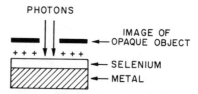

Figure 7.18. Second step of the Xerox process, in which photogenerated carriers in the selenium discharge the illuminated regions, leaving positive surface charge in the dark regions.

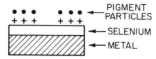

Figure 7.19. Third step of the Xerox process, in which dark pigment particles (●) are attracted to the positively charged regions of the selenium.

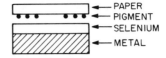

Figure 7.20. Fourth step of Xerox process, in which the dark pigment is transferred to paper and fixed, forming a positive image copy.

laterally. The result of this step is a positive charge distribution on the surface of the selenium which is a replica of the dark portion of the image to be copied. The third step, shown in Figure 7.19, dusts a dark pigment (the toner) onto the selenium, where it is attracted only to the charged regions which were not illuminated in the second step. The result is a positive image of the material to be copied, in which the dark parts of the original appear dark in the image. The fourth step, shown in Figure 7.20, is the transfer of the dark pigment image onto a piece of ordinary paper. The pigment is then fixed onto the paper by heat, forming a permanent positive image which is the final copy. A recent development is the use of amorphous[54] silicon as the photoconductor in the electrophotographic process.

Our next topic is the light-sensitive unit in a television camera; this unit is often a type of tube called a vidicon.[55,56] The vidicon tube contains a thin layer of a photoconductor sensitive in the visible region of the spectrum. Examples are As_2S_3 and PbO, whose energy gaps[52] are, respectively, approximately 2.5 and 2.3 eV. The photoconductor is prepared in polycrystalline form by evaporation in a poor vacuum, resulting in a photoconductive layer with a very high resistivity of the order of 10^{15} Ω cm. Figure 7.21 shows the arrangement of the photoconductive layer in the vidicon tube. Photons are incident on various parts of the photoconductive layer, passing first through a transparent conducting electrode. The photoconductor is scanned from the rear by an electron beam (which moves sequentially over the rear surface), "depositing" negative charges which do

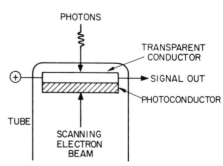

Figure 7.21. Schematic view of a vidicon tube.

not leak off due to the high resistivity of the photoconductive layer. If photons are incident on some part of the front surface of the photoconductive layer, intrinsic electron–hole pairs are generated, increasing the conductivity of the illuminated region. This increased conductivity allows the stored negative charge on the rear surface to be conducted away by an applied positive bias. As a result, when the electron beam returns on its next scan, it "deposits" more electrons in order to return that part of the photoconductive layer to its original condition. The current in the electron beam is thus momentarily increased, giving a pulse of beam current which is essentially information on the light intensity falling on that particular part of the photoconductive layer. In this way, the output signal of the vidicon tube contains information concerning the light falling on the tube. With proper electronic processing, this light pattern information is reproduced on the cathode ray tube of the television receiver.

Another type[55] of vidicon involves a different type of photosensitive surface. The surface is a wafer of single-crystal silicon on which are fabricated about 10^6 p^+–n junctions using integrated circuit technology. The configuration used in this silicon vidicon is shown in Figure 7.22; photons pass through a thin n^+ layer which serves as an electrode. The scanning electron beam reverse-biases each p^+n junction, extending the space charge region into the more lightly doped n-type layer. If photons are incident on the diode between scans by the electron beams, the depletion region is filled with photogenerated carriers, thus reducing the reverse bias on the junction. On the next scan, the electron beam "deposits" additional electrons, resulting in a pulse of current containing video information. The silicon vidicon exhibits high reliability and response into the infrared because of the relatively small energy gap of silicon.

The third application we will consider is photography. The usual photographic process depends on intrinsic photon absorption in the silver halides AgCl and AgBr. These compounds are ionic semiconductors with thresholds for photon absorption[57] at about 2.5 eV at room temperature. A photographic emulsion[58] consists of silver halide microcrystals in a binding

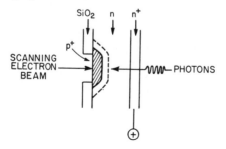

Figure 7.22. Schematic view of one p^+–n junction in a silicon vidicon. The p^+ region is shown shaded, the SiO₂ is an insulator, and the photons pass through the thin n^+ electrode. The dotted line represents the depletion layer of the junction, almost all of which is in the relatively lightly doped n region.

material. A current model[59,59a] of the photographic process in a silver halide AgX (where X is a halogen) is as follows. A photon of sufficient energy is absorbed by an AgX crystal, producing an electron (e^-) and a hole. The electron is then trapped, while the hole can combine with an X^- ion to form a neutral X atom. The trapped electron the combines with a mobile Ag^+ ion,

$$Ag^+ + e^- \to Ag \qquad (7.16)$$

to form a silver atom. A second photon then produces another electron–hole pair. The hole produces a second X atom, which reacts with the first X atom to produce an X_2 molecule, which escapes. The electron is trapped by the silver atom above,

$$Ag + e^- \to Ag^- \qquad (7.17)$$

forming an Ag^- ion. An Ag^+ ion then combines with the Ag^- ion,

$$Ag^+ + Ag^- \to Ag_2 \qquad (7.18)$$

to form a stable two-atom molecule of silver. This Ag_2 is the basic stable unit of silver, and, on repetition of the process by additional photons, becomes a speck of metallic silver. This silver speck, present in an AgX silver halide microcrystal which has been exposed to light, is called the latent image.

When the latent image is developed, it is exposed to a chemical reagent which converts AgX to silver, in such a way that AgX grains that have been exposed to light are converted at a higher rate[60] than unexposed grains. The result is that the latent image is converted into a metallic silver image, with the silver density highest in the regions that had the highest illumination level.

Summary on Semiconductor Photon Detectors

It is useful to summarize how the physics of the semiconductors involved yields useful photosensitive devices. First, the existence of an energy gap between valence and conduction bands is important from the device standpoint. The photon absorption spectrum for intrinsic excitation has a sharp threshold which can be chosen to match the spectral region of interest. If the energy gap is sufficiently large relative to $k_B T$, the density of thermally excited intrinsic carriers will be small. This allows the photo-generated carriers to change the electrical conductivity appreciably. Second, the appreciable lifetime of carriers in semiconductors allows the maintenance of significant concentrations of photogenerated carriers at usefully low light intensities. Third, the existence of discrete impurity levels, with ionization energies smaller than the band gap energy, yields photoconductivity at low photon energies. Finally, the semiconductor p–n junction permits a variety of useful devices to be made (e.g., photovoltaic detectors) which depend directly on the properties of the junction (e.g., the built-in electric field) for their operation.

Emission of Photons in Semiconductors

We now turn to the use of semiconductors as generators of electromagnetic radiation, discussing some processes whereby an electron and a hole recombine with the spontaneous emission of a photon. These processes, collectively called luminescence, are the basis of several useful light sources.

We consider first the recombination process in which a free electron in the conduction band makes a transition to an empty (i.e., hole) state in the valence band, emitting a photon. This process is called band-to-band recombination[61,62] of the electron and hole, and can occur in both direct-gap and indirect-gap semiconductors, as shown in Figures 7.23 and 7.24. In these figures, the energy difference ΔE is defined as

$$\Delta E = E_i - E_f \tag{7.19}$$

where E_f and E_i are, respectively, the energies of the final and initial states

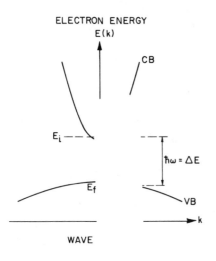

Figure 7.23. Band-to-band recombination of an electron (●) and a hole (○) in a direct-gap semiconductor with the emission of a photon of energy ΔE. The initial and final energies of the electron are, respectively, E_i and E_f. The conduction and valence bands are labeled CB and VB, respectively.

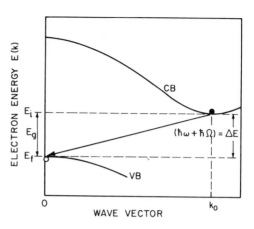

Figure 7.24. Band-to-band recombination of an electron (●) and a hole (○) in an indirect-gap semiconductor, with the emission of a photon of energy $\hbar\omega = \Delta E - \hbar\Omega$ and of a phonon of energy $\hbar\Omega$. The quantity $\Delta E = E_i - E_f$, where E_i and E_f are, respectively, the initial and final energies of the electron. The conduction and valence bands are labeled CB and VB, respectively.

of the electron. Figure 7.23 shows a band-to-band transition of an electron with no change of wave vector (if the photon wave vector is neglected) and the emission of a photon of energy $\hbar\omega = \Delta E$. Since the minimum energy gap is direct, the energy spectrum of the emitted photons will have a low-energy threshold at the band gap energy. The transition shown in Figure 7.23 can take place with emission of a photon of energy ΔE somewhat larger than the energy of the gap if the electron and hole involved have energies above those of their respective band edges.

Figure 7.24 shows a band-to-band transition in an indirect-gap semiconductor. In this process an electron of initial energy E_i and initial wave vector k_i makes a transition to a final state of energy E_f and wave vector k_f with the emission or absorption of a phonon of energy $\hbar\Omega$ and wave vector k'. Conservation of energy requires that

$$E_i = E_f \pm \hbar\Omega + \hbar\omega \qquad (7.20)$$

where $\hbar\omega$ is the energy of the emitted photon. The plus sign refers to the process in which a phonon is emitted and the minus sign to that in which a phonon is absorbed. Conservation of wave vector requires that

$$k_i = k_f \pm k' \qquad (7.21)$$

where the plus and minus signs have the same meaning as in equation (7.20). From equation (7.20), the energy spectrum of emitted photons will have a low-energy threshold when $E_i - E_f = E_g$, the gap energy; the lowest-energy photon emitted will have an energy

$$\hbar\omega = E_g - \hbar\Omega \qquad (7.22)$$

for the transition in which a phonon of energy $\hbar\Omega$ is emitted. Figure 7.24 shows the emission of the minimum-energy photon of energy $E_g - \hbar\Omega$ with the simultaneous emission of a phonon of wave vector k_0. As described earlier for indirect absorption of photons, the indirect band-to-band emission of photons with phonon participation is a three-body process. As such, indirect recombination is less probable than direct electron–hole recombination. An estimate[63] of the electron lifetime with respect to band-to-band recombination in a direct-gap semiconductor is 10^{-6} sec. A similar estimate[63] for an indirect-gap semiconductor is of the order of 0.5 sec, showing that the indirect process is much less probable than the direct process.

From the point of view of applications, the main conclusion is that band-to-band recombination of electrons and holes in a semiconductor results in the emission of photons. In both the direct- and indirect-gap cases, there is a continuous spectrum of photon energies above a low-energy threshold at approximately the energy of the band gap. The specific shape and width of the spectrum of emitted photons will depend on the details of the transition probability, band densities of states, impurity content, etc. Figure 7.25 shows the luminescence[64] from n-type InAs (a direct-gap semiconductor), with an electron density of 9×10^{16} cm^{-3}, at 77 K. The photon emission is due to direct radiative recombination of free electrons and free holes. The peak of the emission spectrum at 0.41 eV agrees well with the value[65] of the energy gap; the emission at lower photon energies is due to other processes.[66]

Photon emission in semiconductors can also take place by recombination transitions between a band state and an impurity level, and between two impurity levels. The former type of transition is known[67] to occur in both n- and p-type GaAs. Figure 7.26 shows schematically the recombination of an electron in the conduction band of GaAs with a hole bound to an acceptor level about 0.03 eV above the valence band edge. The emitted photon energy $\hbar\omega$ is thus less than the band gap energy by about 0.03 eV.

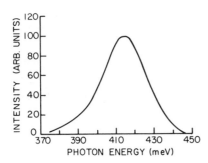

Figure 7.25. Direct band-to-band recombination luminescence in InAs at 77 K. (After Mooradian and Fan.[64])

Figure 7.26. Schematic recombination of an electron (●) in the conduction band (CB) with a hole (○) bound to acceptor slightly above the valence band (VB). The energy $\hbar\omega$ of the emitted photon is less than the band gap energy E_g. (The acceptor ionization energy is exaggerated for clarity.)

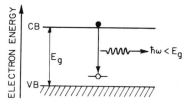

The latter type[68] of transition is also known to occur in various other semiconductors.

p–n Junction Luminescence

One of the most important applications of luminescence in semiconductors is in a *p–n* junction, known as a light-emitting diode[69–73] or LED. In our earlier discussion of photon emission in semiconductors we did not discuss the way in which the recombining electrons and holes were produced. In this section, we consider a technologically important device in which the free electrons and holes are injected across a *p–n* junction. Radiative recombination of electrons and holes then produces photons.

Consider the *p–n* junction at equilibrium shown in Figure 7.27, in which both the *p* and *n* sides are degenerate. If a large forward bias is applied to the junction, the energy barrier to the flow of electrons and holes is reduced as shown in Figure 7.28. Electrons are injected from the *n* side to the *p* side, where they recombine with holes, emitting a photon whose energy $\hbar\omega$ is approximately the gap energy of the semiconductor. A similar process takes place with holes injected into the *n* side, also producing photons. In this way, photon emission takes place in the vicinity of the

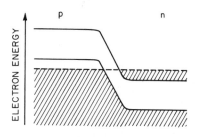

Figure 7.27. Band diagram of a degenerate *p–n* junction at equilibrium. The filled electron states in the conduction band on the *n* side are shown shaded; the empty (hole) states in the valence band on the *p* side are unshaded.

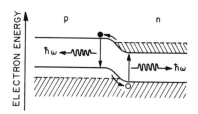

Figure 7.28. Band diagram of a degenerate *p–n* junction under forward bias. Electrons (●) are injected from *n* to *p*, and holes (○) from *p* to *n*, with emission of photons by radiative recombination.

junction. (Actually, the injection of electrons from the n side to the p side is more complicated than the simplified description given above. At forward bias values smaller than that which produces a significant injection current of holes and electrons, electrons may tunnel from n to p, either directly or via an intermediate state, with emission of a photon.[74-76])

From the point of view of applications of light-emitting diodes, the main interest is in the emission of photons in the visible region of the spectrum. For this reason, interest has been concentrated on semiconductors whose band gaps are greater than about 1.5 eV; examples[69,73] are III-V semiconductors and their solid solutions. As discussed earlier, the radiative recombination of electrons and holes is more probable in a direct-gap semiconductor.[77] For this reason, semiconductors, such as GaAs and various III–V solid solutions, with direct gaps have been the object of considerable research.

Light Amplification by Stimulated Emission of Radiation

In our discussion of photon emission from semiconductors, we have been considering only spontaneous emission. We now consider the stimulated emission[78] of photons. Stimulated emission is the transition of an electron from a state of energy E_2 to a state of lower energy E_1 under the influence of an incident photon of energy $\hbar\omega_{21}$, where

$$\hbar\omega_{21} = E_2 - E_1 \qquad (7.23)$$

The incident photon of energy $\hbar\omega_{21}$ stimulates the electronic transition and thereby the emission of a photon whose energy is also $\hbar\omega_{21}$. In certain solid state systems, n_i "input" photons of a particular energy incident on the system will produce the stimulated emission of n_o "output" photons of the same energy. If n_o is larger than n_i, then the process is one of amplification of the number of photons by the process of stimulated emission. The word LASER is an acronym for *l*ight *a*mplification by *s*timulated *e*mission of *r*adiation, that is, the process described above.

We begin by discussing some general ideas[79-83] concerning lasers. Consider a quantum system (as yet unspecified) in which there are three electronic energy levels in a solid, as shown schematically in Figure 7.29. The three electronic energy levels are E_0 (ground state), E_1, and E_2, and various possible electronic transitions in this level system are indicated by arrows. Transition A is from level E_0 to level E_2 and may be excited by the absorption of a photon of energy $E_2 - E_0$ by an electron in the ground

Figure 7.29. Energy level diagram with three electronic levels. The ground state at energy E_0 has an electron population N_0, the intermediate state at energy E_1 has an electron population N_1, and the upper state at energy E_2 has an electron population N_2. Transition A is excited by the absorption of a photon of energy $E_2 - E_0$ by an electron in the ground state. Transition B is radiationless; phonons are given off. Transition C emits photons of energy $E_1 - E_0$.

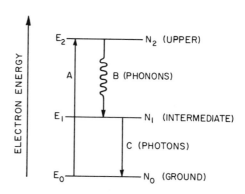

state. Transition B from level E_2 to level E_1 is radiationless (phonons are given off) and is characterized by a relaxation time t_{21}. Transition C from level E_1 to the gound state emits a photon of energy $\hbar\omega_{10} = E_1 - E_0$ and is described by a relaxation time t_{10}. The electron populations, per unit volume, of the three levels are denoted by N_0, N_1, and N_2.

The transition that emits photons is C, so we want to consider the possibility of the amplification of photons of energy $\hbar\omega_{10}$ by the use of this transition. In the amplification process an "input" photon of energy $\hbar\omega_{10}$ will stimulate the emission of additional photons of the same energy. Considering the two levels E_1 and E_0, with electron populations, at some instant, N_1 and N_0, an incident photon of energy $\hbar\omega_{10}$ is just as likely to be absorbed in stimulating an $E_0 \rightarrow E_1$ transition as it is to stimulate the emission of an additional photon by an $E_1 \rightarrow E_0$ transition. For amplification to occur, it is necessary that more electrons in the system be able to emit an $\hbar\omega_{10}$ photon than are able to absorb such a photon. This will be true if there are more electrons in the E_1 level than in the E_0 level. In other words, we want the electron population N_1 of the E_1 level to be larger than the population N_0 of the ground state.

We can see in more detail that this is so as follows; in this argument, statistical degeneracies are neglected for simplicity. The energy given off, per unit time per unit volume, due to stimulated emission from transitions from E_1 to E_0 is given by

$$\hbar\omega_{10}N_1P_{10} \tag{7.24}$$

where P_{10} is the transition probability[84] per unit time of the stimulated emission transition from $E_1 \rightarrow E_0$. Similarly, the energy absorbed, per unit time per unit volume, due to absorption of photons of energy $\hbar\omega_{10}$ is equal to

$$\hbar\omega_{10}N_0P_{01} \tag{7.25}$$

where P_{01} is the transition probability per unit time for absorption. The net energy given off, per unit volume per unit time, due to absorption and stimulated emission is just the difference

$$\hbar\omega_{10}P_{10}N_1 - \hbar\omega_{10}P_{01}N_0 \tag{7.26}$$

Since the transition probabilities P_{10} and P_{01} are equal,[85] the net energy given off is

$$\hbar\omega_{10}P_{10}(N_1 - N_0) \tag{7.27}$$

If a population inversion exists, N_1 is larger than N_0, net energy is given off due to stimulated emission, and photon amplification can take place. (In the discussion above, spontaneous emission was not included since stimulated emission is the faster process and so dominates spontaneous emission.)

Statistical mechanics tells us that, at equilibrium, the ratio N_1/N_0 is less than unity, and the ground state has the larger electron population. If N_1 should be made, in some way, larger than N_0, then the system would not be in equilibrium, and we would have a situation called a population inversion. We conclude that a population inversion such that N_1 is larger than N_0 is necessary if the $E_1 \rightarrow E_0$ transition is to be used for the amplification of $\hbar\omega_{10}$ photons.

Under certain circumstances, the three-level system in Figure 7.29 allows the achievement of the desired population inversion. Consider the excitation, say optically, of electrons from the ground state to the upper state. Suppose the rate at which electrons in the E_2 (upper) level decay to the E_1 (intermediate) level is very fast compared to the rate at which electrons in the E_1 level decay to the E_0 (ground) state. This is equivalent to the statement that the relaxation times t_{21} and t_{10} are such that

$$t_{10} \gg t_{21} \tag{7.28}$$

Equation (7.28) is a statement that the electron lifetime in the intermediate state is long compared to the electron lifetime in the upper state. If the condition (7.28) is satisfied, then, as electrons are excited into the upper state, they will decay rapidly into the intermediate state via transition B, and then, much more slowly, into the ground state via transition C. The result will be an accumulation of electrons in the intermediate state relative to the ground state, or, in other words, a population inversion in which N_1 is larger than N_0.

It is clear that the existence of the intermediate state in the three-level system is essential to the attainment of the desired population inversion. If there were only the two levels E_2 and E_0, then incident photons of energy $E_2 - E_0$ could produce, at most, a 50% population of the E_2 level. This is because such a photon is just as likely to stimulate an $E_0 \rightarrow E_2$ absorption transition as it is to stimulate an $E_2 \rightarrow E_0$ emission transition. The existence of the intermediate state, coupled with the difference in lifetime expressed in equation (7.28), allows the achievement of the population inversion between the intermediate and ground states in the three-level system.

Once the population inversion is achieved, and N_1 is larger than N_0, then spontaneously emitted photons of energy $\hbar\omega_{10}$ can cause stimulated emission of additional $\hbar\omega_{10}$ photons with concurrent amplification. This process is what is called laser action or lasing.

In the three-level system in Figure 7.29, we can see that, in order to obtain a population inversion, more than half of the electrons originally in the ground state must be excited to the intermediate state (via the upper state). This means that considerable energy must be put into the exciting transition A in order to achieve the population inversion, and leads to a relatively inefficient[79] laser when, as in the three-level case, the final state of the emission transition is also the gound state. The four-level system[79] shown in Figure 7.30 is one that avoids this disadvantage. Considering such a set of energy levels in a solid, electrons are excited by photons from the ground state E_0 to the upper state E_3 in transition A. A fast radiationless transition B from level E_3 to level E_2 is characterized by a relaxation time t_{32}. The slow transition C from level E_2 to level E_1 emits a photon of energy $\hbar\omega_{21} = E_2 - E_1$ and is described by the relaxation time t_{21}. Finally, a fast radiationless transition D, with relaxation time t_{10}, returns the electron to the ground state. If the relaxation times are such that

$$t_{21} \gg t_{32}, t_{10} \qquad (7.29)$$

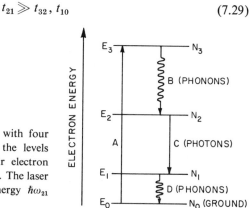

Figure 7.30. Energy level diagram with four electronic levels. The energies of the levels are $E_0, E_1, E_2,$ and E_3, and their electron populations are $N_0, N_1, N_2,$ and N_3. The laser transition C emits photons of energy $\hbar\omega_{21} = E_2 - E_1$.

a population inversion with N_2 larger than N_1 may be achieved. Laser action, with the stimulated emission of photons of energy $\hbar\omega_{21}$, is obtained by transition C, in which the final state E_1 is relatively unpopulated as compared to the ground state E_0 in the three-level scheme.

It should not, however, be inferred that a net output of stimulated photons takes place as soon as a population inversion is achieved. Emission with amplification will occur only if the rate of stimulated emission of photons is greater than the rate of loss of photons due to factors[86] such as absorption, scattering, and diffraction, in the laser system. For this reason, there exists a threshold intensity of pump (exciting) radiation which is necessary to achieve a critical or threshold value[87] of the population inversion. For pump intensity values greater than the threshold value, the population inversion is large enough to give a rate of stimulated emission able to overcome losses and so produce amplification and laser action.

Solid State Lasers

A very important class of laser systems, which exhibit three- and four-level schemes similar to Figures 7.29 and 7.30, is that of solids containing small amounts of an impurity atom. The host solid may be a crystal (such as Al_2O_3) or a glass, and the impurity atom is generally one in which electronic transitions can take place between states of inner, incompletely filled electron shells. An example of such a three-level scheme is that of the Cr^{+3} ion in Al_2O_3 (ruby); a four-level system is that of the Nd^{+3} ion in

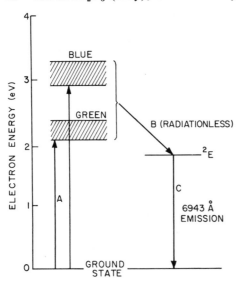

Figure 7.31. Simplified energy level scheme of the Cr^{3+} ion in Al_2O_3 (ruby). The ^2E state is actually a closely spaced (3.6 meV) pair; the shaded regions are broad bands of electron states; arrows represent electronic transitions.

yttrium aluminum garnet (YAG).

The energy level scheme[88] of the Cr^{+3} ion in Al_2O_3, the ruby laser, is shown in simplified form in Figure 7.31. At just less than 2 eV above the ground state is a very closely spaced pair of ^2E states, shown as a single level in Figure 7.31. Above ^2E there are two broad bands of states, referred to as "blue" and "green." In the operation of the ruby laser, electrons are optically excited (transitions A) by pumping from the ground state to the blue and green bands by a broad-spectrum photon source like a xenon lamp. The lifetime of an electron in these excited states is about 10^{-7} sec, after which they decay via radiationless transition B to the intermediate ^2E state with the emission of phonons. Considering the ^2E states as a single level, the lifetime of an electron in the ^2E state is relatively long, about 5×10^{-3} sec. Electrons accumulate in the intermediate state as pumping continues, and a population inversion of the intermediate state relative to the ground state is produced. Laser action takes place, via transition C in Figure 7.31, with the emission of a photon of wavelength 6943 Å, the familiar red line of the ruby laser.

In practice, single crystals of ruby in a cylindrical shape are constructed so that the two ends are plane and parallel, and silvered on the outside to act as mirrors, as shown schematically in Figure 7.32. (In practice, external mirrors are more commonly used.) Assume that the pump radiation has effected a population inversion via transitions A and B in Figure 7.31. There will eventually (in a small fraction of a second) be spontaneous emission of photons due to transition C in Figure 7.31. Many of these photons will leave the cylinder through the sides, but some will travel along the axis of the rod, being reflected at the ends and causing stimulated emission of photons of wavelenght 6943 Å via transition C. As these photons traverse the rod (which is still being pumped), there is an "avalanche" of stimulated emission of photons, and amplification takes place. The photons are extracted from the crystal rod by having one mirror slightly transparent.

While ruby is still an important solid state laser because of its visible output, there are several other solid state lasers, generally four-level systems with outputs in the near infrared. Discussions of various solid state lasers are to be found in the literature.[89]

Next, we mention some of the solid state physics factors[90] relevant to laser action. Considering a crystal in which the active impurity resides in a passive host, the host crystal must be transparent to both the exciting pump radiation and the laser emission. This means that the host crystal

Figure 7.32. Schematic view of cylindrical laser crystal. The mirrored surfaces are shown shaded.

must be transparent from the visible through the near infrared. Suitable materials are generally ionic insulators with large energy gaps, like Al_2O_3, $CaWO_4$, and yttrium aluminum garnet (YAG). Various glasses form another such group of hosts. (The energy level scheme of the active impurity ion must also be suitable, but that is more a problem in atomic physics than in solid state physics even though, of course, the energy levels of the free ion will be modified by interaction with the host crystal potential.) Last, the phonon spectrum of the host crystal will influence the details of the nonradiative transitions in the energy level scheme used for laser action.

Semiconductor Injection Lasers

A semiconductor injection laser[91-95] also involves an inversion of the electron population between two bands of energy levels. However, in this case, the population inversion is accomplished by the injection of electrons into the p-type region of a p–n junction, rather than with the optical excitation used in, for example, the ruby laser.

Consider a degenerate p–n junction, whose band diagram at equilibrium is shown in Figure 7.33. The energy gap of the semiconductor is E_g. The conduction band on the n side is filled with electrons to an energy Δ_n above the band edge, and the valence band on the p side is filled with holes to an energy Δ_p below the band edge. If a forward-bias voltage V_a is applied to the junction, the electron energies on the n side will be increased an amount eV_a relative to the p side. For an applied voltage V_a approximately equal to E_g/e, the situation shown in Figure 7.34 obtains, with electrons being injected into the p region. In this situation, there is an effective population inversion established in a narrow region near the junction. There are filled electron states in the conduction band at a higher energy than empty states in the valence band. Electrons in the active region containing the inverted

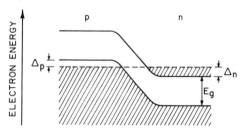

Figure 7.33. Equilibrium band diagram of a degenerate p–n junction. Filled electron states are shown shaded. The energy gap of the semiconductor is E_g. The conduction band on the n side is filled with electrons to an energy Δ_n above the band edge; the valence band on the p side is filled with holes to an energy Δ_p below the band edge. The figure is drawn for Δ_n larger than Δ_p.

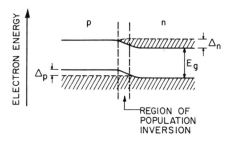

Figure 7.34. Band diagram of the degenerate *p–n* junction under an applied forward bias V_a approximately equal to E_g/e, where E_g is the energy gap of the semiconductor. Injection of electrons into the *p* side forms a region of population inversion near the junction.

population can combine with holes with the emission of photons. As the forward current through the junction is increased, a threshold value of the current is reached. Above this threshold, the photon emission is stimulated; below threshold, the emission is spontaneous junction luminescence. The process shown in Figure 7.34 is typical of GaAs junction lasers,[96] in which electrons are injected into the *p* side, so recombination and photon emission take place in the *p* region of the junction.

If we approximate the energy width of the filled and empty states in the region of population inversion in Figure 7.34 by Δ_n and Δ_p, respectively, we will have the band structure shown in Figure 7.35. This figure is drawn for a direct-gap semiconductor like GaAs, and shows that the electron population is inverted with respect to the emission of photons with energies as large as $E_g + \Delta_n + \Delta_p$ and as small as E_g, if we consider only the process of band-to-band recombination of electrons and holes. Actually, Figure 7.35 is not correct[97] because the high impurity densities necessary for forming a degenerate junction will introduce a hole impurity band in the *p* region where the population inversion exists. It is believed[98] that the lasing transition in GaAs junctions is from the electron distribution in the conduction band to an acceptor state, so the energy of the emitted photon may be less than the gap energy. In any case, we conclude that stimulated

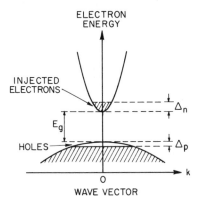

Figure 7.35. Semiconductor band structure within the region of population inversion. The direct energy gap is E_g, and the energy widths of the inverted populations of injected electrons and of holes are Δ_n and Δ_p, respectively.

emission is possible for a range of photon energies near E_g.

A semiconductor junction laser is generally fabricated so the structure forms an optical resonant cavity of length L, such that there are an integral number m of half-wavelengths contained in the cavity length. If λ is the wavelength in the semiconductor and λ_0 is the wavelength in vacuum, then

$$\lambda = \lambda_0/n \qquad (7.30)$$

where n is the refractive index. Then the cavity condition is

$$L = m(\lambda/2) = m(\lambda_0/2n) \qquad (7.31)$$

where m is an integer. The vacuum wavelengths λ_0 satisfying equation (7.31) are given by

$$\lambda_0 = 2nL/m \qquad (7.32)$$

for integral values of m. The stimulated emission spectrum above the threshold current will be at a wavelength or wavelengths which are the modes (7.32) of the optical resonant cavity. The laser emission is thus "selected" by the resonant cavity from the range of possible transitions in the neighborhood of the gap energy E_g.

Junction lasers have been made in a number[99] of direct-gap semiconductors. The emitted photon energy due to a single mode of emission is usually approximately equal to the energy gap when considering the simple and more probable case of direct recombination (i.e., no phonon participation) of electrons and holes. However, since a variety of emission transitions[100] with slightly different energies (due to impurity levels) is known, the emission is only approximately at the gap energy. Junction lasers using narrow gap semiconductors[101] have been operated at emission wavelengths as long as 32 μm.

As a final point on semiconductor injection lasers, we consider heterojunction structures.[102–104] A heterojunction is a junction between two semiconductors with different energy gaps. An example is a heterostructure

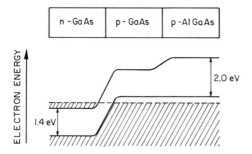

Figure 7.36. Band diagram of a heterojunction structure at equilibrium. Filled electron states are shown as shaded.

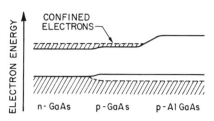

Figure 7.37. Band diagram of a hetero-junction structure at high forward bias, showing confinement of injected electrons. Filled electron states are shown as shaded.

with a junction between p-type GaAs ($E_g = 1.4$ eV) and a p type solid solution of AlAs and GaAs. The latter has a variable energy gap depending on the composition of the solid solution; in the case shown in Figure 7.36, the energy gap of the AlGaAs is 2.0 eV. The doping of the p-type GaAs and the p-type AlGaAs is the same, so the equilibrium band diagram is as shown in Figure 7.36. Under high forward bias, electrons are injected from the n-type GaAs into the p-type GaAs, giving an inverted population as shown in the band diagram in Figure 7.37. The effect of the p-GaAs–p-AlGaAs heterojunction is to provide an energy barrier which helps prevent the diffusion of the electrons injected into the p-type GaAs. This confinement of the inverted electron population results in the achievement of stimulated emission at lower values of the threshold current.

Summary on Solid State Lasers

It is useful to summarize how some aspects of the physics of the solids considered give them their utility as lasers. First, the energy levels of various ions (e.g., Cr^{+3}, Nd^{+3}) in transparent host crystals or glasses provide the three- or four-level systems which make a population inversion possible. Second, in these three- or four-level schemes, differences in electron lifetime in upper and intermediate states allow the achievement of an inverted population. Third, in semiconductor injection lasers, a population inversion can be achieved because of the following. Injection of electrons produces a large number of electrons on the p side of the junction. The fact that the p side can be doped degenerately gives a large number of vacant final states, thereby contributing to the population inversion.

Problems

7.1. *Direct Transition.* Using the band structure of InSb in Figure 1.6, consider a direct transition at an electron wave vector k (say 0.1×10^7 cm^{-1}) near the zone center. Estimate the photon energy involved, and show that the wave vector of this photon is negligible compared to that of the electron.

7.2. *Indirect Transitions with Phonon Emission.* Using equations for the conservation of energy and wave vector, show that equation (7.7) is true for indirect transitions involving the emission of phonons of energy $\hbar\Omega$ in a semiconductor whose minimum (indirect) energy gap is E_g.

7.3. *Automobile Exhaust Monitor.* It is desired to monitor the CO_2 concentration in auto exhaust gas by using the CO_2 molecular absorption line whose wavelength is 4.25 μm. Discuss the design of a photon source (either a laser or a light-emitting diode) and a detector suitable for this spectral region. Comment on the factors that must be taken into account if the system is to be used "on board" an automobile.

References and Comments

1. T. S. Moss, G. J. Burrell, and B. Ellis, *Semiconductor Opto-Electronics*, John Wiley, New York (1973), Chapter 3.
2. J. I. Pankove, *Optical Processes in Semiconductors*, Prentice-Hall, New York (1971), Chapter 3.
3. C. Kittel, *Introduction to Solid State Physics*, Sixth Edition, John Wiley, New York (1986), pages 183–187.
4. See, for example, J. M. Stone, *Radiation and Optics*, McGraw-Hill, New York (1963). pages 376–383.
5. T. S. Moss *et al.*, Reference 1, pages 55–69.
6. See, for example, the data on germanium of W. C. Dash and R. Newman, *Physical Review*, **99**, 1151 (1955).
7. See, T. S. Moss *et al.*, Reference 1, pages 68–69, for examples.
8. T. S. Moss *et al.*, Reference 1, pages 90–94; J. I. Pankove, Reference 2, pages 62–66.
9. See H. M. Rosenberg, *Low Temperature Solid State Physics*, Oxford University Press, New York (1963), pages 249–252 for examples.
10. For impurity energy levels in germanium and silicon, see P. R. Bratt, "Impurity Germanium and Silicon Infrared Detectors," in *Semiconductors and Semimetals*, R. K. Willardson and A. C. Beer (editors), Academic Press, New York (1977), Volume 12, pages 44–45.
11. D. Long, *Energy Bands in Semiconductors*, John Wiley, New York (1968), pages 123–134.
12. D. L. Greenaway and G. Harbeke, *Optical Properties and Band Structure of Semiconductors*, Pergamon Press, Oxford (1968), pages 88–95.
13. Adapted from D. Long, Reference 11, page 129, Figure 7.8.
14. D. Long, Reference 11, page 197.
15. E. Gutsche, J. Voight, and E. Ost, in *Proceedings of the Third International Conference on Photoconductivity*, 1969, E. M. Pell (editor), Pergamon Press, Oxford (1971), page 106.
16. H. Levinstein, *Physics Today*, **30**, 23–28 (November 1977).
17. C. T. Elliott, "Infrared Detectors," in *Handbook on Semiconductors* (T. S. Moss, series editor), Volume 4, *Device Physics* (C. Hilsum, volume editor), pages 727–798, North-Holland, Amsterdam (1981).
18. R. Dalven, in *Solid State Physics*, F. Seitz, D. Turnbull, and H. Ehrenreich (editors), Academic Press, New York (1973), Volume 28, pages 179–224.
19. C. T. Elliott, Reference 17, pages 741–746.

20. I. Melngailis and T. C. Harman, in *Semiconductors and Semimetals*, R. K. Willardson and A. C. Beer (editors), Academic Press, New York (1970), Volume 5, pages 111–174.
21. T. C. Harman, in *The Physics of Semimetals and Narrow-Gap Semiconductors*, D. L. Carter and R. T. Bate (editors), Pergamon Press, Oxford (1971), pages 363–382.
22. T. C. Harman and I. Melngailis, in *Applied Solid State Science*, R. Wolfe (editor), Academic Press, New York (1974), Volume 4, pages 1–94.
23. D. Long and J. L. Schmidt, in Reference 20, pages 175–255.
24. T. S. Moss *et al.*, Reference 1, pages 290–296.
25. A. G. Milnes, *Deep Impurities in Semiconductors*, John Wiley, New York (1973), pages 175–177.
26. P. R. Bratt, "Impurity Germanium and Silicon Infrared Detectors," in *Semiconductors and Semimetals*, R. K. Willardson and A. C. Beer (editors), Academic Press, New York (1977), Volume 12, pages 39–142, especially pages 108–113.
27. R. H. Bube, in *Photoconductivity and Related Phenomena*, J. Mort and D. M. Pai (editors), Elsevier, Amsterdam (1976), pages 117–153.
28. Y. Marfaing, "Photoconductivity and Photoelectric Effects," in *Handbook on Semiconductors* (T. S. Moss, series editor), Volume 4, *Device Physics* (C. Hilsum, volume editor), pages 417–495, North-Holland, Amsterdam (1981).
29. C. Kittel, *Introduction to Solid State Physics*, Fourth Edition, John Wiley, New York (1971), pages 628–632.
30. R. H. Bube, *Photoconductivity in Solids*, John Wiley, New York (1960), pages 59–60, 74–77.
31. T. S. Moss *et al.*, Reference 1, pages 191–192.
32. T. S. Moss *et al.*, Reference 1, pages 168–169.
33. P. W. Kruse, "The Photon Detection Process," in *Optical and Infrared Detectors*, R. J. Keyes (editor), Springer-Verlag, New York (1977), pages 42–47.
34. I Melngailis and T. C. Harman, *Applied Physics Letters*, **13**, 180–183 (1968), Figure 2.
35. B. G. Streetman, *Solid State Electronic Devices*, Second Edition, Prentice-Hall, New York (1980), pages 216–218.
36. S. M. Sze, *Physics of Semiconductor Devices*, Second Edition, John Wiley, New York (1980), pages 754–760.
37. T. P. Pearsall and M. A. Pollack, "Compound Semiconductor Photodiodes," in *Semiconductors and Semimetals*, R. K. Willardson and A. C. Beer (editors), Volume 22D, Academic Press, Orlando (1985), pages 173–245.
38. T. S. Moss *et al.*, Reference 1, pages 153–158; J. I. Pankove, Reference 2, pages 302–312.
39. I. Melngailis and T. C. Harman, Reference 20, page 158.
40. H. J. Hovel, "Solar Cells," in *Semiconductors and Semimetals*, R. K. Willardson and A. C. Beer (editors), Academic Press, New York (1975), Volume 11.
41. T. S. Moss *et al.*, Reference 1, pages 192–197.
42. T. S. Moss *et al.*, Reference 1, page 193.
43. S. Wagner, "The Status of Photovoltaic Devices," in *Solid State Devices* 1982, A. Goetzberger and M. Zerbst (editors), Physik-Verlag, Weinheim (1983), pages 1–24; H. M. Hubbard, "Photovoltaics Today and Tomorrow," *Science*, **244**, 297–304 (21 April 1989).
44. D. E. Carlson and C. R. Wronski, "Amorphous Silicon Solar Cells," in *Amorphous Semiconductors*, M. H. Brodsky (editor), Second Edition, Springer-Verlag, Berlin (1985), pages 287–329.

45. A. Madan *et al.*, editors, *Amorphous Silicon—Pure and Hydrogenated*, Materials Research Society, Pittsburgh (1987), Part VII, pages 497–548.

46. A. Madan, "Amorphous Silicon: From Promise to Practice," *IEEE Spectrum*, Volume 23, No. 9, 38–43 (September 1986).

47. K. Takahashi and M. Konagai, *Amorphous Silicon Solar Cells*, John Wiley, New York (1986).

48. D. Burland and L. Schein, "Physics of Electrophotography," *Physics Today*, **39**, No. 5, 46–53 (May 1986).

49. D. M. Pai and A. R. Melnyk, "Photoconductors in Electrophotography," in *Amorphous Semiconductors for Microelectronics*, D. Adler (editor), SPIE, Bellingham, Washington (1986), pages 82–94.

49a. L. B. Schein, *Electrophotography and Development Physics*, Springer-Verlag, Berlin (1988), Section 2.1.

50. M. D. Tabak, S. W. Ing, and M. E. Scharfe, *IEEE Transactions on Electronic Devices*, ED-**20**, 132–139 (1970).

51. F. W. Schmidlin, in Reference 27, pages 421–478.

52. R. H. Bube, Reference 30, pages 233–234.

53. W. R. Beam, *Electronics of Solids*, McGraw-Hill, New York (1965), page 202.

54. D. M. Pai and A. R. Melnyk, Reference 49, page 91.

55. B. G. Streetman, *Solid State Electronic Devices*, First Edition, Prentice-Hall, New York (1972), pages 244–246.

56. W. R. Beam, Reference 53, pages 202–205.

57. C. E. K. Mees and T. H. James, *The Theory of Photographic Process*, Third Edition, Macmillan, New York (1966), pages 19–30.

58. See, for example, *Color as Seen and Photographed*, Second Edition, Eastman Kodak Co., Rochester (1972), pages 30–39.

59. C. E. K. Mees and T. H. James, Reference 57, pages 103–111.

59a. L. M. Slifkin, "The Photographic Latent Image," *Phys. Bull.*, **39**, 274–277 (1988).

60. C. E. K. Mees and T. H. James, Reference 57, page 278.

61. J. I. Pankove, Reference 2, pages 124–131.

62. T. S. Moss *et al.*, Reference 1, pages 198–200; 202–206.

63. S. Wang, *Solid State Electronics*, McGraw-Hill, New York (1966), page 278.

64. A. Mooradian and H. Y. Fan, *Radiative Recombination in Semiconductors* (Seventh International Conference on the Physics of Semiconductors, Paris, 1964), Academic Press, New York (1965), pages 39–46.

65. D. Long, Reference 11, page 111.

66. J. I. Pankove, Reference 2, pages 125–126.

67. J. I. Pankove, Reference 2, pages 132–136.

68. T. S. Moss *et al.*, Reference 1, pages 210–216.

69. C. J. Neuse, H. Kressel, and I. Ladany, *IEEE Spectrum*, **9**, 28–38 (May 1972).

70. B. G. Streetman, Reference 55, pages 239–244.

71. W. V. Smith, in *Topics in Solid State and Quantum Electronics*, W. D. Hershberger (editor), John Wiley, New York (1972), pages 264–269.

72. A. A. Bergh and P. J. Dean, *Light Emitting Diodes*, Oxford University Press (1976).

73. M. H. Pilkuhn, "Light Emitting Diodes," in *Handbook on Semiconductors* (T. S. Moss, series editor), Volume 4, *Device Physics* (C. Hilsum, volume editor), pages 539–616, North-Holland, Amsterdam (1981).

74. T. S. Moss *et al.*, Reference 1, pages 224–228.

75. J. I. Pankove, Reference 2, pages 177–193.

76. P. J. Dean, in *Applied Solid State Science*, R. Wolfe (editor), Academic Press, New York (1969), Volume 1, pages 24–29.

77. M. H. Pilkuhn, Reference 73, Table 1, page 545, gives values of the radiative lifetime in several semiconductors with direct and with indirect gaps.

78. L. I. Schiff, *Quantum Mechanics*, Third Edition, McGraw-Hill, New York (1968), pages 403–404.

79. W. Koechner, *Solid State Laser Engineering*, Springer-Verlag, New York (1976), Chapter 1.

80. W. V. Smith, Reference 71, pages 241–249.

81. C. Kittel, Reference 3, pages 486–488.

82. W. V. Smith and P. P. Sorokin, *The Laser*, McGraw-Hill, New York (1966), pages 5–7.

83. A. Yariv, *Optical Electronics*, Third Edition, Holt, Rinehart, and Winston, New York (1985), pages 148–157.

84. L. I. Schiff, Reference 78, pages 404–405 and 414.

85. See, for example, H. Eyring, J. Walter, and G. E. Kimball, *Quantum Chemistry*, John Wiley, New York (1944), pages 113–114; also L. I. Schiff, Reference 78.

86. A Yariv, Reference 83, pages 106–108.

87. A. Yariv, Reference 83, pages 151, 157.

88. W. Koechner, Reference 79, page 46; C. Kittel, Reference 3, pages 488–489.

89. W. Koechner, Reference 79, Chapter 2; W. V. Smith, Reference 71, Table 1, pages 254–255; A. Yariv, Reference 83, pages 198–209.

90. W. V. Smith, Reference 71, pages 249–253; W. Koechner, Reference 79, pages 32–44.

91. A. Yariv, *Quantum Electronics*, Second Edition, John Wiley, New York (1975), pages 219–238.

92. W. V. Smith, Reference 71, pages 269–274.

93. B. G. Streetman, Reference 35, pages 389–399.

94. G. H. B. Thompson, *Physics of Semiconductor Laser Devices*, John Wiley, New York (1980).

95. A. Yariv, Reference 83, Chapter 15.

96. W. V. Smith, Reference 71, page 270.

97. See, for example, B. Lax, *Science*, **141**, 1247–1255 (1963).

98. A. Yariv, Reference 91, pages 230–231.

99. A. Yariv, Reference 83, gives a number of examples in Chapter 15.

100. W. V. Smith, Reference 71, page 267.

101. T. C. Harman and I. Melngailis, Reference 22, pages 1–94, Section 13.

102. T. S. Moss *et al.*, Reference 1, pages 241–242.

103. M. B. Panish and I. Hayashi, in *Applied Solid State Science*, R. Wolfe (editor), Academic Press, New York (1974), Volume 4, pages 235–328.

104. Y. Suematsu, "Advances in Semiconductor Lasers," *Physics Today*, **38**, 32–39 (May 1985).

Suggested Reading

T. S. Moss, G. J. Burrell, and B. Ellis, *Semiconductor Opto-Electronics*, John Wiley, New York (1973). A good book on the physics of the interaction of radiation with semiconductors, covering both basic and applied topics.

J. I. Pankove, *Optical Processes in Semiconductors*, Prentice-Hall, New York (1971). This book is more of a research monograph than the book above, but covers many of the same topics.

W. Koechner, *Solid State Laser Engineering*, Springer-Verlag, New York (1976). Chapter 1 of this monograph offers a brief introduction to optical amplification for the non-specialist, while Chapter 2 describes solid state laser materials (excluding semi-conductors) in some detail.

W. V. Smith, in *Topics in Solid State and Quantum Electronics*, W. D. Hershberger (editor), John Wiley, New York (1972). This collection of articles contains a chapter by Smith discussing both optically pumped solid state lasers and semiconductor injection lasers.

A. Yariv, *Quantum Electronics*, Second Edition, John Wiley, New York (1975). Chapters 9 and 10 of this advanced textbook discuss many aspects of laser physics, including non-solid-state lasers.

A. Yariv, *Optical Electronics*, Third Edition, Holt, Rinehart, and Winston, New York (1985). This text covers many of the topics discussed in Yariv's *Quantum Electronics*, but at a more introductory level.

A. A. Bergh and P. J. Dean, *Light Emitting Diodes*, Oxford University Press (1976). This treatise covers all aspects of the physics, design, and technology of light-emitting diodes.

C. Hilsum, editor, *Device Physics* (Volume 4 of *Handbook on Semiconductors*, edited by T. S. Moss), North-Holland, Amsterdam (1981). Chapters 5A, by M. H. Pilkuhn, 5B by H. Kressel, 6A by R. H. Bube, and 6B by C. T. Elliott cover, respectively, light-emitting diodes, semiconductor lasers, solar cells, and infrared detectors. The level is advanced and the treatments are authoritative.

D. E. Carlson, and C. R. Wronski, "Amorphous Silicon Solar Cells," in *Amorphous Semiconductors*, M. H. Brodsky (editor), Second Edition, Springer-Verlag, Berlin (1985), pages 287–329. This review article covers all aspects of its topic.

8

Superconductive Devices and Materials

Introduction

This chapter discusses some of the applications of superconductivity. After a brief review, the wave function for a condensed phase of Cooper pairs is introduced and used to discuss the Josephson effects. The physics of the DC and AC Josephson effects is developed, and current–voltage plots are described as preparation for a treatment of the effect of electromagnetic radiation on Josephson junctions. Quantization of magnetic flux in a superconducting ring leads to the idea of superconducting quantum interference and devices (DC SQUIDs) based thereon. Finally, the chapter concludes with a discussion of superconducting materials, with emphasis on the factors determining the magnitudes of the transition temperature, the critical magnetic field, and, briefly, the critical current density.

Review of Some Aspects of Superconductivity

We recall[1–4] a few points concerning superconductivity that will be useful in discussing some superconducting devices utilizing the Josephson effect. (Other aspects connected with superconductive materials will be introduced as needed in later sections.)

We remember that the BCS ground state of a superconductor differs from the ground state of a normal metal. In a normal metal at 0 K, the electrons fill the Fermi sphere up to the Fermi energy E_F. Since this free-electron model of the normal metal includes no interaction between the electrons, the energy of an excited electronic state may be arbitrarily small. In a superconductor, however, there is an attractive electron–electron interaction which leads to the existence of an energy gap $E_g = 2\Delta$ (at the Fermi energy) between the ground state of the superconductor and the lowest excited state. The attractive electron–electron interaction leads to the formation of electron pairs (Cooper pairs), in which the electrons occupy states of opposite wave vector and spin. Since each electron is a fermion, the pair will behave (approximately) as a boson since exchanging both electrons multiplies the wave function by $(-1)^2$. We will consider the Cooper pair as a "particle" that behaves like a boson.

We will focus our attention on pairs because we will want to discuss Josephson effect devices that involve pairs in their operation. Since we are considering pairs as bosons, we expect qualitatively that, if the temperature is low enough, almost all of the pairs will be in their lowest quantum state. This would be a Bose-Einstein condensation, in momentum space, in which all of the pairs are in the same quantum state and have the same wave function. The assembly of pairs, all in the same state, is often referred to as the condensed phase or condensate of pairs.

Wave Function of the Condensed Phase of Pairs

We next discuss the wave function for an assembly of pairs, following the macroscopic approach of Feynman.[5,6] If the wave function for one particle in three dimensions is $\psi(\mathbf{r})$, then, ignoring the time dependence, we recall[7] that the probability $P(\mathbf{r})$, where

$$P(\mathbf{r}) = \psi^*(\mathbf{r})\psi(\mathbf{r}) \tag{8.1}$$

is the probability that the particle will be in a unit volume located at $\mathbf{r}$. There is[7] an equation of continuity for P such that

$$\nabla \cdot \mathbf{S} + \partial P/\partial t = 0 \tag{8.2}$$

In equation (8.2), $\mathbf{S}$ is the probability current density given[7] by

$$\mathbf{S} = (1/2m)[\psi \hat{p}^*\psi^* + \psi^* \hat{p}\psi] \tag{8.3}$$

where $\hat{p} \equiv -jh\nabla$ is the momentum operator in the absence of a magnetic field.

We know also, from equation (8.1), that, for one particle of wave function $\psi(\mathbf{r})$, the quantity

$$P(\mathbf{r})\,dV = \psi^*(\mathbf{r})\psi(\mathbf{r})\,dV \tag{8.4}$$

is the probability that the particle will be found in a volume elemend dV located at $\mathbf{r}$. If we consider a simple case for which $\psi^*\psi$ is constant in space, we have (assuming normalization of ψ) that

$$\int \psi^*\psi\,dV = 1 \tag{8.5}$$

in this case. Since $\psi^*\psi$ is constant, we obtain

$$\psi^*\psi = 1/V \tag{8.6}$$

where V is the total volume. Equation (8.6) suggests that we may, in a crude way, interpret the quantity $\psi^*\psi$ as the number of particles (in this case, one) per unit volume. In this sense, $\psi^*\psi$ is the "particle density" for one particle described by the wave function ψ. Consider next a large number N of identical particles, all is the same quantum state. These N particles all have the same wave function $\psi(\mathbf{r})$, and, as we did for one particle, we interpret

$$\psi^*(\mathbf{r})\psi(\mathbf{r}) = \varrho = N/V \tag{8.7}$$

where ϱ is the particle density. This interpretation is reasonable; the particle density $\varrho = N/V$ should certainly be N times larger for N particles than it is for one particle in the same volume V.

This crude plausibility argument is designed to introduce the idea[5] that, for a large number of particles all in the same state, and thus all with the same wave function $\psi(\mathbf{r})$,

$$\psi^*(\mathbf{r})\psi(\mathbf{r}) = \varrho(\mathbf{r}) \tag{8.8}$$

where $\varrho(\mathbf{r})$ is the particle density in the system at point $\mathbf{r}$. Further, from (8.8), if each particle has an electric charge q,

$$\varrho q = q\psi^*\psi \tag{8.9}$$

is the electric charge density. Since $\psi^*\psi$ is the particle density, the quantity $\mathbf{S}$ given by equation (8.3) is the particle flux, i.e., the number of particles crossing unit area in unit time. Then the electric current density $\mathbf{J}$ is given by

$$\mathbf{J} = q\mathbf{S} = (q/2m)[\psi\hat{p}^*\psi^* + \psi^*\hat{p}\psi] \tag{8.10}$$

Equations (8.9) and (8.10) tell us that, when there is a large number of particles in exactly the same state, the electric charge density and the electric current density can be calculated *directly* from the wave function ψ. Both of the above are *macroscopic* physical quantities, so the wave function has a macroscopic significance for the case of many particles in the same state.

Equation (8.8) suggests that we write the wave function $\psi(\mathbf{r})$ in the form

$$\psi(\mathbf{r}) = [\varrho(\mathbf{r})]^{1/2} \exp[j\theta(\mathbf{r})] \tag{8.11}$$

where $\theta(\mathbf{r})$ is the phase of the wave function, and $\varrho(\mathbf{r})$ is the particle density in the system. If we consider one dimension for simplicity, equation (8.11) reduces to

$$\psi(x) = [\varrho(x)]^{1/2} \exp[j\theta(x)] \tag{8.12}$$

The wave function ψ given by (8.11) or (8.12) involves the particle density ϱ, which certainly has a classical macroscopic physical meaning. We can investigate the meaning of the phase[8,9] $\theta(x)$ of the wave function by calculating the electric current density $\mathbf{J}$ (in one dimension) using equations (8.10) and (8.12). We have

$$\psi \hat{p}^* \psi^* = \hbar\varrho(d\theta/dx) + (1/2)j\hbar(d\varrho/dx) \tag{8.13}$$

$$\psi^* \hat{p} \psi = \hbar\varrho(d\theta/dx) - (1/2)j\hbar(d\varrho/dx) \tag{8.14}$$

which, when substituted in equation (8.10), gives

$$J = (\hbar\varrho q/m)(d\theta/dx) \tag{8.15}$$

Since the magnitude J of the electric current density and the particle density ϱ are both macroscopic variables, $d\theta/dx$, the spatial variation of the phase of the wave function, is also a macroscopic variable or observable. Since

$$J = \varrho q v \tag{8.16}$$

where v is the velocity of particle flow, we have, on combining (8.16) with (8.15), that

$$mv = \hbar(d\theta/dx) \tag{8.17}$$

a result which, extended to three dimensions, becomes

$$m\mathbf{v} = \hbar \nabla \theta \tag{8.18}$$

Equation (8.18) tells us that the particle momentum mv is related to the gradient $\nabla\theta$ of the phase $\theta(\mathbf{r})$ of the wave function. The absolute phase is not observable, but, if $\nabla\theta$ is known everywhere, then θ is known except for a constant. If the phase is defined at one point of the system, then that constant is determined, and the phase is known everywhere in the system.

We next apply these results to an assembly of Cooper pairs at a very low temperature. All of the pairs are in the same quantum state and have the same wave function. We write the pair wave function $\psi(\mathbf{r},\,t)$ as

$$\psi(\mathbf{r},t) = |\psi(\mathbf{r},t)|\exp[j\theta(\mathbf{r},t)] \tag{8.19}$$

where $\theta(\mathbf{r},\,t)$ is the phase of the pair wave function. Since all of the pairs are in the same state, the density $\varrho(\mathbf{r},\,t)$ of pairs is given by

$$\varrho(\mathbf{r},t) = \psi^*(\mathbf{r},t)\,\psi(\mathbf{r},t) \tag{8.20}$$

so, from (8.19)

$$\varrho(\mathbf{r},t) = |\psi(\mathbf{r},t)|^2 \tag{8.21}$$

and we may write (8.19) in the form

$$\psi(\mathbf{r},t) = [\varrho(\mathbf{r},t)]^{1/2} \exp[j\theta(\mathbf{r},t)] \tag{8.22}$$

We regard the wave function (8.22) as the wave function of the *entire* assembly of pairs. The phase $\theta(\mathbf{r},\,t)$ in (8.22) is[10] the phase of the entire condensate of pairs and is, as discussed above, a physical observable of the system.

Consider next a superconductor in which no magnetic field is present and in which there is no center-of-mass motion of the pairs. In such a case, there is no net velocity of the pairs, so no supercurrent flows. From equation (8.17), we have, in one dimension,

$$d\theta/dx = 0 \tag{8.23}$$

since the pair velocity v vanishes in this case. The conclusion from equation (8.23) is that the phase θ of the condensate of pairs is a constant in space in a superconductor in which no current flows.

We note that the fact that there is no center-of-mass motion in this case is in agreement with the fact that the electrons of the pair have opposite wave vectors and spin, so the net momentum equals zero. If a current does flow in the superconductor, then the pair velocity v is a nonzero constant, so equation (8.17) becomes

$$d\theta/dx = C \tag{8.24}$$

$$\theta = Cx + \theta_0 \tag{8.25}$$

where C and θ_0 are constants. When a supercurrent flows, the phase θ varies linearly in space for a one-dimensional time-independent situation.

We may summarize the results of this section as follows. The superconductor is regarded as an assembly of Cooper pairs, all in the same state. The wave function ψ of the entire assembly of pairs is given by equation (8.22), where $\psi^*\psi = \varrho$, the pair density. The phase θ of the wave function is a macroscopic observable quantity related, through equation (8.15), to the current density J of pairs in the superconductor. If no current flows, the phase is everywhere constant in the superconductor.

The Josephson Effects

We now apply the ideas developed in the previous sections to the physics of the Josephson effect,[6,11–13] a term applied to the effects arising from the tunneling of pairs from one superconductor, through an insulating barrier, into a second superconductor. (The tunneling of pairs is to be distinguished from single-electron tunneling.) The experimental arrangement is shown schematically in Figure 8.1. There are two Josephson effects. The DC Josephson effect is the fact that, if a DC current is passed through the junction from an external source, no voltage is observed across the junction. The AC Josephson effect is the observation that a DC voltage across the junction causes high-frequency current oscillations across the junction. We will discuss the DC and AC effects in that order.

Physics of the DC Josephson Effect

The approach[11–13] to the DC Josephson effect will be as follows. We consider a Josephson junction in which sides (1) and (2) of Figure 8.1 are the same superconductor. We will set up equations describing the pair wave function ψ on the two sides of the junction, and include the existence of tunneling by pairs. We will then consider a DC current passed through the junction and observe its effect, via the phase, on the wave function.

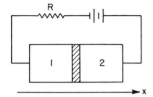

Figure 8.1. Schematic view of a Josephson junction, composed of two superconductors separated by an insulating barrier, which is shown shaded. The barrier, usually a metallic oxide, is of the order of 10 Å thick; its width is exaggerated in the drawing. The direction x is through the junction, as shown.

Let ψ_1 and ψ_2 be the wave functions on the sides (1) and (2) of the Josephson junction. The time-dependent Schrödinger equation must hold on both sides, so

$$j\hbar\,(\partial\psi_1/\partial t) = U_1\psi_1 \tag{8.26}$$

$$j\hbar\,(\partial\psi_2/\partial t) = U_2\psi_2 \tag{8.27}$$

where U_1 and U_2 are the energies of lowest states of the superconductors on sides (1) and (2). Next, we include tunneling in equations (8.26) and (8.27). We recall[14] that, in one dimension, tunneling through an energy barrier of width d, as shown in Figure 8.2, is the existence of a nonzero wave funcion in the region of space beyond d. Figure 8.2 shows an incident particle of plane-wave wave function $A\,\exp\,(jkx)$ incident on an energy barrier of height V_0 and width d; the energy of the incident particle is less than V_0. Physically, we may think of tunneling as the "leaking" of the wave function $A\,\exp\,(jkx)$ through the energy barrier, resulting in a nonzero transmitted wave function $C\,\exp(jkx)$ on the other side ($x > d$) of the barrier.

We may incorporate tunneling into equations (8.26) and (8.27) as follows. Assume that the tunneling of pairs from side (2) to side (1) increases the amplitude ψ_1 of the pair wave function on side (1). Further, assume that the time rate of increase of ψ_1 is proportional to ψ_2, the amplitude of the pair wave function on side (2). We write the time rate of chance of ψ_1 in the form[11]

$$\hbar T\psi_2 \tag{8.28}$$

Figure 8.2. Plot of energy $V(x)$ as a function of distance x showing an energy barrier height V_0 and width d. The wave functions of the incident, reflected, and transmitted particles are, respectively, $A\,\exp(jkx)$, $B\,\exp(-jkx)$, and $C\,\exp(jkx)$.

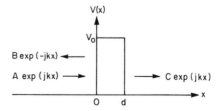

where the constant T is characteristic of the junction and is a measure of the transfer of pairs from side (2) to side (1). The dimension of T will be that of a frequency. We add the expression (8.28) to equation (8.26) for $\partial \psi_1/\partial t$ and obtain

$$j\hbar \, (\partial \psi_1/\partial t) = U_1 \psi_1 + \hbar T \psi_2 \qquad (8.29)$$

Equation (8.29) neglects any decrease in ψ_1 due to pairs tunneling back from side (1) to side (2). In the same way, we add a term $\hbar T \psi_1$ to equation (8.27) to represent tunneling of pairs from side (1) to side (2), obtaining

$$j\hbar \, (\partial \psi_2/\partial t) = U_2 \psi_2 + \hbar T \psi_1 \qquad (8.30)$$

In obtaining (8.30), we have assumed that the constant T is the same for tunneling in both directions.

Since T is a measure of tunneling through the barrier, $T = 0$ when the barrier is very thick, and there is no coupling via tunneling between the two pieces of superconductor. The two sides act like separate superconductors with different values of the phase θ since the phase θ_1 in side (1) is independent of the phase θ_2 in side (2). If the barrier is very thin, and approaches zero in thickness, the coupling between the two pieces of superconductor is large, and the properties of the system change continuously from those of two isolated superconductors to those of a single superconductor. In particular, as the width d of the barrier goes to zero, the phases θ_1 and θ_2 will become equal, as long as no supercurrent is flowing, in accordance with equation (8.23).

We next consider equations (8.29) and (8.30) for the case of appreciable tunneling. For simplicity, we consider the situation in which the superconductors on sides (1) and (2) of the junction are the same. Then, the lowest-state energies U_1 and U_2 are the same, so we set

$$U_1 = U_2 \equiv U \qquad (8.31)$$

We define the energy U as the zero of energy in the problem, so $U \equiv 0$ and it vanishes from equations (8.29) and (8.30). We obtain

$$j\hbar \, (\partial \psi_1/\partial t) = \hbar T \psi_2 \qquad (8.32)$$

$$j\hbar \, (\partial \psi_2/\partial t) = \hbar T \psi_1 \qquad (8.33)$$

Equation (8.32) and (8.33) are the time-dependent Schrödinger equations for the pair wave functions ψ_1 and ψ_2 on the two sides of the junction, including the effect of tunneling. We note that, if there is no tunneling and $T = 0$, then both ψ_1 and ψ_2 are constant in time, as is reasonable.

We now introduce complex wave functions of the form (8.12) for the superconductor on sides (1) and (2) of the junction. We have

$$\psi_1 = [\varrho_1]^{1/2} \exp(j\theta_1) \qquad (8.34)$$

$$\psi_2 = [\varrho_2]^{1/2} \exp(j\theta_2) \qquad (8.35)$$

where the subscripts refer to sides (1) and (2), and we are considering the one-dimensional case, so the ψ's, ϱ's, and θ's are all functions of x and of the time. The direction of x is through the junction, as shown in Figure 8.1. Our next step will be to substitute the wave functions (8.34) and (8.35) into the coupled Schrödinger equations (8.32) and (8.33), thereby obtaining equations relating ϱ_1, ϱ_2, θ_1, and θ_2. We then connect the phases θ_2 and θ_1 with the current density J flowing through the junction.

However, before making those substitutions, let us see if we can get a qualitative idea of the result we expect when a current density flows through the junction. Consider a current density J, determined by the battery and resistor, flowing through the junction in Figure 8.1. From equation (8.15), since J is not zero, we expect $d\theta/dx$ to be nonzero, and

$$\frac{d\theta}{dx} = \frac{mJ}{\hbar \varrho q} \qquad (8.36)$$

From equation (8.36), we expect that there will be a gradient $d\theta/dx$ in the phase of the wave function between the two sides of the junction when a current is flowing. Further, we expect that the phase gradient will be almost entirely across the oxide barrier. That this is so can be seen from equation (8.36), which says that $d\theta/dx$ is inversely proportional to the pair density ϱ for a given value of J. In the superconductor itself, the density of pairs is large, of the order of 10^{22} cm^{-3}, assuming a very low temperature so pairing is essentially complete.[15] In the oxide barrier layer, the density of tunneling pairs is small, typically of the order[16] of 10^{10} cm^{-3}. Thus, $d\theta/dx$ is small in the superconductor and large in the oxide barrier layer. We will assume that, approximately, the phase θ is constant in space in the two superconducting sides of the junction. This is shown in Figure 8.3, where $\theta = \theta_1$ in side (1), $\theta = \theta_2$ in side (2), and the width of the oxide barrier is Δx. Then $d\theta/dx = 0$ in the superconductors, and the entire gradient of phase

$$d\theta/dx = (\theta_2 - \theta_1)/\Delta x \equiv \Delta\theta/\Delta x \qquad (8.37)$$

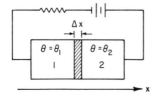

Figure 8.3. Current flowing through a Josephson junction, with a phase gradient $\Delta\theta/\Delta x = (\theta_2 - \theta_1)/\Delta x$ across the oxide barrier layer (shaded). The phases on sides (1) and (2) are θ_1 and θ_2, respectively.

appears across the oxide barrier. Our central physical result is that the flow of a current through the junction produces a change of phase $\Delta\theta \equiv \theta_2 - \theta_1$ across the oxide barrier layer.

We now substitute the wave functions (8.34) and (8.35) into the coupled Schrödinger equations (8.32) and (8.33). Since we are assuming that the phase is constant in space within the superconductor itself, we have $\theta = \theta(t)$ in the superconductor on both sides of the junction. Substituting the wave functions ψ_1 and ψ_2 into the Schrödinger equations yields

$$j\hbar \frac{\partial}{\partial t}(\varrho_1^{1/2} e^{j\theta_1}) = \hbar T \varrho_2^{1/2} e^{j\theta_2} \tag{8.38}$$

$$j\hbar \frac{\partial}{\partial t}(\varrho_2^{1/2} e^{j\theta_2}) = \hbar T \varrho_1^{1/2} e^{j\theta_1} \tag{8.39}$$

Performing the differentiation, and equating[11] real and imaginary parts of the resulting two equations, yields

$$\dot{\varrho}_1 = 2T(\varrho_1\varrho_2)^{1/2} \sin{(\Delta\theta)} \tag{8.40}$$

$$\dot{\varrho}_2 = -2T(\varrho_1\varrho_2)^{1/2} \sin{(\Delta\theta)} \tag{8.41}$$

$$\dot{\theta}_1 = -T(\varrho_2/\varrho_1)^{1/2} \cos{(\Delta\theta)} \tag{8.42}$$

$$\dot{\theta}_2 = -T(\varrho_1/\varrho_2)^{1/2} \cos{(\Delta\theta)} \tag{8.43}$$

where $\Delta\theta \equiv \theta_2 - \theta_1$ and $\dot{\varrho}_1 = \partial\varrho_1/\partial t, \dot{\theta}_1 = \partial\theta_1/\partial t$, etc. Since the superconductors on sides (1) and (2) of the junction are the same, we set the pair densities equal on both sides of the junction, so

$$\varrho_1 = \varrho_2 \tag{8.44}$$

The condition (8.44) leads, from equations (8.42) and (8.43), to the result, since $\dot{\theta}_1 = \dot{\theta}_2$, that

$$\frac{\partial}{\partial t}(\theta_2 - \theta_1) \equiv \frac{\partial}{\partial t}(\Delta\theta) = 0 . \tag{8.45}$$

Equation (8.45) says that the phase difference $\Delta\theta$ across the junction is constant in time. The condition (8.44) applied to equations (8.40) and (8.41) yields

$$\dot{\varrho}_2 = -\dot{\varrho}_1 \qquad (8.46)$$

an equation that says the time rate of increase of the pair density ϱ_2 is equal to the time rate of decrease of the pair density ϱ_1. The magnitude J of the current density flowing from side (1) to side (2) is proportional to $\partial\varrho_2/\partial t$, so we conclude, from (8.40) or (8.41), that J is proportional to $\sin(\Delta\theta)$. We write

$$J = J_0 \sin(\Delta\theta) \qquad (8.47)$$

where the constant J_0 is a function[17] of the properties of the barrier, including the constant T, and the temperature. Since the magnitude of $\sin(\Delta\theta)$ varies between zero and one, we see from equation (8.47) that the magnitude J varies from zero to J_0. The quantity J_0 is called the critical current density[6] and is the largest DC supercurrent that can flow through the junction at zero voltage in the DC Josephson effect. (Currents exceeding this magnitude must, as discussed below in the AC Josephson effect, flow by some other process, which involves a nonzero voltage,[18] such as single-particle tunneling.) The values[19] of the corresponding critical currents may vary from a fraction of a microampere to tens of milliamperes. The relation given by (8.47) is thus valid for values of J less than the critical current density J_0.

We note especially that no electric field appears in equation (8.47) for the DC supercurrent density J. If an experiment is set up, as shown in Figure 8.3, in which a current density J, smaller than J_0, is passed through the junction, the current flows as a direct supercurrent and *no voltage* appears across the junction. This is the DC Josephson effect. When J is increased until it equals the critical current density J_0, the junction can no longer sustain a supercurrent. (A discussion in terms of the energy difference (the coupling energy) between the two sides of the junction is given by Clarke.[6]) While no voltage is developed across the junction for values of the current density below the critical value, a phase difference $\Delta\theta$, given by

$$\Delta\theta = \sin^{-1}(J/J_0) \qquad (8.48)$$

is produced across the junction. As the current density J through the junction (determined by the external circuit in Figure 8.3) increases from zero to a value less than J_0, the phase difference $\Delta\theta$ is developed according to equation

(8.48). As a final point, we note that, for the junction, the current density is related to the phase *difference* $\Delta\theta$ between the two sides of the junction, while from equation (8.15), the current density in a single superconductor is related to the phase *gradient* $d\theta/dx$.

To summarize our results for the DC Josephson effect, the DC supercurrent density J through the junction is described by equation (8.47), where $\Delta\theta$ is the difference in the phase of the wave function between the two sides of the junction. Equation (8.47) is valid for values of J less than the critical value J_0, the maximum supercurrent density the junction can sustain. For values of J smaller than J_0, no voltage is observed across the junction.

Physics of the AC Josephson Effect

Suppose the current J through a Josephson junction has a value larger than the critical value J_0. In this situation, a new effect, the AC Josephson effect,[6,11–13] is observed, in which a DC voltage is produced across the junction, resulting in high-frequency current oscillations.

If J is larger than J_0, the junction can no longer maintain a DC supercurrent. Some of the current is carried by single normal electrons tunneling through the barrier, and a DC voltage V appears across the junction. If a potential difference V exists between the two sides of the junction,[†] then a particle of electric charge q will change its potential energy by qV on passing through the barrier. For a Cooper pair, $q = -2e$, and the change in potential energy per pair is $-2eV$. This is equivalent to saying that a pair on one side of the junction is at potential energy $-eV$, and a pair on the other side is at potential energy $+eV$.

We now modify equations (8.26) and (8.27) to include these potential energy terms in the Hamiltonian, obtaining

$$jh(\partial\psi_1/\partial t) = U_1\psi_1 + eV\psi_1 \tag{8.49}$$

$$jh(\partial\psi_2/\partial t) = U_2\psi_2 - eV\psi_2 \tag{8.50}$$

Setting, as before, $U_1 = U_2 = U \equiv 0$, and including, again as before, the tunneling terms $\hbar T\psi_2$ and $\hbar T\psi_1$ in (8.49) and (8.50), respectively, yields

$$jh(\partial\psi_1/\partial t) = \hbar T\psi_2 + eV\psi_1 \tag{8.51}$$

$$jh(\partial\psi_2/\partial t) = \hbar T\psi_1 - eV\psi_2 \tag{8.52}$$

---

† Strictly speaking, this argument should be given in terms of the chemical potential.[20,22]

as the equations into which we again substitute the wave functions (8.34) and (8.35). From (8.51) we obtain

$$jh\left(\tfrac{1}{2}\varrho_1^{-1/2}\dot\varrho_1 e^{j\theta_1} + j\varrho_1^{1/2}e^{j\theta_1}\dot\theta_1\right) = \hbar T\varrho_2^{1/2}e^{j\theta_2} + eV\varrho_1^{1/2}e^{j\theta_1} \qquad (8.53)$$

Multiplying by $\varrho_1^{1/2}\exp(-j\theta_1)$, gives, with $\Delta\theta \equiv \theta_2 - \theta_1$, the result

$$jh\left(\tfrac{1}{2}\dot\varrho_1 + j\varrho_1\dot\theta_1\right) - eV\varrho_1 = \hbar T(\varrho_1\varrho_2)^{1/2}\left[\cos(\Delta\theta) + j\sin(\Delta\theta)\right] \qquad (8.54)$$

Equating real and imaginary parts of (8.54) gives

$$\dot\varrho_1 = 2T(\varrho_1\varrho_2)^{1/2}\sin(\Delta\theta) \qquad (8.55)$$

$$\dot\theta_1 = (-e/\hbar)V - T(\varrho_2/\varrho_1)^{1/2}\cos(\Delta\theta) \qquad (8.56)$$

Equation (8.55) is the same as (8.40) found for the DC Josephson effect. Equation (8.56), when compared with (8.42), shows that the term $(e/\hbar)V$ has entered the expression for $\dot\theta_1$, the time rate of change of the phase θ_1. In a similar manner, from equation (8.52), one obtains

$$\dot\varrho_2 = -2T(\varrho_1\varrho_2)^{1/2}\sin(\Delta\theta) \qquad (8.57)$$

$$\dot\theta_2 = (e/\hbar)V - T(\varrho_1/\varrho_2)^{1/2}\cos(\Delta\theta) \qquad (8.58)$$

where equations (8.57) and (8.58) may be compared with (8.41) and (8.43).

We again set, as we did in equation (8.44), the pair densities ϱ_1 and ϱ_2 equal, so (8.56) and (8.58) become

$$\dot\theta_1 = (-e/\hbar)V - T\cos(\Delta\theta) \qquad (8.59)$$

$$\dot\theta_2 = (e/\hbar)V - T\cos(\Delta\theta) \qquad (8.60)$$

which, in turn, lead to the result

$$\frac{d}{dt}(\Delta\theta) \equiv \dot\theta_2 - \dot\theta_1 = 2eV/\hbar \qquad (8.61)$$

We note that, if $V = 0$, equation (8.61) reduces (as it should) to equation (8.45) describing the DC Josephson effect.

Integrating equation (8.61) gives

$$\Delta\theta(t) = (\Delta\theta)_0 + (2e/\hbar)\int_0^t V(t)\,dt \qquad (8.62)$$

where $V(t)$ is the (generally time-dependent) voltage across the junction and $(\Delta\theta)_0$ is the value of the phase difference $\Delta\theta$ at time $t = 0$. Equation (8.62) gives the time-dependent phase difference $\Delta\theta(t)$ in terms of the voltage $V(t)$ across the junction. Equation (8.47) for the supercurrent density J is still valid,[23] but now, with V not zero, the phase difference $\Delta\theta$ is the function of time given by equation (8.62). Substituting (8.62) into (8.47) gives[24]

$$J(t) = J_0 \sin\left[(\Delta\theta)_0 + (2e/\hbar) \int_0^t V(t)\,dt \right] \tag{8.63}$$

We now consider the case $V(t) = \text{constant} = V$, corresponding to a DC voltage across the junction. In this case, equation (8.63) becomes

$$J(t) = J_0 \sin\left[(\Delta\theta)_0 + (2eV/\hbar)t \right]$$

$$J(t) = J_0 \sin\left[(\Delta\theta)_0 + \omega t \right] \tag{8.64}$$

where we have defined

$$\omega \equiv 2eV/\hbar$$

Equation (8.64) says that the supercurrent J through the junction varies sinusoidally with time with a frequency ω which depends on the value V of the nonzero DC voltage across the junction. (Note that, so far, we have been dealing only with the supercurrent J, given by (8.64). To obtain the total current through the junction, one must add[25] the current due to the tunneling of single normal electrons.)

The physical situation on the AC Josephson effect is as follows. When the total current through the junction is larger than the critical current J_0, part of the total current must be carried by normal electrons (because J_0 is the largest supercurrent that the junction can sustain at zero voltage in the DC Josephson effect). These normal electrons pass through the junction, producing a nonzero voltage V across the junction. The nonzero voltage across the junction results in a time-dependent phase difference $\Delta\theta(t)$ between the two sides of the junction. This time-dependent phase difference produces, from (8.63), a time-dependent supercurrent $J(t)$ through the junction. For the case in which V is constant, the supercurrent oscillates sinusoidally with time, corresponding to pairs tunneling back and forth between the two sides of the junction, which are now separated in energy by an amount $2eV$. As the pairs tunnel back and forth, they emit and reabsorb photons of energy $\hbar\omega = 2eV$. (These photons can sometimes be radiated out of the junction,[26] but with very low efficiency.)

To summarize, the result of the AC Josephson effect is that a current greater that the critical value produces a nonzero voltage V across the junction. This voltage results in the time-dependent supercurrent $J(t)$ given by (8.63). For a constant voltage V, the supercurrent oscillates sinusoidally, with a frequency $\omega = 2eV/\hbar$, through the junction. A voltage $V = 10^{-6}$ volts corresponds to a frequency of approximately 484 MHz.

Voltage–Current Curves for Josephson Junctions

We now discuss a plot[27,28] of the voltage V across a tunnel junction of the kind indicated in Figure 8.1, as a function of the DC current density J through the junction. We consider the situation in which the impedance R of the current source is large compared to that of the junction. As the current is increased from zero, no voltage appears across the junction for currents less than the critical current J_0. This is the DC Josephson effect and is shown as curve (1) on the voltage–current plot in Figure 8.4. Curve (2) in that figure is the current due to the tunneling[29] of single normal electrons; this begins to increase sharply at a voltage approximately equal to $2\Delta/e$, where 2Δ is the energy gap of the superconductor. The normal electron tunnel current is very small for voltages less than $2\Delta/e$. When the critical current is exceeded, there is a discontinuous jump from zero voltage to a voltage value on the normal electron current curve (2). This is shown in Figure 8.4 as the horizontal dashed line connecting curves (1) and (2). There is now a nonzero voltage across the junction. As the current is increased further, the voltage across the junction follows curve (2), and the DC current through the junction is due to

Figure 8.4. Schematic plot of voltage V across a Josephson tunnel junction as a function of the current density J. The dashed line represents the discontinuous increase in V when the critical current J_0 is exceeded; 2Δ is the energy gap of the superconductor. Curve (1) is the DC supercurrent due to pair tunneling. Curve (2) shows the DC current, for nonzero voltages, due to single-electron tunneling. On curve (2), the supercurrent is the oscillatory supercurrent of the AC Josephson effect.

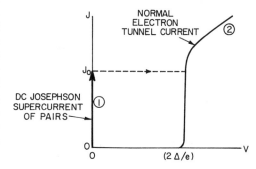

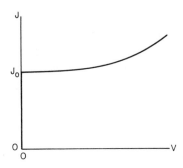

Figure 8.5. Schematic current–voltage characteristic of a point contact Josephson junction, showing the critical current J_0 of the device.

normal electron tunneling. The supercurrent through the junction is now oscillatory in time, as given by equation (8.63) for the AC Josephson effect. If the current is decreased to zero, hysteresis effects[28,30] may be observed. Finally, current sources with an impedence less than that of the tunnel junction produce somewhat different current–voltage curves.[28]

It should also be pointed out that there are several other types[28,31,32] of junctions in addition to the oxide tunnel junction shown schematically in Figure 8.1 Among these is the point contact junction in which a superconductor, sharpened to a radius of a few micrometers, is pressed against a block of superconducting material. The current–voltage characteristic of these other types of junctions differs[19,28] from that shown in Figure 8.4, and looks approximately like that in Figure 8.5, which is a schematic curve for a point contact.

Effect of Electromagnetic Radiation on the Junction

We now consider the effect[31-35] of high-frequency (microwave and far infrared) electromagnetic radiation on a Josephson junction. It will be found that changes are observed in the curve of DC current as a function of voltage. These changes enable the junction to be used as a detector[19,32,34,36] of electromagnetic radiation.

Consider a junction across which a constant DC voltage V_0 is produced, resulting in the alternating supercurrent of the AC Josephson effect. The frequency of the alternating supercurrent will depend, from equations (8.63) and (8.64), on the DC voltage. Suppose further that microwave electromagnetic radiation of frequency ω_0 is incident on the junction, inducing an AC voltage

$$V_1 \cos \omega_0 t \qquad (8.65)$$

across the junction. The radiation given by (8.65) is mixed by the junction with the alternating Josephson supercurrent, resulting in frequency modulation of the latter. The experimental result is the alteration, referred to above, of the DC current–voltage characteristic of the junction.

Following Clarke;[33] we may calculate the effect of the applied radiation on the junction on the assumption that both the DC current and the incident radiation have low source impedances. The total voltage V across the junction is the sum of the applied DC and induced AC voltages, so

$$V = V_0 + V_1 \cos \omega_0 t \tag{8.66}$$

From equation (8.61), the time rate of change $d(\Delta\theta)/dt$ of the phase difference $\Delta\theta$ across the junction is

$$d(\Delta\theta)/dt = 2eV/\hbar = (2e/\hbar)(V_0 + V_1 \cos \omega_0 t) \tag{8.67}$$

since the total voltage V across the junction in given by (8.66). Integrating (8.67) gives

$$\Delta\theta(t) = (\Delta\theta)_0 + \left(\frac{2eV_0}{\hbar}\right)t + \left(\frac{2eV_1}{\hbar\omega_0}\right)\sin \omega_0 t \tag{8.68}$$

for the time dependence $\Delta\theta(t)$ of the phase difference, and where V_0, V_1, and ω_0 are constant; $(\Delta\theta)_0$ is the value of the phase difference at time $t = 0$. The supercurrent density J is, as before, obtained by substituting the phase difference (8.68) into equation (8.47) for J, yielding

$$J(t) = J_0 \sin\left[(\Delta\theta)_0 + \left(\frac{2eV_0}{\hbar}\right)t + \left(\frac{2eV_1}{\hbar\omega_0}\right)\sin \omega_0 t\right] \tag{8.69}$$

showing that the supercurrent is a function of time. Equation (8.69) can be expanded in a Fourier–Bessel series[37] using the relations[38]

$$\sin(X \sin a) = \sum_{n=-\infty}^{\infty} B_n(X) \sin(na) \tag{8.70}$$

$$\cos(X \sin a) = \sum_{n=-\infty}^{\infty} B_n(X) \cos(na) \tag{8.71}$$

where, to avoid, confusion with the current density J, the symbol $B_n(X)$ has been used for the Bessel function of order n of argument X. Using relations (8.70) and (8.71), equation (8.69) may be expanded to give[32]

$$J(t) = J_0 \sum_{n=-\infty}^{\infty} \left\{ B_n\left(\frac{2eV_1}{\hbar\omega_0}\right) \sin\left[\left(n\omega_0 + \frac{2eV_0}{\hbar}\right)t + (\Delta\theta)_0\right] \right\} \quad (8.72)$$

for the time-dependent supercurrent density $J(t)$. A discussion of the initial phase difference $(\Delta\theta)_0$ may be found in the literature.[39]

We may examine the properties of the supercurrent through the junction by considering equation (8.72). The equation is a sum of components at infinitely many frequencies

$$n\omega_0 + (2eV_0/\hbar) \quad (8.73)$$

The amplitude of the component at frequency $n\omega_0 + (2eV_0/\hbar)$ is

$$J_0[B_n(2eV_1/\hbar\omega_0)]\sin(\Delta\theta)_0 \quad (8.74)$$

In particular, when the frequency $n\omega_0 + (2eV_0/\hbar)$ vanishes, so

$$n\hbar\omega_0 = -2eV_0 \quad (8.75)$$

where $n = 0, \pm 1, \pm 2, ...$, there will be a component of the supercurrent density (8.72) at zero frequency. A DC supercurrent will therefore be present in the junction whenever the DC voltage V_0 across the junction has one of the values

$$V_0 = n(\hbar\omega_0/2e) \quad (8.76)$$

where n is any integer. These zero-frequency supercurrents correspond to the appearance of a DC Josephson effect at a nonzero voltage V_0 across the junction when the junction is irradiated with electromagnetic radiation of frequency ω_0. These results predict a series of spikes of DC supercurrent superimposed on the DC current–voltage characteristic of the junction. The spikes occur, from equation (8.76), at voltages $\hbar\omega_0/2e$, $2\hbar\omega_0/2e$, $3\hbar\omega_0/2e$, etc. The voltage separation between adjacent spikes is $\hbar\omega_0/2e$ and is determined by the radiation frequency ω_0; a frequency of approximately 484 MHz will correspond to a spike separation of 1 μV. In practice the spike structure has not been observed.[40] Instead, for the usual experimental arrangement of a high-impedance source of DC current, current *steps*[33] are produced in the DC junction characteristic. The voltage separation of the steps is also equal

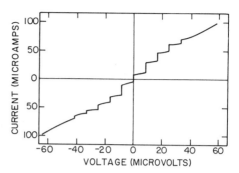

Figure 8.6. Steps induced in the DC current–voltage characteristic of an Sn–SnO–Sn tunnel junction by 4 GHz electromagnetic radiation. The spacing of the steps is approximately 8.5 μV (after Clarke[41]).

to $\hbar\omega_0/2e$. Figure 8.6 shows[41] the steps induced in a Sn–SnO–Sn tunnel junction by 4 GHz electromagnetic radiation.

From equation (8.74), the amplitude J^n_{DC} of step n of the DC supercurrent, induced by the radiation, is proportional to the Bessel function of order n and argument $2eV_1/\hbar\omega_0$, so

$$J^n_{DC} \propto B_n(2eV_1/\hbar\omega_0) \qquad (8.77)$$

The step n of amplitude J^n_{DC} appears on the DC current–voltage characteristic at the voltage $n\hbar\omega_0/2e$ determined by equation (8.76). If we consider the step for $n = 0$ at zero voltage, the amplitude of the zero-voltage step is

$$J^0_{DC} \propto B_0(2eV_1/\hbar\omega_0) \qquad (8.78)$$

where, since the voltage $n\hbar\omega_0/2e$ equals zero for $n = 0$, J^0_{DC} given by (8.78) is the critical current of the junction. From the properties[42] of the Bessel function of order zero, $B_0(x)$ decreases monotonically with increasing x, for x less than approximately 2.4. The amplitude J^0_{DC} of the $n = 0$ step responds to incident radiation (at low values of V_1 where $2eV_1$ is less than about $2\hbar\omega_0$) by decreasing with increasing V_1. The critical current of the junction therefore decreases[19] with increasing microwave power incident on the device.

On the other hand, the amplitude J^1_{DC} of the $n = 1$ step, appearing at the voltage $\hbar\omega_0/2e$ on the DC current–voltage characteristic, is

$$J^1_{DC} \propto B_1(2eV_1/\hbar\omega_0) \qquad (8.79)$$

From the properties[42] of the Bessel function of order unity, $B_1(x)$ increases with increasing x, for x less than about 1.8. The amplitude J^1_{DC} of the $n = 1$ step increases with increasing V_1 (for low values of V_1, such that $2eV_1 \lesssim 1.8\hbar\omega_0$).

Finally, we note that equation (8.76) is satisfied, for $n = 0$, by *any* incident frequency ω_0. If broadband radiation, containing many frequencies, is applied to a junction, the critical current will respond to, and be modified by, each frequency component incident. Since,[43] for small values of x, $B_0(x)$ may be expressed as the series

$$B_0(x) = 1 - x^2/4 + \cdots \tag{8.80}$$

equation (8.78) shows that J^0_{DC} is proportional to $V_1{}^2/\omega_0{}^2$ for small values of V_1/ω_0. This result leads to the conclusion[19] that the junction will act as a broadband square-law detector, and becomes less sensitive with increasing incident frequency ω_0 of the radiation.

Based on their response to incident electromagnetic radiation, different types of Josephson junctions can be used as detectors in a number of different ways. Several review articles[19,32,44] discuss the field, with particular emphasis on more realistic models of the practical junctions used.

Quantization of Magnetic Flux in a Superconducting Ring

In preparation for a discussion of the physics of superconductive quantum interference devices, we discuss the quantization of magnetic flux in a superconducting ring.[45] First, we return to our discussion of the phase θ of the wave function of a superconductor, and include the effect of a magnetic field **B**. It is well known[46] that the total momentum **p** of a particle, of mass m and electric charge q, moving with velocity **v** in field of magnetic induction **B** is

$$\mathbf{p} = m\mathbf{v} + (q/c)\mathbf{A} \tag{8.81}$$

where the magnetic vector potential **A** is given by $\mathbf{B} = \nabla \times \mathbf{A}$. The quantity $(q/c)\mathbf{A}$ is called the field momentum and $m\mathbf{v}$ is the kinetic momentum. The relation (8.18),

$$m\mathbf{v} = \hbar \nabla \theta \tag{8.18}$$

between the kinetic momentum $m\mathbf{v}$ and the gradient $\nabla\theta$ of the phase of the wave function, was obtained for the situation in which no magnetic field, and hence no field momentum, were present. To include the effect of a non-zero magnetic field, we add the field momentum $(q/c)\mathbf{A}$ to the left-hand side of (8.18), obtaining for the total momentum **p** the result

$$\mathbf{p} = m\mathbf{v} + (q/c)\mathbf{A} = \hbar\nabla\theta \tag{8.82}$$

which we rewrite as

$$m\mathbf{v} = \hbar\nabla\theta - (q/c)\mathbf{A} \tag{8.83}$$

Next, we derive a relation for the current density J in a magnetic field. From (8.16),

$$\mathbf{J} = \varrho q\mathbf{v} \tag{8.84}$$

is the current density of particles of charge q and density ϱ. Substituting the velocity **v** from (8.83) into (8.84) gives us

$$\mathbf{J} = (\varrho q/m)[\hbar\nabla\theta - (q/c)\mathbf{A}] \tag{8.85}$$

for the current density **J** in a magnetic field described by the vector potential **A**, in terms of the gradient of the phase of the wave function.

We now want to consider **J** in the interior of a superconducting ring, shown schematically in Figure 8.7, in which C is a curve deep in the bulk of the superconductor. It is well known[47] that currents in a superconductor must flow in the surface. This means that, inside the superconductor (in the region of the curve C), no current flows, and **J** = 0. We thus have, deep in the super-conductor, that, using (8.85),

$$0 = \mathbf{J} = (\varrho q/m)[\hbar\nabla\theta - (q/c)\mathbf{A}] \tag{8.86}$$

leading to

$$\hbar\nabla\theta = (q/c)\mathbf{A} \tag{8.87}$$

inside the superconducting ring.

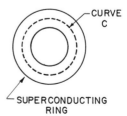

Figure 8.7. Superconducting ring, showing the (dashed) curve C used as the path of integration in the line integral in equation (8.88).

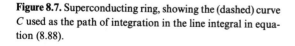

Consider next the line integral of $\nabla\theta$ around a path C well inside the superconducting ring, as shown in Figure 8.7. The line integral is

$$\oint_C \nabla\theta \cdot d\mathbf{l} = \Delta\theta \tag{8.88}$$

where $\Delta\theta$ is the change in the phase of the wave function on going around the path C around the ring. On integration around the superconducting ring, we require that the wave function be single-valued, meaning that the phase angle must come back to the same value, plus some integral multiple of 2π. From equation (8.11), we see that this means that the change in phase $\Delta\theta$ must be an integral multiple of 2π since $\exp(j2\pi) = 1$. Our requirement that the wave function be single valued leads to the condition

$$\Delta\theta = 2\pi s \tag{8.89}$$

where s is an integer, on the change of phase $\Delta\theta$.

We examine next the line integral of $\mathbf{A}$ around the curve C in the ring, and, using Stokes' theorem, obtain

$$\oint_C \mathbf{A} \cdot d\mathbf{l} = \int_{S'} (\nabla \times \mathbf{A}) \cdot \mathbf{n}\, dS \tag{8.90}$$

where $\mathbf{n}$ is the unit normal to the element of area dS of the surface S' bounded by curve C. Since $\mathbf{B} = \nabla \times \mathbf{A}$, equation (8.90) becomes

$$\oint_C \mathbf{A} \cdot d\mathbf{l} = \int_{S'} \mathbf{B} \cdot \mathbf{n}\, dS = \Phi \tag{8.91}$$

where Φ is the magnetic flux passing through the curve C. Since, from (8.87),

$$\mathbf{A} = (\hbar c/q)\nabla\theta \tag{8.92}$$

we have also

$$\oint_C \mathbf{A} \cdot d\mathbf{l} = (\hbar c/q) \oint_C \nabla\theta \cdot d\mathbf{l} = (\hbar c/q)\Delta\theta = 2\pi s(\hbar c/q) \tag{8.93}$$

on using (8.88) and (8.89). Comparing (8.93) with (8.91) gives the result

$$\Phi = s(hc/q) \tag{8.94}$$

where the particle charge $q = |-2e| = 2e$ for Cooper pairs. We rewrite (8.94) as

$$\Phi = s\Phi_0 \qquad (8.95)$$

giving the flux Φ through the superconducting ring as an integral multiple of

$$\Phi_0 \equiv (hc/2e) \qquad (8.96)$$

where Φ_0 is called the flux quantum and has the approximate value 2×10^{-7} G cm^2.

Our conclusion is that the magnetic flux Φ through a superconducting ring is quantized in multiples of the flux quantum or fluxoid Φ_0. The flux Φ is the sum of the flux Φ_e from the external sources and the flux Φ_s from supercurrents flowing in the surface of the ring. Since there is no quantization imposed on Φ_e, the flux Φ_s must adjust itself[45] in order that the total flux Φ take on a quantized value that is a multiple of Φ_0.

Superconducting Quantum Interference

To begin a discussion of superconducting quantum interference, [48-50] we consider the effect of a magnetic field on the DC supercurrent flowing through a Josephson junction. From equations (8.89) and (8.94), we have

$$\Delta\theta = (q/\hbar c)\Phi \qquad (8.97)$$

for a superconducting ring, where Φ is the quantized magnetic flux enclosed by the ring, and $\Delta\theta$ is the change in the phase of the wave function on performing the line integral in equation (8.88). Equation (8.97) shows that there is a connection between the phase θ of the superconducting wave function and the magnetic flux Φ.

We now want to find a relation between Φ and the DC supercurrent J flowing in a Josephson junction. We consider the junction shown schematically in Figure 8.8, where a and b are the end points of the junction. From equation (8.82), the total momentum $\mathbf{p}$ of a particle in a magnetic field is given by

$$\mathbf{p} = m\mathbf{v} + (q/c)\mathbf{A} = \hbar\nabla\theta \qquad (8.82)$$

where $\mathbf{v}$ is the velocity, m the mass, q the charge, $\mathbf{A}$ the magnetic vector poten-

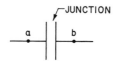

Figure 8.8. Schematic Josephson junction with end points a and b.

tial, and $\nabla\theta$ the gradient of the phase. We apply equation (8.82) to the junction in Figure 8.8, obtaining

$$\hbar\,\Delta\theta = \hbar\,(\theta_b - \theta_a) = \hbar \int_a^b \nabla\theta \cdot d\mathbf{l} = \int_a^b \mathbf{p} \cdot d\mathbf{l} \tag{8.98}$$

which becomes

$$\hbar\,\Delta\theta = \int_a^b m\mathbf{v} \cdot d\mathbf{l} + (q/c)\int_a^b \mathbf{A} \cdot d\mathbf{l} \tag{8.99}$$

where $\Delta\theta$ is the phase difference across the junction, and the line integrals are taken between the end points b and a.

We now use equation (8.99) to consider a superconducting loop L containing two Josephson junctions, (1) and (2), as shown in Figure 8.9. No voltage is applied. The phase differences $\Delta\theta_1$ and $\Delta\theta_2$ across the two junctions are

$$\hbar\,\Delta\theta_1 = \int_a^b m\mathbf{v} \cdot d\mathbf{l} + (q/c)\int_a^b \mathbf{A} \cdot d\mathbf{l} \tag{8.100}$$

$$\hbar\,\Delta\theta_2 = \int_a^b m\mathbf{v} \cdot d\mathbf{l} + (q/c)\int_a^b \mathbf{A} \cdot d\mathbf{l} \tag{8.101}$$

where $\mathbf{A}$ is the vector potential of the magnetic field through the loop, and $\mathbf{v}$ is the particle (pair) velocity in the supercurrent flowing in the loop. Rewriting equation (8.100) as

$$-\hbar\,\Delta\theta_1 = \int_b^a m\mathbf{v} \cdot d\mathbf{l} + (q/c)\int_b^a \mathbf{A} \cdot d\mathbf{l} \tag{8.102}$$

Figure 8.9. Superconducting loop L containing two Josephson junctions (1) and (2), where a and b are points on the loop; J_T is the total supercurrent through the loop device.

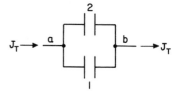

and adding (8.102) to (8.101) gives

$$\hbar(\Delta\theta_2 - \Delta\theta_1) = \oint_L m\mathbf{v} \cdot d\mathbf{l} + (q/c) \oint_L \mathbf{A} \cdot d\mathbf{l} \qquad (8.103)$$

where, in equation (8.103), the line integrals are around a closed curve in the superconducting loop L containing the two Josephson junctions. The integrand $\mathbf{v} \cdot d\mathbf{l}$ in the first integral in equation (8.103) is[51] proportional to the supercurrent, which flows in the surface of the superconductor. The path of integration may be chosen in a region of zero current, so the first integral in (8.103) vanishes. Thus equation (8.103) becomes

$$\hbar(\Delta\theta_2 - \Delta\theta_1) = (q/c) \oint_L \mathbf{A} \cdot d\mathbf{l} \qquad (8.104)$$

Next, using equation (8.91), we have

$$(q/c) \oint_L \mathbf{A} \cdot d\mathbf{l} = (q/c)\Phi \qquad (8.105)$$

where Φ is the total magnetic flux through the loop L. Equations (8.104) and (8.105) give

$$\Delta\theta_2 - \Delta\theta_1 = (2e/\hbar c)\Phi \qquad (8.106)$$

on putting $|\,q\,| = 2e$ for pairs. Equation (8.106) connects the phase differences $\Delta\theta_2$ and $\Delta\theta_1$ across the two junctions with the total magnetic flux Φ through the superconducting loop containing the junctions. We note that $\Delta\theta_2$ and $\Delta\theta_1$ are equal if the flux Φ is zero.

Equation (8.106) is satisfied if

$$\Delta\theta_2 = \theta_0 + (e/\hbar c)\Phi \qquad (8.107)$$

$$\Delta\theta_1 = \theta_0 - (e/\hbar c)\Phi \qquad (8.108)$$

where θ_0 is an introduced constant whose significance will be discussed below. Since (8.107) and (8.108) give the phase differences across the two junctions, we may use equation (8.47) to find the DC supercurrents J_1 and J_2 through junctions (1) and (2). We obtain

$$J_1 = J_{01} \sin [\theta_0 - (e/\hbar c)\Phi] \tag{8.109}$$

$$J_2 = J_{02} \sin [\theta_0 + (e/\hbar c)\Phi] \tag{8.110}$$

where J_{01} and J_{02} are, respectively, the maximum DC supercurrents through junctions (1) and (2). Since the individual supercurrents J_1 and J_2 are, from (8.109) and (8.110), functions only of the flux Φ, the total supercurrent $J_T \equiv J_1 + J_2$ through the entire loop device is given by

$$J_T = J_{01} \sin [\theta_0 - (e/\hbar c)\Phi] + J_{02} \sin [\theta_0 + (e/\hbar c)\Phi] \tag{8.111}$$

If we assume that junctions (1) and (2) are identical and have the same maximum current J_0, where $J_{01} = J_{02} \equiv J_0$, we obtain

$$J_T = J_0 \{\sin [\theta_0 + (e/\hbar c)\Phi] + \sin [\theta_0 - (e/\hbar c)\Phi]\} \tag{8.112}$$

which can be written as

$$J_T = 2J_0 (\sin \theta_0) \cos (e\Phi/\hbar c) \tag{8.113}$$

showing that the total supercurrent J_T varies harmonically with $e\Phi/\hbar c$, and hence with the total magnetic flux through the loop containing the junctions.

Equation (8.113) represents a total supercurrent J_T due to interference between the supercurrents J_1 and J_2, given by (8.109) and (8.110), through the individual junctions (1) and (2). From equations (8.107) and (8.108), the phase differences $\Delta\theta_1$ and $\Delta\theta_2$ across the two junctions are functions of the total magnetic flux Φ through the loop. As Φ then is varied by changing the applied magnetic field, the currents J_1 and J_2 change also. The result is the interference effect expressed in equation (8.113).

The quantity θ_0 is, from equations (8.107) and (8.108), the value of the phase difference $\Delta\theta_2 = \Delta\theta_1$ across both junctions when the magnetic flux $\Phi = 0$. We may therefore regard[50] $\sin \theta_0$ as a quantity that is free to adjust to the current through the loop. Its maximum value is unity. Further, we have lumped into J_0 the effect[52] (which we did not discuss) of the magnetic flux on the maximum supercurrents of the individual junctions. (This leads to the long-period diffraction effects[49,50] observed experimentally.) Equation (8.113) shows that the total supercurrent J_T has maxima when

$$e\Phi/\hbar c = s\pi \tag{8.114}$$

where s is an integer, or, equivalently, when

$$\Phi = s\Phi_0$$

where $\Phi_0 \equiv (hc/2e)$ is the flux quantum defined in (8.96).

Physically, we expect that a plot of total current J_T as a function of magnetic field will show oscillations. There will be maxima in the total current whenever the value of the magnetic field is such that an integral number of flux quanta pass through the superconducting loop. The effect has been observed experimentally,[49,50] and is termed superconducting quantum interference.

The Superconducting Quantum Interference Device (SQUID)

We now discuss a device, the DC SQUID,[53,54] based on superconducting quantum interference. We consider, as in the previous section, the two identical junctions on a superconducting ring, as shown schematically in Figure 8.9. The total supercurrent through the device is given by equation (8.113), rewritten as

$$J_T = 2J_0 (\sin \theta_0) \cos (\pi\Phi/\Phi_0) \qquad (8.115)$$

as a function of magnetic flux Φ through the loop of the device. Since the maximum value of the quantity $\sin \theta_0$ is unity, the total supercurrent will have a maximum magnitude $| J_{max} |$, for a given magnetic flux Φ, equal to

$$| J_{max} | = 2J_0 | \cos (\pi\Phi/\Phi_0)| \qquad (8.116)$$

where this current is composed of superconducting electron pairs passing through the entire device of two junctions on a superconducting loop. Equation (8.116) is correct[55] if Φ is the total flux from both applied external sources and from supercurrents in the device induced by external magnetic fields. Since $| J_{max} |$ is the maximum current the SQUID can carry at a given value of the flux Φ, we identify $| J_{max} |$ with the critical current J_c (the current at which a voltage first appears across the junction). From (8.116), then,

$$J_c = 2J_0 | \cos (\pi\Phi/\Phi_0)| \qquad (8.117)$$

where Φ is the total magnetic flux through the loop of the SQUID. Figure (8.10) is a plot of equation (8.117), showing the periodic variation of J_c with the flux Φ.

One is generally interested in the variation of the SQUID critical current as a function of the external *applied* magnetic flux Φ_e. The *induced* flux is Φ_i, where

$$\Phi_i \cong LI \tag{8.118}$$

for a SQUID loop of self-inductance L carrying a current I. If L is small, such that

$$LI \ll \Phi_0 \tag{8.119}$$

then the induced flux Φ_i will be small, and the total flux Φ, where

$$\Phi = \Phi_e + \Phi_i \tag{8.120}$$

will be given by

$$\Phi \cong \Phi_e \tag{8.121}$$

Then equation (8.117) for the SQUID critical current becomes

$$J_c = 2J_0 |\cos(\pi\Phi_e/\Phi_0)| \tag{8.122}$$

where Φ_e is the external applied magnetic flux through the loop. The variation of J_c as a function of Φ_e given by equation (8.122) is the same as that shown in Figure 8.10, and is the case for the limit in which the condition (8.119) holds. Equation (8.122) shows that the critical current of the SQUID

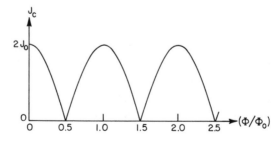

Figure 8.10. Plot of critical current J_c as a function of total flux Φ, as given by equation (8.117).

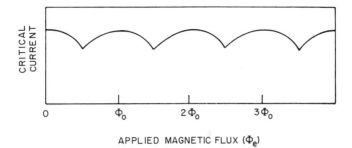

CRITICAL CURRENT

0   Φ_0 $2\,\Phi_0$ $3\,\Phi_0$

APPLIED MAGNETIC FLUX (Φ_e)

Figure 8.11. Plot of critical current J_c as a function of applied flux Φ_e, in the case of a DC SQUID for which the self-inductance is not small. The positions of the maxima and minima are the same as in Figure 8.10, but the minimum value of J_c is not zero (after Clarke[20]).

is a periodic function of the applied magnetic flux Φ_e, and that the minimum critical current is zero. This is approached if the condition (8.119) holds for the device. However, in practical devices,[56] (8.119) is not usually satisfied, and the self-inductance L is of magnitude

$$L \approx \Phi_0/J_0 \tag{8.123}$$

where J_0 is the critical current of each junction. Since the inductance L given by (8.123) is larger than that required to satisfy (8.119), the maximum decrease in the critical current will be smaller than the value $2J_0$ given by equation (8.112), and the critical current will not go to zero in real devices. What is observed in a plot of critical current J_c as a function of applied flux Φ_e is shown schematically in Figure 8.11. There are still maxima and minima in J_c, but the minimum critical current is not zero as it was for the idealized case of very small inductance on which equation (8.122) is based.

The two-junction DC SQUID is thus a device in which the critical current is a function of the magnetic flux passing through the loop. The SQUID can therefore act as a magnetometer[20,57,58] by measuring the critical current. Since the maxima in the critical current in Figure 8.11 are Φ_0 apart on the Φ_e axis, the device can measure magnetic fields much smaller[20] than 10^{-7} G (assuming a maximum device area of about 1 cm²). Other techniques [20,57,60] have been used to obtain even higher sensitivities with DC SQUIDs. Finally, we mention the superconducting voltmeter,[20,57,58,59] the idea behind which is shown schematically in Figure 8.12. In this device, the voltage V to be measured generates a current $I = V/R$ in a superconducting loop of self-inductance L_i, producing a magnetic flux $\Phi = MI$; the mutual inductance

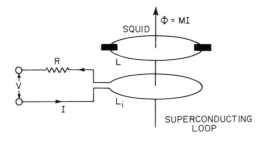

Figure 8.12. Superconducting voltmeter (schematic) where L is the self-inductance of the SQUID, L_i is the self-inductance of the input coil, and M is the mutual inductance (after Clarke[20,21]).

$M = a(LL_i)^{1/2}$, where L is the self-inductance of the SQUID and a is a coupling constant. The flux is then measured with a SQUID magnetometer, yielding a noise-limited voltage sensitivity of the order of 10^{-15} volts. (In practice, the flux change is not measured directly.[59]) There are other types of SQUID devices and many other interesting applications based on the physics we have discussed. The interested reader is referred to the literature.[54,60-63]

Superconducting Materials

We now turn our attention to superconducting materials, examining some of the solid state factors affecting the magnitude of the transition temperature, the critical magnetic field, and, very briefly, the critical current density. As of this writing (late 1988) this field is in explosive ferment owing to the discovery of superconductors with very high critical temperatures. However, these new high-temperature superconductors are not yet well understood. For this reason, and to furnish a background for future developments in this area, this section of the chapter will emphasize the older, lower-temperature superconductors, about which the physical picture seems clear, before attempting an outline of what appears to be reasonably well established about the new materials.

We begin by considering the magnitude of the critical temperature T_c in the BCS microscopic theory.[64-68] The critical temperature of a superconductor is extremely important for applications. The highest value of T_c so far reported (late 1988) is about 125 K,[69] in the copper–oxygen compound $Tl_2Ca_2Ba_2Cu_3O_x$, thereby opening the possibility of superconductive applications above the temperature of liquid nitrogen.

To begin our discussion of the magnitude of T_c, we recall that, in the BCS picture, there is an attractive electron–electron interaction arising from the interaction of the electrons with the lattice phonons of the solid. This attraction results in the formation of Cooper pairs of electrons of opposite momenta and spin. We first discuss the BCS picture and describe its application to older, lower-temperature superconductors in which the electron–electron attraction is phonon-mediated. The attractive interaction between electrons is described[70] by a potential energy $V_{k,k'}$ of interaction between two electrons with wave vectors k and k', and with energies E_k and $E_{k'}$. The simplest form of $V_{k,k'}$ is the BCS "one square well" model in which

$$V_{k,k'} = -V \qquad (8.124)$$

where V is a positive constant, for electron energies E_k and $E_{k'}$ such that

$$|E_k - E_F| < \hbar\omega_D, \qquad |E_{k'} - E_F| < \hbar\omega_D \qquad (8.125)$$

where ω_D is the Debye frequency and E_F is the Fermi energy. For electron energies E_k and $E_{k'}$ other than those given by (8.125), $V_{k,k'} \equiv 0$. The model described by equations (8.124) and (8.125) is thus one with a net attractive potential energy of interaction, in which the phonon-mediated attraction overcomes the Coulomb repulsion, for electrons with energies within $\hbar\omega_D$ of the Fermi energy. For other electron energies, $V_{k,k'}$ is zero, and there is no attraction. The criterion[71] for an attractive interaction, and hence for superconductivity in this model, is simply that V be positive. Then the net attraction will result in the formation of Cooper pairs.

Using the BCS theory with the model of (8.124) and (8.125), one may obtain[68] the following results. The critical transition temperature T_c is given by

$$k_B T_c = 1.13\hbar\omega_D \exp[-1/N(0)V] \qquad (8.126)$$

where $N(0)$ is the density of states (of one spin) at the Fermi energy E_F. Equation (8.126) is correct within this model for zero magnetic field, and for values of T_c less than $\theta_D = (\hbar\omega_D/k_B)$, the Debye temperature. An important restriction on the conditions under which (8.126) holds is that it must also be true that

$$N(0)V \ll 1 \qquad (8.127)$$

This condition is called the weak-coupling limit. For most elemental superconductors, $N(0)V$ is less than[72] about 0.3. The magnitude of the superconducting energy gap $E_g(0) \equiv 2\Delta(0)$, at $T = 0$K, is given by

$$\Delta(0) = 2\hbar\omega_D \exp[-1/N(0)V] \qquad (8.128)$$

again in the weak coupling limit (8.127). The quantity $\Delta(0)$ is the energy gap parameter at 0 K; combining (8.126) and (8.128) gives the relation

$$E_g(0) \equiv 2\Delta(0) = 3.52 k_B T_c \qquad (8.129)$$

Equation (8.129) relates the energy E_g, necessary to break up a Cooper pair, to the transition temperature T_c. A discussion of the comparison of experimental results with the various predictions of the BCS theory may be found in the literature. [72]

From the point of view of applications, we would now like to consider some of the solid state factors influencing, through equation (8.126), the magnitude of T_c. From that equation, we see that, as $N(0)V$ increases, T_c increases, so we expect that large values of the attractive interaction V and/or of the density of states $N(0)$ at the Fermi energy will favor relatively large values of the transition temperature. Also, large values of Θ_D favor high values of T_c, but the exponential in $[-1/N(0)V]$ is the dominant factor. The fact that Θ_D is the less important parameter is seen in the experimental values[1] of $T_c = 7.2$ K for lead, for which $\Theta_D = 105$ K, while $T_c = 1.2$ K for aluminum, whose value of Θ_D is 428 K.

We consider first the quantity V in equation (8.126), subject to the weak-coupling limit. Since the attraction between electrons is due to the electron-phonon interaction, we would expect V to be large if the electron–phonon interaction is, in some sense, large. Since it is the scattering of electrons by phonons that is primarily responsible for the resistivity of a metal at room temperature, we would expect[1] metals with a high resistivity at room temperature to be more likely to be superconductors at low temperatures. This idea may be illustrated by writing[73]

$$\varrho = m^*/ne^2\tau \qquad (8.130)$$

where ϱ is the resistivity, m^* is the effective mass, n the electron density, e the charge on the proton, and τ is the mean time between electron–phonon collisions. If the electron–phonon interaction V is large, we might expect, roughly, that τ will be small, τ^{-1} will be large, and ϱ will be large. Table 8.1 shows such data for a few metallic elements, with values of τ^{-1} calculated[74] from equation (8.130); the elements are arranged in order of increasing room-temperature resistivity. The values of τ^{-1} increase as the resistivity increases, and these values of τ^{-1} may be compared with values[75] of V which may be obtained from experimental data as follows. The electronic specific heat C_e of a normal metal is given[76] by

$$C_e = \frac{1}{3}\pi^2 k_B^2 D(E_F)T \equiv \gamma T \qquad (8.131)$$

where $D(E_F)$ is the density of states of both spins at the Fermi energy E_F. Then

$$D(E_F) = 2N(0) \qquad (8.132)$$

and we obtain

$$N(0) = (3\gamma/2\pi^2 k_B^2) \qquad (8.133)$$

relating the density of states $N(0)$ of one spin at the Fermi energy to the experimental coefficient γ determined from heat capacity measurements. Combining equation (8.133) for $N(0)$ with the BCS equation (8.126) for T_c allows a calculation[75] of V from experimental values of γ, Θ_D, and T_c. From the values of V in Table 8.1, we see that τ^{-1} generally increases as V increases, agreeing with the intuitive idea that high resistivity suggests a large electron–phonon interaction. Even though V represents only very roughly the net attraction between electrons, these illustrative results indicate the correlation between resistivity and the attractive interaction. The values of τ^{-1} and V in Table 8.1 are plotted in Figure 8.13. Silver is omitted from the figure because it is not superconductive, but is included in the table to exhibit its low value of τ^{-1}.

Table 8.1. Resistivity and Superconductivity of Several Elements

	ϱ (295 K, Ω cm)[a]	n (cm^{-3})[b]	$(m^*/m)^c$	τ^{-1} (sec^{-1})	V (eV cm^3)[d]
Ag[e]	1.6×10^{-6}	5.86×10^{22}	1.00	2.6×10^{13}	$< 1.2 \times 10^{-23}$
Al	2.74×10^{-6}	18.1×10^{22}	1.48	9.3×10^{13}	0.97×10^{-23}
W	5.3×10^{-6}	12.6×10^{22}	1.69*	1.0×10^{14}	0.59×10^{-23}
Cd	7.27×10^{-6}	9.27×10^{22}	0.73	2.6×10^{14}	2.5×10^{-23}
In	8.75×10^{-6}	11.5×10^{22}	1.37	2.0×10^{14}	2.1×10^{-23}
Nb	14.5×10^{-6}	5.56×10^{22}	11.7*	1.9×10^{13}	0.34×10^{-23}
Pb	21.0×10^{-6}	13.2×10^{22}	1.97	3.9×10^{14}	1.7×10^{-23}

[a] C. Kittel, Reference 1, page 144, Table 3.
[b] N. W. Ashcroft and N. D. Mermin, Reference 2, page 5, Table 1.
[c] Thermal values from C. Kittel, Reference 1, page 141, Table 2; m is the free electron mass.   The values marked with an asterisk were calculated by the author from the data in Table 2.
[d] G. Gladstone, M. A. Jensen, and J. R. Schrieffer, Reference 75, Table VI.
[e] Not superconductive.

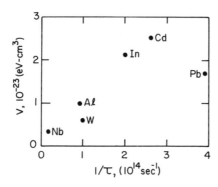

Figure 8.13. Electron–electron attractive interaction V as a function of τ^{-1}, where τ is the electron–phonon collision time in equation (8.130). Data and calculations of Table 8.1.

It is interesting to plot values[75] of V for several different metallic elements as a function of the number Z of valence electrons per atom; such a graph is shown in Figure 8.14. This plot shows a number of interesting qualitative trends. First, V is very small for the nonsuperconducting metals Lu, Y, Sc, Rh, Pt, and Pd. Second, the nontransition metals of valence 2, 3, 4 have relatively high values of V, and they superconduct at relatively high (above 0.5 K) temperatures. Third, almost all of the transition metals[75] superconduct in spite of their relatively low values of V. The value of V is seen to be fairly small and roughly constant for values of Z between 4 and 8, i.e., for most of the transition metals.

We may obtain some additional information by considering the variation of the density of states $N(0)$ as a function of Z. From equation (8.133), one may calculate[75] values† of $N(0)$ from the experimental values of γ; these values of the density of states are plotted against Z in Figure 8.15. The values of $N(0)$ are relatively small for the monovalent metals and for the alkaline earths, which, except for beryllium,[79] are not superconductors. The density of states varies greatly with Z for the transition metals, with peaks in $N(0)$ for $Z = 3$ (nonsuperconducting Sc, Lu, Y, and superconducting La), for $Z = 5$ (superconducting V, Nb, and Ta), and for $Z = 9$ and 10 (nonsuperconducting Rh, Pt, and Pd).

Finally, Figure 8.16 shows the transition temperatures[80] T_c of several superconductors (and upper limits[75] on T_c for nonsuperconductors) plotted as a function of the number Z of valence electrons per atom. The non-

---

† The density of states calculated from γ using equation (8.133) is not the "bare" or unrenormalized density of states $N(0)$. However, since the behavior[77,78] of γ is generally indicative of the behavior of $N(0)$, in the interest of simplicity we will use the symbol $N(0)$ for the density of states calculated from γ.

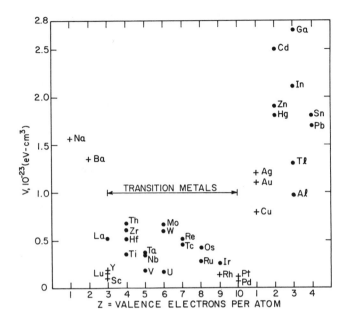

Figure 8.14. Electron–electron attractive interaction V (in eV cm³) as a function of the number Z of valence electrons per atom. Superconductors are indicated by dots, nonsuperconductors (at atmospheric pressure) by crosses, in which cases the value of V is an upper limit.

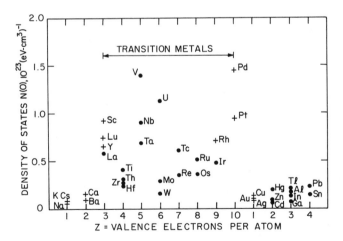

Figure 8.15. Density of states $N(0)$ calculated[75] from equation (8.133), plotted as a function of the number Z of valence electrons per atom. Superconductors are indicated by dots, nonsuperconductors (at atmospheric pressure) by crosses.

superconductivity of the alkali metals correlates with their small density of states as seen from Figure 8.15; the same is true of the alkaline earth and noble metals. The superconductivity of the nontransition metals of valence 2, 3, 4 at relatively high temperatures is in accordance with their large values of the electron–phonon interaction V (Figure 8.14) and is presumably in spite of their small values of the density of states $N(0)$ as seen in Figure 8.15. The values of T_c for the transition metals show complex behavior[81], with peaks in T_c at $Z = 5$ and $Z = 7$ and a relative minimum (excepting uranium) at $Z = 6$. These results may be correlated with the maximum in $N(0)$ at $Z = 5$ and the minimum in $N(0)$ at $Z = 6$; both effects are seen in Figure 8.15. The maximum in T_c at $Z = 7$ correlates with the moderately large values of $N(0)$ and of V for rhenium and technetium seen in Figures 8.14 and 8.15. These conclusions are in accord with the empirical results of Matthias,[82,83] often referred to as "Matthias' rules." These rules include the observation of maxima in T_c at $Z = 5$ and $Z = 7$, and a minimum in T_c at $Z = 6$, results that

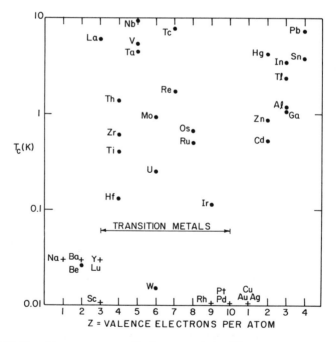

Figure 8.16. Transition temperatures T_c of superconductors[80] (dots) and upper limits[75] on T_c for nonsuperconductors (crosses), plotted as a function of the number Z of valence electrons. Note that T_c is plotted on a logarithmic scale.

correlate[84] well with the description in terms of $N(0)$ and V given above. Further discussion of the systematics of superconductivity may be found in the literature.[85]

To continue discussing the factors affecting the magnitude of T_c, it is necessary to drop the weak-coupling requirement $N(0)V \ll 1$ in the BCS theory. The strong-coupling regime is one in which $N(0)V$ is not small compared to unity, so the result (8.126) is no longer valid. An outline[86] of the results of the application of the strong-coupling regime to the BCS theory is as follows. In the theory, there is a well-known equation, the BCS gap equation, which, in the weak-coupling limit with the attractive interaction V given by (8.124) and (8.125), leads to the well-known results (8.126) and (8.128) for T_c and Δ. In the strong-coupling regime, the BCS gap equation becomes a set of coupled integral equations called the Eliashberg[86] equations. Approximate solution[86a] of the Eliashberg equations will give an expression for T_c, in terms of the density of states at the Fermi energy and the electron–phonon interaction, valid in the strong-coupling regime. However, the results of the strong-coupling treatment are generally expressed, not in terms of $N(0)$ and V directly, but as a function of new parameters, which we now introduce.

The quantity V in equation (8.124) is the *net* attractive interaction between electrons due to the electron–phonon interaction, so

$$V = V_{ph} - V_c \qquad (8.134)$$

where V_{ph} represents the attractive interaction and V_c the repulsive Coulomb interaction. Consider the quantity λ defined by

$$\lambda \equiv N(0)V_{ph} \qquad (8.135)$$

where $N(0)$ is the usual density of electron states of one spin at the Fermi energy. The parameter λ is thus a measure of the attractive electron–phonon interaction; its value[87] ranges from 0.1 to 1.6 in the older materials. Introducing a second parameter μ defined by

$$\mu \equiv N(0)V_c \qquad (8.136)$$

we see that μ is a measure of the Coulomb repulsion between the electrons forming a pair. Our original variable $N(0)V$ is then given by

$$N(0)V = \lambda - \mu \qquad (8.137)$$

However, consideration of the retardation[87] of the phonon-mediated attraction relative to the much more rapid Coulomb repulsion suggests that (8.137) be modified to

$$N(0)V = \lambda - \mu^*$$ (8.138)

where the quantity μ^* is related to μ by the relation

$$\mu^* = \mu/[1 + \mu \ln(E_F/k_B \Theta_D)]$$ (8.139)

where E_F is the Fermi energy, and Θ_D is the Debye temperature. In terms of λ and μ^*, the weak-coupling result (8.126) for T_c becomes[88]

$$T_c \propto \Theta_D \exp\left[-1/(\lambda - \mu^*)\right]$$ (8.140)

For strong coupling, the electron–phonon interaction is large enough that $N(0)V_{ph}$, and hence λ, are close to unity. The McMillan solution[86a] of the Eliashberg equations in the strong-coupling regime results in the approximate relation

$$T_c \cong \frac{\Theta_D}{1.45} \exp\left[-\frac{1.04(1+\lambda)}{\lambda - \mu^*(1 + 0.62\lambda)}\right]$$ (8.141)

Equation (8.141), giving T_c as a function of λ, μ^*, and Θ_D, is usually called the McMillan equation, and values of T_c calculated[89] using it agree quite well with the experimental values. However, calculated values of T_c for the strong coupling superconductors lead and mercury ($\lambda = 1.55$ and 1.60, respectively) are less accurate than those for materials with smaller values of λ.

Our objective will be to use equation (8.141) to discuss some factors influencing the magnitude of T_c. Clearly, as λ increases in (8.141), T_c increases, so an increase in the electron–phonon interaction strength λ will tend to increase the transition temperature. The quantity μ^* may be determined[90] from experimental data on the isotope shift coefficient of a superconductor; most values of μ^* lie near 0.13. Regarding μ^* as a constant in the McMillan equation (8.141) allows us to regard T_c as a function of the electron–phonon interaction strength λ (and, of course, of Θ_D). We now consider the solid state factors which influence the magnitude of λ. It can be shown[91] that λ may be expressed as

$$\lambda = \frac{N(0)\langle I^2 \rangle}{M\langle \omega^2 \rangle}$$ (8.142)

where M is the ionic isotopic mass in the superconductor and $\langle I^2 \rangle$ is the average of the square of the matrix element I of the electron–phonon interaction. The quantity $\langle \omega^2 \rangle$ is the average of the square of the phonon frequency ω, taken over the phonon distribution for the superconductor. As before, a large value of the density of states $N(0)$ favors a large λ, as does a large value of the electron–phonon matrix element I. A small atomic mass M also increases λ and favors a relatively high T_c. (This may be the reason[79] beryllium alone among the alkaline earth metals is a superconductor at atmospheric pressure.) Finally, a decrease in $\langle \omega^2 \rangle$ will increase λ; the resultant expected decrease in $\langle \omega \rangle$, and hence in Θ_D, in (8.141) will not be as important as the increase in λ in the exponential term.

It may also be noted that the McMillan equation (8.141) is an accurate representation of T_c as a function of λ only for values of λ less than about 1.25. For larger values of λ, it has been shown[92,93] that

$$T_C \propto (\langle \omega^2 \rangle \lambda)^{1/2} \tag{8.143}$$

suggesting that T_c increases monotonically with increasing values of the electron–phonon interaction λ. This would mean that, within the BCS theory with a phonon-mediated electron–electron interaction, there is no natural maximum[94-97] in T_c. In this context, "natural" refers to a maximum arising from the structure of the equations alone. There may be maxima in transition temperatures due to lattice instabilities or other external factors.

Considering the transition temperatures of real nonelemental materials, the field has been revolutionized by the discovery of the copper-oxide superconductors with values of T_c as high as 125 K. While these new materials are the main focus of current research, it is still useful to discuss some of the older, lower-temperature compound superconductors. The A-15 compounds[98-100] are of current practical importance for applications, such as superconducting electromagnets involving high magnetic fields and high current densities. (The A-15 structure is shown in Figure 8.25, and is discussed later in this chapter in the context of high critical magnetic fields.) Examples of superconductors with the A-15 structure which are important for superconducting magnets[101] are Nb_3Sn ($T_c = 18$ K) and V_3Ga ($T_c = 16$ K). These values of T_c are typical of the higher figures reached before the discovery of the copper-oxide superconductors. (The maximum observed T_c for an A-15 compound is 23 K for Nb_3Ge.[102]) While low by today's standard, these values of T_c allow the construction of practical superconducting magnets which operate at about 4 K, well below the critical temperature of the superconductor used. For Nb_3Sn, the electron–phonon interaction parameter λ is in the range[103] from 1.65 to 1.95, so Nb_3Sn is a strong-coupling superconductor. While the A_3B superconductors of the A-15 structure have been extensively studied, the reasons for their relatively high critical temperatures, say in terms of the factors in equation (8.142), are still not fully understood.[104–106] Values of T_c, $2\Delta(0)$, and λ are given in Table 8.2 for several superconductors. (The other columns of the table are discussed later in this chapter.) We see from their values of λ of unity or larger that the elements lead and niobium, and the A-15 compound Nb_3Sn, are all strong-coupling superconductors, with energy gaps $E_g(0) \equiv 2\Delta(0)$ of a few milli-electron volts at 0 K. On the other hand, aluminum, with its smaller value of λ, is a weak-coupling superconductor with a smaller energy gap.

On turning to the new high-temperature superconductors, it is difficult, as of this writing (late 1988), to do more than given an incomplete outline of a picture that is yet to become clear. One important class of the copper oxide type has the formula $RM_2Cu_3O_7$, where R is a rare earth atom and M is an alkaline earth atom.[107–109] An example is $YBa_2Cu_3O_7$, which has $T_c \cong 93$ K and has been perhaps the most thoroughly studied to date. The crystal structure of $YBa_2Cu_3O_7$ is orthorhombic, as shown in Figure 8.17, which exhibits

Table 8.2. Properties of Some Superconductors

Material	T_c (K)	$2\Delta(0)$ (meV)	λ	κ	Type	ξ_0 (Å)
Al	1.1^a	0.34^b	0.4^f	0.01^c	I	16000^h
Pb	7.2^a	2.73^b	1.55^e	0.45^c	I	830^h
Nb	9.5^a	3.05^b	0.98^e	1.02^c	II	380^h
Nb_3Sn	18^f	6.5^g	1.8^g	34^d	II	$\sim35^i$
$YBa_2Cu_3O_7$	93^j	$\sim30^k$	?	$60\text{-}100^l$	II	$\sim10^{m,n}$

[a] C. Kittel, Reference 1, page 320, Table 1.
[b] C. Kittel, Reference 1, page 328, Table 3.
[c] C. Kittel, Reference 1, page 337, Table 5.
[d] A. C. Rose-Innes and E. H. Rhoderick, Reference 4, page 195, Table 12.1.
[e] R. M. White and T.H. Geballe, Reference 77, page 228, Table VI.3.
[f] P. B. Allen and B. Mitrovic, Reference 92, page 3, Table I.
[g] E. L. Wolf et al., Phys. Rev. B 22, 1214 (1980).
[h] R. Meservy and B.B. Schwartz, Reference 72, page 174, Table VI.
[i] Calculated by the author using the value of $H_{c2}(0)$ given in Table 8.3 and equation (8.188).
[j] See, for example, C. K. N. Patel and R. C. Dynes, Reference 107, page 4946.
[k] Calculated by the author using $[2\Delta(0)/k_BT_c] \cong 3.5$, quoted by Tinkham and Lobb, Reference 108, page 131.
[l] J. Bardeen et al., in Novel Superconductivity, S. A. Wolf and V. Z. Kresin (editors), Plenum Press, New York (1987), pages 333–339.
[m] M. Tinkham and C.J. Lobb, Reference 108, page 92.
[n] The coherence length ξ_0 in $YBa_2Cu_3O_7$ appears to be very anisotropic; this is discussed later in this chapter in the context of the critical magnetic field.

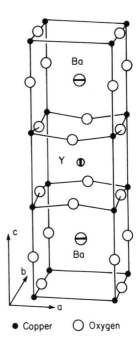

Ba

Y

Ba

Figure 8.17. Unit cell of orthorhombic $YBa_2Cu_3O_7$. The lattice constants are[110] $a = 3.8265$ Å, $b = 3.8833$ Å, $c = 11.6813$ Å at 293 K, and the two "buckled" copper–oxygen planes, separated by the yttrium atom, are about 3 Å apart.

● Copper   ○ Oxygen

a unit cell with lattice constants[110] $a = 3.8265$ Å, $b = 3.8833$ Å, $c = 11.6813$ Å, at 293 K. The "buckled" copper–oxygen planes at the top and bottom of the unit cell, separated by the yttrium atom, are about 3 Å apart, and should be particularly noted. It is generally thought[111] that the superconductivity involves carriers (holes) in hybridized copper–oxygen bonds in these planes.

Cooper pairs are believed to exist in $YBa_2Cu_3O_7$, but the origin of the attractive electron–electron interaction is still an open question. In the older, lower-temperature superconductors, the source of this attraction is the electron–phonon interaction, evidence for which rests on the experimental observation of the isotope effect. Returning to the BCS equation (8.126) for T_c, in which $\Theta_D \equiv (\hbar\omega_D/k_B)$ is the Debye temperature, we see that T_c is proportional to the Debye phonon frequency ω_D. Since, for a simple harmonic oscillator model, one expects the phonon frequency to be inversely proportional to the atomic mass M, it is to be expected that

$$T_c \propto M^{-a} \tag{8.144}$$

where, in the simplest picture, the exponent $a = 0.5$. The dependence of T_c on

M is known as the isotope effect, and has been seen[112] in many superconductors. However, on substituting O^{18} for O^{16} in $YBa_2Cu_3O_7$, it has been observed[113] that the exponent $\alpha \cong 0.02$, a value much smaller than expected *if* the electron–electron attraction is due to the electron–phonon mechanism. This result suggests that a new mechanism, other than (or in addition to) the electron–phonon interaction, may be responsible for the formation of Cooper pairs. (For this reason, there is no entry under "λ" in Table 8.2.) At the moment there are a number of non-phonon attractive interactions under discussion,[114,115] but the question is still open.

It is also not yet established whether $YBa_2Cu_3O_7$ is a weakly-coupled superconductor. While there are some indications[116,116a] that the quantity $[2\Delta(0)/k_BT_c]$ has the BCS weak-coupling value of 3.5, there are also experimentally observed values of $[2\Delta(0)/k_BT]$ ranging[116b] from 4 to 7.5. If if does turn out to be the case that $YBa_2Cu_3O_7$ is weakly coupled, then the critical temperature T_c might be expected to be described by an equation analogous to (8.126) but appropriate for a non-phonon attractive mechanism. In that case, there would be a cut-off energy[108] prefactor $\hbar\omega_c$ (where the cut-off frequency ω_c is the Debye frequency for the phonon mechanism) in the equation. For a non-phonon mechanism, such a cut-off energy might well be considerably larger than $\hbar\omega_D$, thereby resulting in a high value of T_c even though the exponential term would be small for the case of weak coupling. However, these remarks are rather speculative and reflect the uncertainties inherent in this preliminary and incomplete picture.

As of late 1988, superconductivity with a value of T_c of about 125 K has been observed[69,117] in $Tl_2Ca_2Ba_2Cu_3O_x$, a copper–oxygen compound whic does not contain a rare earth atom. As of this writing, however, the investigation of these thallium copper-oxide superconductors is just beginning.

We now discuss another important aspect of superconducting materials, the critical magnetic field. As is well known,[118–121] the application of a sufficiently strong magnetic field to a superconductor will destroy its superconductivity. The magnitude of this magnetic field, called the critical field, is of great interest for applications of superconducting materials, particularly the construction of superconducting solenoids designed to produce high magnetic fields.[122,123]

Before discussing this topic, we review some background material[119,124,125] on Type I and Type II superconductors. We consider the magnetization curve, a plot of $(-4\pi M)$, where **M** is the magnetization inside the superconductor, as a function of the magnitude of the applied magnetic field[126] B_a.

For some superconductors, known as Type I, the magnetization curve is

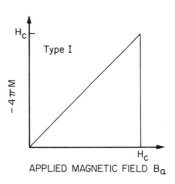

Figure 8.18. Plot of $(-4\pi M)$, where M is the magnitude of the magnetization, as a function of applied magnetic field B_a, for a Type I superconductor. The critical value of the magnetic field $B_{ac} \equiv H_c$. This magnetization curve is for a bulk sample exhibiting a complete Meissner effect, and the temperature is below the critical transition temperature T_c.

as shown in Figure 8.18. Consider applying a magnetic field to the superconductor by increasing it from zero to a value $B_a = H_a$; the result may be viewed in two equivalent ways.[127] In the first, the time-varying magnetic flux induces a surface screening current in the superconductor. This current persists in the superconductor after the applied field has reached a constant value B_a, and (neglecting demagnetizing effects[128]) produces a field $(-B_a)$ inside the sample. The induced field $(-B_a)$ just cancels the applied field B_a, resulting in a net magnetic induction $\mathbf{B} = 0$ within the superconductor. A second way of viewing the process is as follows. Instead of considering the surface screening current directly, one says that the superconductor has magnetic properties such that its magnetic susceptibility $\chi \equiv (M/H)$ has the value $\chi = (-1/4\pi)$. Then, since

$$\mathbf{B} = \mathbf{H} + 4\pi\mathbf{M} = \mathbf{H}(1 + 4\pi\chi) \tag{8.145}$$

we obtain $\mathbf{B} = 0$ inside the superconductor, thus describing the Meissner effect. The situation is referred to as the perfect diamagnetism of the superconductor. Since $\mathbf{B} = 0$, equation (8.145) yields

$$\mathbf{H} = -4\pi\mathbf{M} \tag{8.146}$$

for the magnetization $\mathbf{M}$ of the superconductor. Equation (8.146) gives the magnetization curve in Figure 8.18. As the applied magnetic field[126] $B_a = H_a$ is increased, the magnitude of $(-4\pi\mathbf{M})$ increases linearly until the critical field H_c is reached, and the specimen reverts to the normal state. Thus, a Type I conductor with a diamagnet susceptibility $\chi = (-1/4\pi)$ is referred to as a perfect diamagnet exhibiting a complete Meissner effect, that is, the total exclusion of magnetic flux from the superconductor.

The second class of superconductors, known as Type II, exhibits a

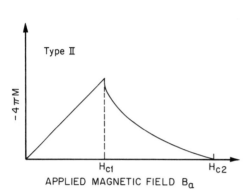

Figure 8.19. Plot of $(-4\pi M)$ as a function of applied magnetic field B_a for a Type II superconductor. (The sample is assumed to be a long cylinder parallel to the applied field.) The sample is superconducting ($B = 0$) when the applied field is less than the lower critical field H_{c1}, and is in the mixed or vortex state ($B \neq 0$) when the applied field is between H_{c1} and the upper critical field H_{c2}. For fields larger than H_{c2}, the superconductivity is destroyed and the sample is a normal conductor.

magnetization curve of the kind shown in Figure 8.19. At values of the applied magnetic field below a value H_{c1}, the specimen (assumed to be a long cylinder parallel to $\mathbf{B}_a$) is superconducting, and magnetic flux is completely excluded. The value H_{c1} is called the lower critical field, and is the value at which flux begins to penetrate the sample. For values of the applied field between H_{c1} and the upper critical field H_{c2}, the flux density B is not zero within the superconductor and the Meissner effect is said to be incomplete. The sample is threaded by flux lines and is said to be in the mixed state or the vortex state; this state contains both superconducting and normal regions. (The reason for the term vortex state will be given later, in a discussion of the microscopic structure of the mixed state.) Finally, when the applied field exceeds H_{c2}, the sample becomes a normal conductor. It will turn out that all superconductors with critical fields large enough to be of interest for applications are of Type II.

The next question we want to address is why the application of a sufficiently strong magnetic field destroys superconductivity. Why is there a critical field? We may approach this question on the macroscopic level by considering the thermodynamics[129–132] of the superconducting and normal states in a magnetic field.

Consider a Type I superconductor of unit volume at a constant temperature T below T_c. If U is the internal energy and S the entropy of the sample, then

$$F = U - TS \qquad (8.147)$$

defines the Helmholtz free energy density. The sample is assumed to be a long thin cylinder parallel to the applied external magnetic field $\mathbf{H}_a \equiv \mathbf{B}_a$; the sample volume is constant. Then the change dF in the free energy density due to changes dT in temperature and dH in the magnetic intensity inside the sample is given[132] by

$$dF = -S\,dT - M\,dH \tag{8.148}$$

where M is the magnetization. Since T is constant,

$$dF = -M\,dH \tag{8.149}$$

gives the free energy density change dF in terms of the magnetization, which, from (8.146), is

$$M = (-1/4\pi)H \tag{8.150}$$

and the change dH in the magnetic intensity H inside the superconductor. From the continuity of the tangential component of $\mathbf{H}$, in this geometry we have[132] $H = H_a = B_a$. Thus (8.149) becomes

$$dF = (+1/4\pi)B_a\,dB_a \tag{8.151}$$

where B_a is the magnitude of the applied magnetic field.

We may use (8.151) to calculate the free energy density $F_S(B_a, T)$ of the superconductor at temperature T in the applied external field B_a. We have

$$F_S(B_a,T) - F_S(0,T) = \int_0^{B_a} dF = \int_0^{B_a} (+1/4\pi)B_a\,dB_a \tag{8.152}$$

where $F_S(0, T)$ is the free energy density of the superconductor in zero field. The result is

$$F_S(B_a, T) - F_S(0, T) = \frac{B_a^2}{8\pi} \tag{8.153}$$

showing that the application of an external magnetic field increases the free energy density of the superconductor.

We may make the same calculation for a normal metal, using

$$M = \chi H \tag{8.154}$$

for the magnetization in place of (8.150). In (8.154), χ is the paramagnetic susceptibility of the normal metal. Typical values[133] are of the order of 10^{-6},

much smaller than the magnitude $(1/4\pi) = 0.08$ of the diamagnetic susceptibility for a Type I superconductor exhibiting perfect diamagnetism. On this
basis, we may neglect[134] the susceptibility of a normal metal and take its free
energy density as independent of the applied magnetic field. We write

$$F_N(B_a,T) = F_N(0,T) \equiv F_N \qquad (8.155)$$

where $F_N(B_a, T)$ and $F_N(0,T)$ are, respectively, the free energy densities of the
normal state in the applied field B_a and in zero field, both at temperature T.
Rewriting (8.153) as

$$F_S(B_a,T) = F_S(0,T) + \frac{B_a^2}{8\pi} \qquad (8.156)$$

it is worth reiterating that the free energy density of the superconductor
increases as the applied magnetic field B_a is increased. When $B_a = B_{ac} \equiv H_c$, the
critical magnetic field, (8.156) becomes

$$F_S(H_c,T) = F_S(0,T) + \frac{H_c^2}{8\pi} \qquad (8.157)$$

Further, the normal and superconducting states coexist in equilibrium at H_c
since their free energy densities are then equal, and we have

$$F_S(H_c,T) = F_N(H_c,T) = F_N \qquad (8.158)$$

when the applied magnetic field has the critical value H_c because we have assumed that the normal state free energy density is independent of the magnetic field.

Figure 8.20 shows the free energy density $F_S(B_a, T)$ of the superconducting state from equation (8.153) and the constant free energy density F_N of
the normal state from equation (8.155), both plotted as functions of the
applied magnetic field B_a. At the critical value $B_a = H_c$, the superconducting
and normal phases are in equilibrium and have the same free energy density,
as expressed by equation (8.158). In the region of the figure to the left of the
line $B_a = H_c$, the sample is superconducting; to the right it is in the normal
state.

We now see, from Figure 8.20, why there is a critical magnetic field H_c
which destroys superconductivity. As the applied magnetic field B_a is turned
on, the free energy density F_S of the superconductor increases, as in equation
(8.156). This expenditure of energy is necessary to set up the surface screening

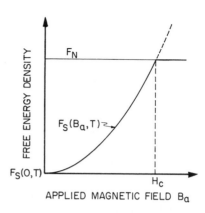

Figure 8.20. Free energy density $F_S(B_a, T)$ for the superconducting state from equation (8.153) and F_N for the normal state from equation (8.155), both plotted as functions of the applied magnetic field B_a. At the critical value $B_a = H_c$, the superconducting and normal phases are in equilibrium, and $F_S(H_c, T) = F_N$. The region to the left of the line $B_a = H_c$ is the superconducting state and the region to the right is the normal state (after Kittel[1]).

current which cancels the applied field within the superconductor and results in the Meissner effect. As the free energy density increases, F_S will, for fields larger than the critical field, become greater than the free energy density F_N of the normal state. When F_S is larger than F_N, the superconducting phase is no longer the stable state, and the sample reverts to the normal phase, allowing the applied magnetic field to penetrate. The content of equation (8.156) may also be restated[135] as follows in terms of the Meissner effect. The expulsion of the magnetic field from the interior of the superconductor can be considered as increasing the energy by an amount equal to the energy of the displaced magnetic field. For a sample of unit volume, that amount is $(B_a^2/8\pi)$, for an applied field of magnitude B_a. When B_a is larger than H_c, the energy of the superconductor has been increased by more than $(H_c^2/8\pi)$, and the sample becomes normal.

Finally, these results may be used to define some common terminology. We note from (8.155) that, since F_N is taken as independent of the applied field, we have, considering temperatures below T_c,

$$F_N(H_c, T) = F_N(0, T) \tag{8.159}$$

If this result is combined with (8.158) and (8.157), we obtain

$$F_N(0, T) = F_S(0, T) + \frac{H_c^2}{8\pi} \tag{8.160}$$

which is usually written as

$$\Delta F \equiv F_N(0, T) - F_S(0, T) = \frac{H_c^2}{8\pi} \tag{8.161}$$

Equation (8.161) defines[136] ΔF, the free energy density difference between the normal and superconducting states in zero applied field; ΔF is called the condensation energy of the superconducting state. Further, if $T = 0$, then, from (8.147), $F = U$, the internal energy. Then

$$\Delta U \equiv U_N(0) - U_S(0) = \frac{[H_c(0)]^2}{8\pi} \qquad (8.162)$$

is the energy difference between the energies $U_N(0)$ and $U_S(0)$ of the normal and superconducting states at absolute zero, and $H_c(0)$ is the magnitude of the critical field at absolute zero. The quantity ΔU is referred to[137] as the stabilization energy density of the superconducting state at $T = 0$.

Given the existence of a critical magnetic field, we now want to discuss the factors determining its magnitude. For applications of superconducting materials in electromagnets,[138,139] a high critical field is desirable. It turns out, as we will see, that Type I superconductors have inherently low critical fields, so our main interest will be in Type II materials. To discuss Type II superconductors, we must review the concepts of penetration depth and coherence length.

We recall[140–142] that a gas of quasi-bosons of charge q and mass m, all in the same state, and with number density ϱ, obeys the London equation

$$\mathbf{\nabla} \times \mathbf{J} = (-c/4\pi\lambda_L^2)\mathbf{B} \qquad (8.163)$$

where $\mathbf{J}$ is the current density and

$$\lambda_L^2 \equiv (mc^2/4\pi\varrho q^2) \qquad (8.164)$$

Equation (8.163) is equivalent to

$$\mathbf{J} = (-c/4\pi\lambda_L^2)\mathbf{A} \qquad (8.165)$$

stating that the current density $\mathbf{J}$ is proportional to the magnetic vector potential $\mathbf{A}$. As is well known, the physical significance of λ_L is that of the depth to which the magnetic field penetrates a superconductor; λ_L is the distance in which B decreases by a factor $(1/e)$. Typical values[143] λ_L are a few hundred Angstroms in pure elemental superconductors, most of which, as we will see, are Type I.

We discuss next the coherence length or range of coherence. The BCS coherence length ξ_0 is defined[144] by

$$\xi_0 \equiv \hbar v_F / \pi \Delta(0) \tag{8.166}$$

where v_F is the Fermi velocity and $\Delta(0)$ is the gap parameter at $T = 0$ K. From (8.166), it is seen that ξ_0 is a characteristic of a given superconductor; ξ_0 is also temperature independent. The physical interpretation[145–148] of ξ_0 is the length over which the electrons in a Cooper pair remain correlated, so ξ_0 may be considered as a kind of microscopic dimension of a pair. It may also be thought of as the range of the pair wavefunction. Values[149] of ξ_0 are several hundred to several thousand Angstroms in pure superconductors.

Another important quantity is the Ginzburg–Landau temperature-dependent coherence length,[150] denoted by $\xi(T)$. The physical significance of $\xi(T)$ may be discussed in terms of the number density ϱ of pairs, given by

$$\varrho = |\psi|^2 \tag{8.21}$$

where ψ is the pair wave function.[151] One may show[152] that $\xi(T)$ is the characteristic length over which a slightly disturbed pair wave function ψ will return to the value ψ_∞ it would have deep in the interior of a pure superconductor. We may also think of $\xi(T)$ as the shortest distance within which there may be an appreciable variation in ψ and, hence, in ϱ. In summary,[153] the coherence length $\xi(T)$ provides the characteristic scale of length for spatial variations of the superconducting wave function or of the density of pairs in the superconductor.

So far, we have been considering the coherence length in pure materials. We now discuss the effect of impurities in terms of the mean free path l of electrons in the normal state of the superconductor. As the material becomes less pure, we expect l to become smaller as the electrons collide with more numerous impurities. The superconducting state is a highly ordered state, and one would expect the range of coherence to be largest in a pure superconductor. It is intuitively reasonable to think of impurities as decreasing the order and coherence of the superconducting state. For this reason, one might expect $\xi(T)$ to vary with the impurity content of the superconductor, while ξ_0 is an intrinsic characteristic of the material. The derivation of the dependence of $\xi(T)$ on the electron mean free path l may be found in the literature[154]; the results may be summarized as follows. In a pure superconductor, at a temperature near T_c, in the "clean limit" in which $l \gg \xi_0$, one finds that

$$\xi(T) = 0.74\ \xi_0/(1-t)^{1/2} \qquad (8.167)$$

where $t \equiv (T/T_c)$. From (8.167), $\xi(T)$ does not depend on l in a pure superconductor in the clean limit, but is, reasonably, related to the BCS coherence length ξ_0. In an impure superconductor, also near T_c, in the "dirty limit" in which $l \ll \xi_0$,

$$\xi(T) = 0.86\,(\xi_0 l)^{1/2}/(1-t)^{1/2} \qquad (8.168)$$

The dependence of $\xi(T)$ on l in the dirty limit, given by equation (8.168), will be of considerable interest to us later.

We now want to consider the dependence of the penetration depth λ on the electron mean free path l. The result[155] found is that, as l decreases, the penetration depth increases. The reason may be inferred from the fact that equation (8.165) giving the current density $\mathbf{J}$ as a function of the vector potential $\mathbf{A}$ is essentially a nonlocal relation. By this is meant that, in order to calculate $\mathbf{J}(\mathbf{r})$ at a point $\mathbf{r}$, one should take into account the vector potential $\mathbf{A}$ in a volume of radius roughly ξ_P about the point $\mathbf{r}$. The quantity ξ_P is called the electromagnetic or Pippard coherence length,[156] and serves as a characteristic length for nonlocal electromagnetic effects in superconductors. The Pippard length ξ_P is related to the electron mean free path l by the relation

$$\xi_P^{-1} = l^{-1} + \xi_0^{-1} \qquad (8.169)$$

where the BCS coherence length ξ_0 is defined by equation (8.166). In an impure sample, l will be small and, if $l \ll \xi_0$, we see from (8.169) that ξ_P will be small. If the Pippard length is thus decreased by impurities, the supercurrent response $\mathbf{J}$ to an external field $\mathbf{A}$ is weakened, allowing increased penetration by the external magnetic field. For the dirty limit in which $l \ll \xi_0$, it is found that[157,158]

$$\lambda(T) = \lambda_L(\xi_0/1.33l)^{1/2}[2(1-t)]^{-1/2} \qquad (8.170)$$

where λ_L is defined by (8.164) and, again, $t \equiv T/T_c$. Equation (8.170) gives the dependence of the penetration depth $\lambda(T)$ on the electron mean free path l in the dirty limit. Last, we note that,[157–158] in the clean limit, a pure superconductor will have a penetration depth $\lambda(T)$ given by

$$\lambda(T) = \lambda_L[2(1-t)]^{-1/2} \qquad (8.171)$$

Finally, we introduce the widely used Ginzburg–Landau parameter κ defined by the equation

$$\kappa \equiv \lambda(T)/\xi(T) \qquad (8.172)$$

where $\lambda(T)$ is the penetration depth and $\xi(T)$ is the Ginzburg–Landau temperature-dependent coherence length. The value of κ is most important since (as we will see) it determines whether a superconductor is Type I or Type II. From equations (8.171) and (8.167), we have[159]

$$\kappa = \frac{\lambda_L[2(1-t)]^{-1/2}}{0.74\xi_0(1-t)^{-1/2}} = 0.96\,\lambda_L\,\xi_0^{-1} \qquad (8.173)$$

in the clean limit of a pure superconductor. In the dirty limit of an impure superconductor, equations (8.170) and (8.168) give[159]

$$\kappa = \frac{\lambda_L\,\xi_0^{1/2}\,l^{-1/2}\,(1.33)^{-1/2}\,[2(1-t)]^{-1/2}}{0.86\xi_0^{1/2}\,l^{1/2}\,(1-t)^{-1/2}} = 0.71\,\lambda_L l^{-1} \qquad (8.174)$$

The important conclusion from equation (8.174) is that, as l becomes very small in an impure superconductor, the Ginzburg–Landau parameter v becomes large.

We now introduce the idea of the surface energy at a boundary between the normal and superconducting phases. We can have such a boundary only when the two phases coexist, i.e., in an applied magnetic field equal to the critical field. As we shall see below, the sign of the surface energy depends on the value of $\kappa = \lambda/\xi$, and that sign determines whether the superconductor is Type I or Type II.

Consider the boundary, located at $x = 0$, between a superconducting (S) region and a normal (N) region. The normal region is in the region $x < 0$ and the superconductor is in the region $x > 0$. We want to consider the variation of the free energy density[160] as a function of distance x into the superconductor. From equation (8.157), we have

$$F_1 \equiv F_S(H_c,T) - F_S(0,T) = (H_c^2/8\pi) \qquad (8.175)$$

as the increase F_1 in the free energy density of the superconductor in the applied magnetic field $B_a = H_c$. This is the increase in energy due to the expulsion of the magnetic flux in the Meissner effect. Since the magnetic field penetrates a distance λ into the superconductor, we would expect that the positive contribution F_1 to the free energy density would vary with distance x, and be smaller near the N–S boundary because the amount of flux expelled varies with x and is smaller near the boundary.

Consider next the condensation energy, given by equation (8.161), and rewritten as

$$F_2 \equiv F_S(0,T) - F_N(0,T) = (-H_c^2/8\pi) \qquad (8.176)$$

The quantity F_2 is the negative contribution to the free energy density due to the ordering of the electrons when the superconducting phase forms. Again, we would expect the magnitude of F_2 to vary with distance from the N–S boundary because we expect the superconducting pair wave function ψ to have a characteristic coherence length ξ. If it takes a distance ξ from the boundary for the pair density $\varrho = |\psi|^2$ to reach its maximum value, one would expect the magnitude of F_2 to be smaller near the N–S boundary because ψ is smaller near the boundary.

If we consider the sum $F = F_1 + F_2$ of the two contributions to the free energy density, we would also expect F to vary with distance from the boundary. Since, from (8.175) and (8.176), on assuming $F_N(0,T) = F_N(H_c,T)$ in the normal phase from (8.155), we have

$$F = F_1 + F_2 = F_S(H_c,T) - F_N(H_c,T) \qquad (8.177)$$

We would also expect, deep in the superconductor, that $F = 0$, expressing the fact that the N and S phases will be in equilibrium at the critical magnetic field H_c. Since, as we shall see, F is not zero in the region near the interface, we may consider F as a surface free energy density associated with the boundary between the normal and superconducting phases. It will turn out, as shown below, that the sign (positive or negative) of F will depend on whether the coherence length ξ is larger or smaller than the penetration depth λ. We consider now the two contributions F_1 and F_2 to the free energy density F for the cases in which ξ is larger than λ and in which ξ is smaller than λ.

Figure 8.21 shows the situation when ξ is larger than λ. In Figure 8.21(a), we see a plot of the positive term F_1 in the free energy density, shown schematically as a function of distance x. In the region near the interface at $x = 0$, the energy increase due to the expulsion of flux is less than $(H_c^2/8\pi)$ because the

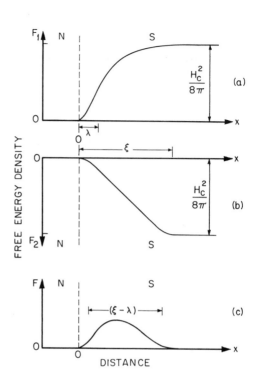

Figure 8.21. Effect of coherence length ξ and penetration depth λ on the surface energy of a Type I superconductor for which $\lambda < \xi$. (a) Variation with distance x of the positive free energy density contribution F_1 due to flux exclusion. (b) Variation (schematic) with distance x of the negative free energy density contribution F_2 due to the condensation energy. (c) Sum $F = F_1 + F_2$ as a function of distance x, showing the positive surface energy. The figure is drawn for $\kappa = (\lambda/\xi) = 0.2$ (after Rose-Innes and Rhoderick[4]).

magnetic field penetrates the superconductor with penetration depth λ. Figure 8.21(b) shows, schematically, a plot of the negative (condensation energy) contribution term F_2, as a function of distance x. The characteristic distance is the coherence length ξ, which is drawn for the case $\xi = 5\lambda$, so $\kappa = (\lambda/\xi) = 0.2$. Last, Figure 8.21(c) shows the free energy density $F = F_1 + F_2$ as a function of distance, and F is seen to be positive in the region near the interface. If we define σ, the surface energy per unit area of the boundary, by the relation[161]

$$\sigma = (H_c^2/8\pi)\delta \qquad (8.178)$$

and identify (σ/δ) with F, the free energy density given by (8.177), we can see that the length δ is roughly equal to $(\xi - \lambda)$. This may be seen[162] most readily by replacing Figures 8.21(a) and (b) by step functions in which the changes in F_1 and F_2 take place, respectively, at $x = \lambda$ and $x = \xi$. Then it is seen that there is then a region of thickness $(\xi - \lambda)$ in which the positive energy of flux expulsion is not cancelled by the negative contribution of the condensation energy.

The result is the positive surface energy shown in Figure 8.21(c) and given in equation (8.178). We conclude that, if ξ is larger than λ, there will be a region of positive surface energy at the normal-superconductor boundary. If the surface energy per unit area is positive, as it is if ξ is larger than λ so $\delta = (\xi - \lambda)$ is also positive, then the free energy will contain a term which increases as the boundary area increases. The free energy will then be minimized by having as small an interface as possible. For this reason, this type of superconductor (with ξ larger than λ) will remain superconducting throughout[163] in a field of magnitude less than the critical field. This class of superconductors is called Type I.

We consider next the situation when ξ is smaller than λ, as shown in Figure 8.22, which is of the same kind as Figure 8.21. Figure 8.22 is drawn for $\xi = 0.5\lambda$, so $\kappa = (\lambda/\xi) = 2$. Just as described for Figure 8.21, the contributions F_1 in Figure 8.22(a) and F_2 in Figure 8.21(b) are added in Figure 8.22(c) to give a region of negative surface energy near the normal-superconducting interface. The conclusion is that, if ξ is smaller than λ, the surface energy per unit area σ will be negative near the boundary, and minimization of the free energy would be favored by the formation of normal-superconducting interfaces. In such a situation, the application of a magnetic field to a supercon-

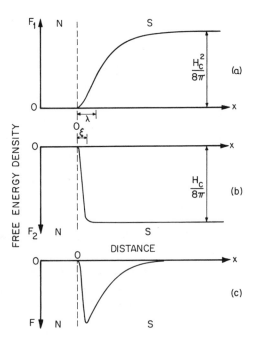

Figure 8.22. Effect of coherence length ξ and penetration depth λ on surface energy of a Type II superconductor, for which $\lambda > \xi$. (a) Variation with distance x of the positive free energy density contribution F_1 due to flux exclusion. (b) Variation (schematic) with distance x of the negative free energy density and contribution F_2 due to the condensation energy. (c) Sum $F = F_1 + F_2$ as a function of distance x showing the negative surface energy. The figure is drawn for $\kappa = (\lambda/\xi) = 2$ (after Rose-Innes and Rhoderick[4]).

ductor of this type produces some fine-scale mixture of normal and superconducting regions called the mixed state, or, for reasons to be described presently, the vortex state. Superconductors for which ξ is smaller than λ, and which exhibit a mixed state, are called Type II superconductors, and are the type in which we will be primarily interested for applications.

We may now see why most elemental superconductors are Type I. From equation (8.173), we have, in the clean limit, that

$$\kappa \propto (1/\xi_0) \tag{8.179}$$

For pure elements, we expect the clean limit to be appropriate and the intrinsic coherence length ξ_0 to be relatively long. Thus κ is expected to be small, so ξ will be larger than λ, leading to Type I behavior. For typical pure superconductors,[164] λ is of the order of 500 Angstroms and ξ_0 is of the order of 3000 Angstroms, so $\kappa \ll 1$. Values of ξ_0 and κ for the Type I superconductors aluminum and lead are given in Table 8.2.

Conversely, most alloys[165] are Type II, as can be seen from equation (8.174) for the dirty limit, which says

$$\kappa \propto l^{-1} \tag{8.180}$$

In an alloy, we expect strong electron scattering, small values of l, and large values of $\kappa \gg 1$, so we will have ξ smaller than λ, indicating Type II behavior. Some typical values of κ for Type II materials are[166] $\kappa = 34$ for Nb_3Sn and $\kappa = 20$ for Ti_2Nb alloys. It might also be noted that some elemental superconductors (e.g., Nb) have intrinsic coherence lengths small enough to give values of κ large enough to produce Type II behavior. Finally, it should be pointed out that the exact value of κ at which the change from Type I to Type II behavior occurs is[164] $\kappa = (1/\sqrt{2}) = 0.707$. From Table 8.2, the value[143] of κ for niobium is 1.02, and its coherence length $\xi_0 = 380$ Å.

Having discussed the distinction between Type I and Type II superconductors, we now discuss the latter in some detail,[167,168] emphasizing the structure of the mixed state. Consider a Type II superconductor (whose value of κ is large compared to unity) in an applied magnetic field; as usual, the sample is assumed to be a long thin cylinder aligned parallel to the field direction. At values of the applied field less than the lower critical field H_{c1} (see Figure 8.19), the Type II material behaves just like a Type I superconductor, exhibiting complete flux exclusion and perfect diamagnetism. At the lower critical field H_{c1}, the mixed state appears, with its mixture of normal (N) and superconducting (S) regions whose formation is favored by the negative surface energy discussed above. The structure of the mixed state is composed[169] of filaments of normal material, oriented parallel to the applied field, interspersed throughout a matrix of superconducting material.

We consider first the structure of the mixed state at an applied magnetic field slightly larger than H_{c1}. Here there are only a few filaments of normal material, and the separation between filaments is large compared to the penetration depth λ. The majority of the sample is still superconducting, with a surface supercurrent flowing around the perimeter of the specimen[168]; the magnetic field due to this supercurrent cancels the applied field in the superconducting regions of the material. Each filament may be thought of, approximately, as a cylinder of normal material with its long axis parallel to the applied magnetic field. The situation is shown in Figure 8.23. This figure[170] shows, schematically, the structure of the mixed state; the normal cylindrical regions are shaded, and the superconducting material is unshaded. Each normal filament is threaded by a magnetic field parallel to the applied magnetic field. The field in each filament is due to a supercurrent[168] circulating about each cylinder, as indicated schematically in the drawing.

The circulating supercurrent around each normal region is called a current vortex, so the mixed state is also called the vortex state. For the cylindrical picture of the normal region used in Figure 8.23, the supercurrent vortex around the normal region is also cylindrical in shape. However, the cylindrical structure described above is not really correct; the normal "core," surrounded by the current vortex does not have a sharply defined radius. The structure of a single vortex is better described by giving the spatial variation of the superconducting pair wave function ψ and of the magnetic field B threading the vortex. (The entire structure of normal core and circulating

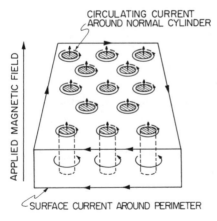

CIRCULATING CURRENT
AROUND NORMAL CYLINDER

APPLIED MAGNETIC FIELD

SURFACE CURRENT AROUND PERIMETER

Figure 8.23. Schematic view of normal (shaded) and superconducting (unshaded) regions in the mixed state of a Type II superconductor in a magnetic field slightly larger than the lower critical field H_{c1}. The circulating supercurrent around each normal cylinder produces the magnetic field parallel to the applied field, threading the normal regions. The perimeter of the specimen contains a circulating supercurrent which cancels the applied magnetic field in the superconducting portion of the sample (after Rose-Innes and Rhoderick[4]).

Figure 8.24. Approximate variation with distance
x of the pair wave function ψ and magnetic field B
in a single vortex in the mixed state of a Type II superconductor. The center of the vortex is at $x = 0$,
ξ is the coherence length, and λ is the penetration
depth. The figure is drawn for $\lambda = 5\xi$, so $\kappa = 5$.
The maximum value of B, at the center of the
vortex, is approximately $2H_{c1}$ (after Tinkham[3]).

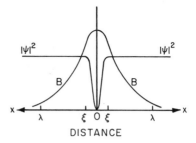

current vortex is generally referred to as a "vortex.") Figure 8.24 shows[171,172]
the approximate variation of $|\psi|^2$ and B with distance x in a single isolated
vortex. The normal core,[165] at $x = 0$, has a radius roughly equal to the coherence length ξ; in the core (which, strictly speaking, is a line), the material is
normal and $|\psi|^2 \cong 0$. Outside the core, for $x > \xi$, the material is fully super-
conducting, and $|\psi|^2 = \varrho$, the density of pairs. The magnetic field $B(x)$ de-
creases[173] with increasing distance from the core, with a characteristic
length equal to the penetration depth λ. (Figure 8.24 is drawn for $\lambda = 5\xi$, so
$\kappa = 5$.) The magnetic flux in the normal core of the vortex is quantized in
units of the flux quantum $\Phi_0 = (hc/2e)$ since we may consider the normal core
and its surrounding current vortex as a supercurrent flowing in a superconducting ring. Since the flux within the core of the vortex is Φ_0, and the field
B falls off with characteristic length λ, we expect B to have roughly[174] the
value

$$B \cong (\Phi_0/\pi\lambda^2) \tag{8.181}$$

within the normal material of the filament. A more accurate[175] result is, at
the center of the vortex,

$$B \cong (\Phi_0/2\pi\lambda^2) \ln \kappa \tag{8.182}$$

where, for typical values of κ from 5 to 35, $\ln \kappa$ ranges from 1.6 to 3.6, so
(8.181) is sufficient for our qualitative purposes. Finally, one would expect
intuitively, since the mixed state appears at the lower critical field H_{c1}, that
H_{c1} should be roughly equal to the value of the magnetic field within the
normal core of the vortex. This turns out to be approximately correct, and a
calculation of H_{c1} gives

$$H_{c1} = (\Phi_0/4\pi\lambda^2) \ln \kappa \tag{8.183}$$

so H_{c1} is approximately one-half the value given by equation (8.182) for the magnetic field at the vortex center.

As the magnetic field applied to the Type II superconductor in the mixed state is increased above H_{c1}, more flux enters the sample. Since there is only one flux quantum Φ_0 in the normal core of each vortex,[176] more flux entering the sample results in the formation of additional vortices. The separation between vortices then decreases, and the interaction[177] between the vortices must be considered. The key result is that two vortices whose fluxes are in the same direction repel each other. As the applied magnetic field on the mixed state is increased and more vortices are formed, the vortices interact and repel each other, raising the energy of the vortex state. It turns out[177,178] that a regular two-dimensional array or lattice of vortices is formed, with a triangular lattice being stable with the lowest energy. The shaded regions in Figure 8.23, representing the normal cores of the vortices, form the stable triangular lattice. While not indicated on Figure 8.23, it can readily be seen that the unit cell[179] of the vortex lattice is hexagonal and contains one vortex at its center.

As the applied magnetic field is increased further, and approaches the upper critical field H_{c2}, more flux penetrates the sample, more vortices are formed, and the distance between vortices decreases. The spacing between vortices approaches the coherence length ξ, the characteristic distance for variation of the pair wave function. Near H_{c2}, the vortices are so tightly packed that their normal cores begin to overlap and, at H_{c2}, the material is essentially full of normal electrons and the density of pairs goes to zero.[180] We may obtain an approximate value of H_{c2} as the flux quantum Φ_0 divided by $\pi\xi^2$, the "area" of the cylindrical normal core, so

$$H_{c2} \cong (\Phi_0/\pi\xi^2) \tag{8.184}$$

An accurate calculation[181] gives

$$H_{c2} \cong (\Phi_0/2\pi\xi^2) \tag{8.185}$$

again in the approximation in which $\kappa = (\lambda/\xi) \gg 1$. Physically, just below H_{c2}, we expect the mixed state to be composed of tightly-packed vortices. At the upper critical field H_{c2}, the normal cores fill the sample, which is then completely normal. Finally, we may note, from (8.183) and (8.185) that

$$(H_{c2}/H_{c1}) = 2(\lambda^2/\xi^2)(\ln\kappa)^{-1} = 2\kappa^2/(\ln\kappa) \cong \kappa^2 \tag{8.186}$$

since $2/(\ln\kappa)$ is of the order of unity for the values of κ for most Type II superconductors. Equation (8.186) tells us that superconductors with large values of κ will have high ratios of H_{c2} to H_{c1}. The upper and lower critical fields H_{c2} and H_{c1} are the physically important magnetic fields for Type II superconductors. (The thermodynamic critical field H_c defined in terms of the condensation energy by equation (8.161) is not physically significant[180] for Type II materials.) As far as applications go, our key result is that the important upper critical field H_{c2} varies, from equation (8.185), inversely with the square of the coherence length ξ. We will return to this result later in discussing superconducting materials for applications requiring high critical fields.

As the final topic in our description of the mixed or vortex state, we may now discuss[182] the origin of the Type II magnetization curve in Figure 8.19. At applied fields below H_{c1}, the material acts like a Type I superconductor, exhibiting perfect diamagnetism and a magnetization $\mathbf{M} = (-1/4\pi)\mathbf{H}$ as given by equation (8.145) with $\mathbf{B} = 0$. At an applied field equal to H_{c1}, magnetic flux penetrates the sample as vortices with normal cores form. This flux in the specimen means B is no longer zero within the sample, so from (8.145), written as

$$-4\pi\mathbf{M} = \mathbf{H} - \mathbf{B} \qquad (8.187)$$

we see that the magnitude of $(-4\pi\mathbf{M})$ begins to decrease as shown in Figure 8.19. As the applied field is increased above H_{c1}, the number of vortices and normal cores increases, the flux density B within the sample increases further, resulting in a continued decrease in the magnitude of $(-4\pi\mathbf{M})$. As the applied field approaches H_{c2}, the vortices are packed more closely together until, at H_{c2}, the magnitude of $(-4\pi\mathbf{M})$ becomes zero as the sample becomes a normal conductor with zero magnetization.

With the information we have now developed, we may discuss the magnitude of the critical field in various types of superconductors. We begin with Type I pure elemental materials, which, as we shall see, have critical fields too small to be of interest for applications. From equation (8.161),

$$F_N(0,T) - F_S(0,T) = (H_c^2/8\pi) \qquad (8.161)$$

the critical field H_c for a pure superconductor is given in terms of the difference in the free energy densities $F_N(0,T)$ and $F_S(0,T)$ of the normal and superconducting states in zero magnetic field and at temperature T. This free energy difference may be calculated[183,184] at $T = 0$ as $(1/2)N(0)[\Delta(0)]^2$, where $N(0)$ is the density of states of one spin at the Fermi energy and $\Delta(0)$ is the energy gap parameter at $T = 0$. Combining results gives

$$V[H_c(0)]^2/8\pi = (1/2)N(0)[\Delta(0)]^2 \tag{8.188}$$

where $H_c(0)$ is the critical field at absolute zero and V is the volume of the sample. Since, from equation (8.133),

$$N(0) = (3\gamma/2\pi^2 k_B^2) \tag{8.133}$$

where γ is the experimentally determined coefficient in the electronic specific heat, equation (8.188) may be rewritten in the useful form[185]

$$V[H_c(0)]^2 = 5.92\gamma T_c^2 \tag{8.189}$$

on using the weak coupling relation (8.129) between $\Delta(0)$ and T_c. Equation (8.189) tells us that, for a Type I superconductor, the critical field $H_c(0)$ is directly proportional to the transition temperature T_c. Equation (8.189) may be used[186] to estimate $H_c(0)$ for a typical Type I superconductor as being of the order of several hundred Gauss. We see that Type I materials are thus superconductors with low critical fields, and we therefore turn our attention to Type II.

As noted above, a few elements (e.g., Nb, V) have intrinsic coherence lengths short enough to make them Type II superconductors. However, their values of H_{c2} are so low (less than 2000 Gauss[187]) that they are not of interest for applications. We therefore now consider alloys and compounds that are Type II materials. We saw, in equation (8.185), that H_{c2} is given by

$$H_{c2} = (\Phi_0/2\pi\xi^2) \tag{8.185}$$

where ξ is the coherence length. For an impure superconductor, in the dirty limit in which $l \ll \xi_0$, we had the result (8.168),

$$\xi(T) = 0.86(\xi_0 l)^{1/2}/(1-t)^{1/2} \tag{8.168}$$

stating that ξ was proportional to $(\xi_0 l)^{1/2}$, where ξ_0 is the intrinsic coherence length. Combining (8.168) and (8.185) yields the important result that

$$H_{c2} \propto (\xi_0 l)^{-1} \tag{8.190}$$

Equation (8.190) may be rewritten in terms of more accessible parameters by using the BCS coherence length

$$\xi_0 = \hbar v_F / \pi \Delta (0) \tag{8.166}$$

given in equation (8.166). Since, from equation (8.129),

$$2\Delta (0) = E_g (0) = 3.52 \, k_B T_c \tag{8.129}$$

in the case of weak coupling, we can rewrite (8.166) as

$$\xi_0 = (2\hbar/3.52 k_B \pi)(v_F/T_c) = 0.18 \hbar v_F / k_B T_c \tag{8.191}$$

Combining equations (8.190) and (8.191) gives

$$H_{c2} \propto (T_c / v_F l) \tag{8.192}$$

a result[188] giving the upper critical field in terms of the solid state parameters T_c, V_F, and l, the electron mean free path in the normal state. It should be noted here that equation (8.168), and hence equation (8.192), really apply in the "dirty limit" in which the intrinsic coherence length ξ_0 is much larger than the mean free path l. While it is not certain that this condition is satisfied in all of the Type II superconductors we are about to discuss, we will still use equation (8.192) to provide qualitative explanations of their large upper critical fields.

Equation (8.192), and its rewritten form

$$(H_{c2}/T_c) \propto (v_F l)^{-1} \tag{8.193}$$

stating that H_{c2}, and (H_{c2}/T_c), are inversely proportional to both v_F and l, allow us to discuss a number of the older, lower-temperature Type II superconductors which are of present or potential interest because of their high critical fields. We will discuss[189] several classes of materials: alloys of the transition metals (such as Nb–Ti), intermetallic compounds (such as Nb_3Sn, V_3Si) with the A-15 structure, and some other materials, including the transition metal dichalcogenides (such as TaS_2) and the Chevrel compounds (such as $PbMo_6S_8$). Values of $H_{c2}(0)$, the upper critical field at 0 K, for several of these Type II superconductors are shown in Table 8.3. Values of $H_{c2}(0)$ are usually obtained by extrapolation; quoted figures are often sample-dependent and should be considered representative.

Table 8.3. Values of $-H_{c2}(0)$ in Kilogauss for Several Type II Superconductors.

$Nb_{38}Ti_{62}$	120^a	TaS_2	$150^{c,d,e}$
Nb_3Sn	240^b	$PbMo_6S_8$	540^f
V_3Si	$\sim 200^b$	$YBa_2Cu_3O_7{}^g$	above 380 at 80 K

^{a} T. H. Geballe and M. R. Beasley, Reference 100, page 519.
^{b} Values taken from J. Muller, *Rep. Prog. Phys.* **43**, 641–687 (1980), Figures 14 and 15.
^{c} Data for TaS_2 intercalated with pyridine.
^{d} R. C. Morris and R. V. Coleman, *Phys. Rev.* B **7**, 991–1001 (1973), Table I.
^{e} H_{c2} measured at 1.4 K with the applied magnetic field parallel to the plane of the layers in TaS_2.
^{f} Ø. Fischer and M. Decroux, *J. Magnetism and Magnetic Materials (Netherlands)* **11**, 164–168 (1979), Figure 2.
^{g} T. Sakakibara, Reference 206. As described in the text this value of H_{c2} is for the magnetic field aligned parallel to the copper–oxygen planes in $YBa_2Cu_3O_7$.

As pointed out earlier, in the discussion of the parameter $\kappa = (\lambda/\xi)$, one expects alloys to have short coherence lengths ξ, short mean free paths l, values of κ greater than unity, and hence in general to be Type II superconductors. This is illustrated[190] in the change of lead from a Type I superconductor to Type II by the addition of several weight percent of indium. The decreased coherence length in the Type II Pb–In alloy results also in an increase of the critical field. The same effect is seen in the technologically important alloys of niobium and titanium; the alloy $Nb_{37}Ti_{63}$ is presently[191] a very widely used material for magnets; it is believed to be a body-centered cubic structure. The separate components have modest critical fields: $H_c = 1980$ Gauss for Nb and $H_c = 100$ Gauss for Ti, both values[187] at 0 K, and the intrinsic coherence length $\xi_0 = 400$ angstroms for Nb.[192] On the other hand, values of $H_{c2} = 120,000$ Gauss are reported[191] for Nb–Ti alloys, for which the coherence length $\xi = 30$–50 angstroms and the mean free path[193] $l = 10$ angstroms. The alloying process presumably decreases the mean free path drastically relative to the pure components, thereby resulting, via equation (8.168), in a small coherence length and a large value of the upper critical field. Other alloy systems of importance[194] because of their high critical fields include Nb–Zr and Mo–Re.

The intermetallic compounds with the A-15 structure are important because of their high critical fields[195] even though their values of T_c are low by today's standards. All of these compounds are Type II superconductors with the crystal structure shown in Figure 8.25. (For Nb_3Sn, as given in Table 8.2, $\kappa = 34$.) The structure for A_3B shows the three nonintersecting chains of A (for example, Nb in Nb_3Sn) atoms in the crystal. Their high values of H_{c2}

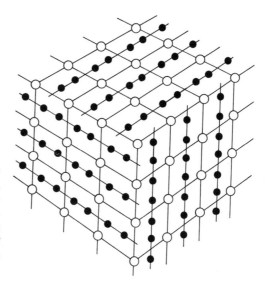

Figure 8.25. Schematic crystal structure of A_3B in the A15 structure, showing the three orthogonal, nonintersecting chains of A atoms (●). (This drawing emphasizes the chain structure, and does not show the B atom at the center of each cubic unit cell.) (After White and Geballe[77].)

certainly reflect a very small value of the coherence length, with values of approximately 30 angstroms being given[192] for V_3Si, Nb_3Sn, and Nb_3Ge. It is not, however, clear on a microscopic basis why the coherence length is so short. It has been proposed[193] that the reason is that superconductivity in the A15 compounds is due to the d-electrons of the A atoms located on the nonintersecting linear chains. These electrons would then be quite localized, leading to a short mean free path and a small coherence length. However, this essentially one-dimensional model might be expected[196] to lead to considerable anisotropy of H_{c2}, an anisotropy which has not been observed. At the moment, then, the question remains under discussion.

Both the transition metal alloys and the A15 intermetallic compounds just discussed are presently important in applications involving superconducting magnets. The remaining two classes of materials, the layer compounds (e.g., TaS_2) and the Chevrel compounds (e.g., $PbMo_6S_8$), are not now technologically important, but merit discussion because of their high values of H_{c2} and their interesting physics.

A number of transition metal dichalcogenides (e.g., TaS_2, $NbSe_2$) have a crystal structure composed of layers of atoms, as shown schematically in Figure 8.26. As such, they are essentially two-dimensional, with rather weak forces between the planes, leading to very anisotropic electronic properties.[197] These materials are Type II superconductors with rather low ($T_c \leq 4$ K) critical temperatures, but very high (about 150 kGauss) values of H_{c2} for magnetic

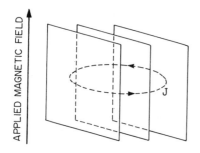

APPLIED MAGNETIC FIELD

Figure 8.26. Schematic view of the structure of a layer compound transition metal dichalcogenide, such as TaS_2. The circle labeled J is the supercurrent in a plane perpendicular to both the applied magnetic field and the plane of the layers; the magnetic field is parallel to the plane of the layers.

fields oriented parallel to the plane of the layers, as shown in Figure 8.26. It has been suggested[193] that, for the applied field parallel to the layers, the coherence length will be very small. The reason is that the screening supercurrent J, which is in a plane normal to the applied field, will, in this configuration, be in a plane normal to the layers, as shown schematically in Figure 8.26, leading to a small mean free path and a small coherence length. Of further interest is the fact that organic molecules (e.g., pyridine) may be intercalated between the layers of these compounds, increasing the interlayer distance while preserving the superconductivity. In such cases,[197] parallel critical fields in the neighborhood of 150 kGauss have been observed. However, in spite of these high values of H_{c2}, their low critical temperatures have kept these materials from being of interest for applications.

Next, we discuss the Chevrel phases,[198–200] which are compounds of the type MMo_6X_8, where M is a metal (e.g., Pb, Sn) and X is either S, Se, or Te. A well-studied member of this group is $PbMo_6S_8$, with a T_c of 12 to 15 K. Their structure contains octahedrons of six molybdenum atoms, and the Mo_6 octahedrons are surrounded by a cube of eight chalcogen atoms.[201] While it is probably not true that the criterion $\xi_0 \gg l$ for the "dirty limit" is satisfied by $PbMo_6S_8$ (the values[202] reported are $\xi_0 \cong 50$ Angstroms and $l \cong 23$ Angstroms), this compound may be thought of, approximately, as a dirty Type II superconductor. The reasons for the enormous critical fields (approaching 600 kGauss in some cases) of the Chevrel compounds are not yet clear. However, it has been suggested[193,199] that the Mo_6 cluster structure leads to low dimensionality and small values of the mean free path l and coherence length ξ. Further, band structure calculations suggest a narrow d-band at the Fermi surface; the narrow band implies a small Fermi velocity V_F. The combination of a small v_F and a small l lead, from equation (8.192), to a high upper critical field. Again, however, the subject is still under discussion.[203]

Finally, we briefly discuss the critical field in $YBa_2Cu_3O_7$, whose value of T_c is about 93 K. This new superconductor is strongly Type II, with (as listed in Table 8.2) the parameter $\kappa \cong 60$–100, possibly the largest value reported to date. As seen in Figure 8.17, the crystal structure of $YBa_2Cu_3O_7$ is quite anisotropic, with the superconducting carriers believed to be those in the copper–oxygen planes. The BCS coherence length is also thought to be very anisotropic[204,205]; at 0 K, $\xi_{\|}(0) \cong 30$ Å for carriers in the copper–oxygen planes. This is the approximate spatial extent of a Cooper pair composed of carriers located in a copper–oxygen plane. However, the coherence length $\xi_{\perp}(0)$ for Cooper pairs in the direction perpendicular to the copper–oxygen planes is about 4 Å.[204,205] Note that $\xi_{\perp}(0)$ is considerably smaller than the distance $c \cong 11.8$ Å, in the $YBa_2Cu_3O_7$ unit cell, normal to the copper–oxygen planes. These results suggest that $H_{c2\perp}$, the upper critical field with the applied field perpendicular to the copper–oxygen planes, and in which the supercurrents flow *in* the copper–oxygen planes, will be given, from equation (8.185), by[205]

$$H_{c2\perp} = \Phi_0/2\pi\,[\xi_{\|}(0)]^2 \qquad (8.194)$$

Similarly, the upper critical field $H_{c2\|}$, when the applied field is parallel to the copper–oxygen planes, is given by

$$H_{c2\|} = \Phi_0/2\pi\,[\xi_{\perp}(0)]^2 \qquad (8.195)$$

This would be the same situation as that indicated schematically in Figure 8.26 for the transition metal dichalcogenide layer compounds. This picture is in agreement with the observed[206] anisotropic upper critical field in $YBa_2Cu_3O_7$. With the applied magnetic field perpendicular to the c-axis, and hence parallel to the copper–oxygen planes, superconductivity at 80 K could not be destroyed by an applied field of 380,000 Gauss, so $H_{c2\|}$ at 80 K is larger than this value. It was also found[206] that $H_{c2\|}$ is larger than $H_{c2\perp}$ at all temperatures (down to about 60 K) at which $H_{c2\perp}$ was measured. It appears clear that the small coherence length $\xi_{\perp}(0)$ in $YBa_2Cu_3O_7$ leads to an upper critical field $H_{c2\|}$ of unparalleled magnitude.

We end the discussion of superconducting materials with a few words on the critical current density[207–212] in superconductors. It is observed experimentally that electrical resistance appears for current densities larger than a critical value J_c; the magnitude of J_c is different for different materials. In a homogeneous Type I superconductor with dimensions larger than the penetration depth, and with no external magnetic field applied, the critical current density J_c is the transport supercurrent density which produces the critical magnetic field H_c at the surface, outside the superconductor. If there is also an external applied magnetic field, the critical field H_c will be obtained at the superconductor surface for smaller values of the transport supercurrent in the sample. The critical current density J_c will therefore decrease as the magnetic field applied to the sample is increased. Further, the observed value of J_c for a given specimen may depend on sample geometry and metallurgical structure and so can be less than the value determined by H_c at the surface. In any event, since Type I superconductors are not of great technological interest because of their small critical fields, we move on to consider briefly the critical current density in Type II superconductors. Here the situation is more complex because of the vortex structure of the mixed state existing at applied fields between H_{c1} and H_{c2}. If there is a transport supercurrent flowing, a Lorentz force will be exerted on the moving charged carriers by the magnetic field threading the normal cores of the vortices. By Newton's Third Law, a force of equal magnitude will be exerted on the vortices, producing a vortex motion usually called flux flow. This dissipative motion results in the appearance of electrical resistance when the Type II superconductor carries a current. However, suppose that the flux flow motion of the vortices is prevented by some mechanism which opposes the Lorentz force. Usually called "pinning," such mechanisms localize the vortices at particular locations, such as grain boundaries and dislocations, in the sample. Pinning of vortices thus prevents flux flow and the appearance of electrical resistance, at current densities up to some critical value J_c, called the critical current density of the Type II superconductor. The stronger the force pinning the vortices, the larger the value of the critical current density. It is clear that, for applications such as superconducting solenoids, a high value of J_c is necessary. As an example,[213] for Nb$_3$Sn, $J_c \cong 10^5$ Ampères per cm^2 in a magnetic field of 150 kiloGauss; as might be expected, the critical current density decreases as the magnetic field on the superconductor increases. Research on the critical currents of the new high-temperature superconductors is just beginning, and obtaining wires of these materials with usefully high values of J_c will be one of the main challenges of the field.

Problems

8.1. *DC Josephson Effect Equations with Reverse Flow.* In the derivation of equation (8.32), the effect on ψ_1 of "reverse flow" tunneling from side (1) to side (2) was neglected. The same is true for tunneling from side (2) to side (1) in equation (8.33). To explore this question, add a term $-\hbar T\psi_1$ to equation (8.32) and a term

$-\hbar T\psi_2$ to equation (8.33). Proceed with the development and obtain the equivalents of equations (8.40)–(8.43). Comment on the physical meaning and reasonableness of your results. (The author thanks B. Black for suggesting this problem.)

8.2. *Voltage Measurements with a DC SQUID.* (a) Ignoring any capacitance, consider measurement of the voltage V in the circuit of Figure 8.12 for a situation in which the ratio (M/R) is about 1 second. Estimate the voltage sensitivity of measurements with this SQUID circuit. (b) Compare your result in (a) with the Nyquist noise voltage (see, for example, C. Kittel, Reference 131, page 403) at 4 K of the resistance R in the circuit of Figure 8.12. (For appropriate values of the parameters in the expression for the Nyquist noise voltage, see J. Clarke, Reference 20, page 35.) This comparison shows that the voltage sensitivity of the SQUID voltmeter is limited by noise.

8.3. *Three-Terminal Superconducting Devices.* There exists a class of *three*-terminal superconducting devices roughly analogous to a bipolar transistor. (One such device is the "quiteron.") While it is not year clear whether this type of device will be important, it is an interesting application of superconductivity. Discuss the fundamental physics of this device by consulting, for example, the review article by W.J. Gallagher, "Three Terminal Superconducting Devices," *IEEE Transactions on Magnetics* MAG-21, 709–716 (March 1985) and the further references therein.

8.4. *Magnitude of Coulomb Term in the McMillan Equation.* Using representative values of the solid state parameters in equation (8.139) defining μ^*, show that μ^* is of the order of 0.1 for a typical metal.

8.5. *Critical Fields for Type I Superconductors.* Using equation (8.189), calculate the critical field $H_c(0)$ for (a) tin ($\kappa \cong 0.15$) and (b) lead ($\kappa \cong 0.45$) which has perhaps the highest value of T_c for a Type I superconductor. (The required values of the material parameters, except for the molar volumes, may be found in Reference 1.) Compare your calculated values of $H_c(0)$ with the experimental values given in Reference 1.

References and Comments

1. C. Kittel, *Introduction to Solid State Physics*, Sixth Edition, John Wiley, New York (1986), Chapter 12.
2. N. W. Ashcroft and N. D. Mermin, *Solid State Physics*, Holt, Rinehart, and Winston, New York (1976), Chapter 34.
3. M. Tinkham, *Introduction to Superconductivity*, McGraw-Hill, New York (1975), Chapter 1.
4. A. C. Rose-Innes and E. H. Rhoderick, *Introduction to Superconductivity*, Second Edition, Pergamon Press, Oxford (1978).
5. R. P. Feynman, R. B. Leighton, and M. Sands, *The Feynman Lectures on Physics*, Addison-Wesley, Reading, Mass. (1965), Volume III, Sections 21-4, 21-5.

6. For an introductory discussion of a microscopic quantum mechanical treatment of the Josephson equations, see J. Clarke, "The Josephson Effect and e/h," *Amer. J. Phys.*, **38**, 1071-1095 (1970), Section II.1.

7. L. I. Schiff, *Quantum Mechanics*, Third Edition, McGraw-Hill, New York (1968).

8. R. P. Feynman, R. B. Leighton, and M. Sands, Reference 5, Sections 21-3, 21-5.

9. D. Bohm, *Quantum Theory*, Prentice-Hall, New York (1951), Sections 4.9, 6.6, 6.7, 6.8.

10. J. Clarke, Reference 6, page 1073.

11. C. Kittel, Reference 1, pages 350–351.

12. R. P. Feynman, R. B. Leighton, and M. Sands, Reference 5, Section 21-9.

13. L. Solymar, *Superconductive Tunneling and Applications*, John Wiley, New York (1972), Chapters 8–10.

14. L. I. Schiff, Reference 7, pages 101–105.

15. B. D. Josephson, "Superconductive Tunneling," in *Superconductivity in Science and Technology*, M. H. Cohen (editor), University of Chicago Press, Chicago (1968), page 20.

16. J. Clarke, private communication.

17. See, for example, M. Tinkham, Reference 3, page 194.

18. J. E. Mercereau, "Macroscopic Quantum Phenomena," in *Superconductivity*, R. D. Parks (editor), Marcel Dekker, New York (1969), Volume 1, Chapter 8, page 400.

19. J. Clarke, "Josephson Junction Detectors," *Science*, **184**, 1235–1242 (1974).

20. J. Clarke, "Electronics with Superconducting Junctions," *Physics Today*, **24**, 30–37 (August 1971).

21. J. Clarke, private communication.

22. M. Tinkham, Reference 3, pages 204–205.

23. J. Clarke, Reference 20, page 31.

24. J. E. Mercereau, "Superconductivity," in *Topics in Solid State and Quantum Electronics*, W. D. Hershberger (editor), John Wiley, New York (1972), page 232; Reference 18, page 402.

25. L. Solymar, Reference 13, page 153; M. Tinkham, Reference 3, pages 195–196.

26. L. Solymar, Reference 13, page 256.

27. M. Tinkham, Reference 3, pages 194–195.

28. J. Clarke, Reference 6, Section II.2.

29. C. Kittel, Reference 1, pages 348–350.

30. L. Solymar, Reference 13, Chapter 11.

31. L. Solymar, Reference 13, Chapters 8, 15.

32. P. L. Richards, "The Josephson Junction as a Detector of Microwave and Far Infrared Radiation," in *Semiconductors and Semimetals*, R. K. Willardson and A. C. Beer (editors), Academic Press, New York (1977), Volume 12, Chapter 6.

33. J. Clarke, Reference 6, Section II.3, pages 1078–1080.

34. J. E. Mercereau, Reference 24, Chapter 5, pages 235–236.

35. L. Solymar, Reference 13, pages 157–164.

36. L. Solymar, Reference 13, Chapter 17, pages 271–275.

37. See, for example, F. B. Hildebrand, *Advanced Calculus for Applications*, Prentice-Hall, New York (1962), pages 226–231.

38. P. L. Richards, Reference 32, page 398.

39. P. L. Richards, Reference 32, page 398; J. E. Mercereau, Reference 18, pages 402, 404.

40. L. Solymar, Reference 13, pages 159–160.

41. J. Clarke, Reference 6, Figure 5.

42. See, for example, E. Butkov, *Mathematical Physics*, Addison-Wesley, Reading, Mass. (1968), Figure 9.2, page 364; E. Jahnke and F. Emde, *Tables of Functions*, Dover, New York (1943), pages 156–157.

43. F. B. Hildebrand, Reference 37, page 144, equation (70).

44. A. H. Silver and J. E. Zimmerman, "Josephson Weak Link Devices," in *Applied Superconductivity*, V. L. Newhouse (editor), Academic Press, New York (1975), Volume 1, Chapter 1, pages 89–96.

45. See, for example, C. Kittel, Reference 1, pages 340–342.

46. See, for example, C. Kittel, *Introduction to Solid State Physics*, Fourth Edition, John Wiley, New York (1971), Advanced Topic I, pages 727–731.

47. A. C. Rose-Innes and E. H. Rhoderick, Reference 4, pages 22 and 36; N. W. Ashcroft and N. D. Mermin, Reference 2, page 739.

48. C. Kittel, Reference 1, pages 353–354.

49. J. E. Mercereau, Reference 18, Chapter 8.

50. R. C. Jaklevic, J. Lambe, J. E. Mercereau, and A. H. Silver, "Macroscopic Quantum Interference in Superconductors," *Phys. Rev.*, **140**, A1628–A1637 (1965).

51. J. Clarke, "Low Frequency Applications of Superconducting Quantum Interference Devices," *Proceedings IEEE*, **61**, 8–19 (1973), Section III.

52. L. Solymar, Reference 13, page 201, equation (13.6).

53. J. Clarke, "Squids, Brains, and Gravity Waves," *Physics Today*, **39**, 36–44 (March 1986).

54. H. D. Hahlbohm and H. Lubbig, *SQUID '85: Superconducting Quantum Interference Devices and Their Applications*, de Gruyter, Berlin (1985).

55. R. C. Jaklevic *et al.*, Reference 50, page A1631.

56. See J. Clarke, Reference 51, for a discussion.

57. J. Clarke, Reference 51, Section IV.

58. L. Solymar, Reference 13, Chapter 18.

59. J. Clarke, "Superconducting Quantum Interference Devices for Low Frequency Measurements," in *Superconductor Applications: Squids and Machines (1977)*, B. B. Schwartz and S. Foner (editors), Plenum Publishing, New York (1978), Chapter 3, pages 67–124.

60. J. Clarke, "The Application of Josephson Junctions to Computer Storage and Logic Elements and to Magnetic Measurements," in *Magnetism and Magnetic Materials—1975*, J. J. Becker, G.H.Lander, and J. J.Rhyne (editors), Conference Proceedings No. 29, American Institute of Physics, New York (1975), pages 20–21; W. Anacker, "Computing at 4 K," *IEEE Spectrum*, **16**, 26 (May 1979); H. Hayakawa, "Josephson Computer Technology," *Physics Today*, **39**, 46–52 (March 1986).

61. L. Solymar, Reference 13, Chapters 14, 16, 19.

62. A. H. Silver and J. E. Zimmerman, Reference 44, pages 67–89; 96–106.

63. *Future Trends in Superconductive Electronics (Charlottesville, 1978)*, B. S. Deaver, Jr., C. M. Falco, J. H. Harris, and S. A. Wolf (editors), Conference Proceedings No. 44, American Institute of Physics, New York (1978).

64. C. Kittel, Reference 1, pages 338–340.

65. N. W. Ashcroft and N. D. Mermin, Reference 2, pages 739–746.

66. A. C. Rose-Innes and E. H. Rhoderick, Reference 4, Chapter 9.

67. D. R. Tilley and J. Tilley, *Superfluidity and Superconductivity*, Second Edition, Adam Hilger, Bristol (1986), Chapter 4.

68. M. Tinkham, Reference 3, Chapter 2.

69. See, for example, S. S. P. Parkin *et. al.*, "Bulk Superconductivity at 125 K in $Tl_2Ca_2Ba_2Cu_3O_x$," *Phys. Rev. Lett.*, **60**, 2539–2542 (1988), for one of a number of reports on superconductivity in this temperature range.

70. N. W. Ashcroft and N. D. Mermin, Reference 2, page 743; M. Tinkham, Reference 3, pages 17–21.

71. C. Kittel, *Quantum Theory of Solide*, John Wiley, New York (1963), page 162.

72. R. Meservey and B. B. Schwartz, "Equilibrium Properties: Comparison of Experimental Results with Predictions of BCS Theory," in *Superconductivity*, R. D. Parks (editor), Marcel Dekker, New York (1969), Volume 1, pages 117–191, Table I, page 122.

73. C. Kittel, Reference 1, page 142, equation (44) with an effective mass m^* in place of the free electron mass m.

74. Calculated by the author.

75. G. Gladstone, M. A. Jensen, and J. R. Schrieffer, "Superconductivity in the Transition Metals: Theory and Experiment," in *Superconductivity*, R. D. Parks (editor), Marcel Dekker, New York (1969), pages 665–816; Table VI, page 734.

76. C. Kittel, Reference 1, pages 135–139, equation (34).

77. R. M. White and T. H. Geballe, *Long Range Order in Solids*, Supplement 15 to *Solid State Physics*, H. Ehrenreich, F. Seitz, and D. Turnbull (editors), Academic Press, New York (1979), page 112, equation (3.46).

78. G. Gladstone *et al.*, Reference 75, pages 682–685.

79. R. M. White and T. H. Geballe, Reference 77, page 221, give $T_c = 0.026$ K for bulk beryllium at atmospheric pressure.

80. R. M. White and T. H. Geballe, Reference 77, front end papers.

81. G. Gladstone *et al.*, Reference 75, page 736, Figure 30.

82. B. T. Matthias, "Superconductivity in the Periodic System," *Progress in Low Temperature Physics*, C. J. Gorter (editor), North-Holland, Amsterdam (1957), Volume II, pages 138–150; Figure 2, page 140.

83. B. T. Matthias, "The Empirical Approach to Superconductivity," in *Applied Solid State Physics*, W. Low and M. Schieber (editors), Plenum Press, New York (1970), pages 179–188; Figure 2, page 184.

84. D. Dew-Hughes, "Practical Superconducting Materials," in *Superconducting Machines and Devices*, S. Foner and B. B. Schwartz (editors), Plenum Press, New York (1974), Chapter 2, pages 91–92.

85. R. M. White and T. H. Geballe, Reference 77, pages 220–246.

86. R. M. White and T. H. Geballe, Reference 77, pages 104–113.

86a. W. L. McMillan, "Transition Temperature of Strong-Coupled Superconductors," *Phys. Rev.*, **167**, 331–334 (1968).

87. R. M. White and T. H. Geballe, Reference 77, pages 92–103; Table VI.3, page 228.

88. D. J. Scalapino, "The Electron–Phonon Interaction and Strong Coupling Superconductors," in *Superconductivity*, R. D. Parks (editor), Marcel Dekker, New York (1969), Volume 1, pages 449–560; equation (135), page 542.

89. R. M. White and T. H. Geballe, Reference 77, page 228.

90. R. M. White and T. H. Geballe, Reference 77, pages 112, 226, and 228.

91. R. M. White and T. H. Geballe, Reference 77, pages 93–98.

92. P. B. Allen and B. Mitrovic, "Theory of Superconducting T_c," in *Solid State Physics*, D. Turnbull and H. Ehrenreich (editors), **37**, 1–92, Academic Press, New York (1982), especially pages 50–56 and Figure 8.

93. P. B. Allen and R. C. Dynes, "Transition Temperatures of Strong-Coupled Supercon-ductors Reanalyzed," *Phys. Rev. B*, **12**, 905–922 (1975).

94. R. M. White and T. H. Geballe, Reference 77, pages 118–121.

95. See, for example, K. M. Ho, M. L. Cohen, and W. E. Pickett, "Maximum Supercon-ducting Transition Temperature in A15 Compounds?," *Phys. Rev. Lett.*, **41**, 815–818 (1978).

96. C. M. Varma, "What Limits T_c?," in *Superconductivity in d- and f-Band Metals, 1982*, W. Buckel and W. Weber (editors), Kernforschungszentrum Karlsruhe GmbH, Karlsruhe (1982), pages 603–613.

97. P. B. Allen and B. Mitrovic, Reference 92, pages 88–89.

98. See, for example, J. Muller, "A-15 Superconductors," *Rep. Prog. Phys.*, **43**, 641–687 (1980).

99. M. R. Beasley and T. H. Geballe, "Superconducting Materials," *Physics Today*, **37**, 60–68 (October 1984), pages 63–65.

100. T. H. Geballe and M. R. Beasley, "Superconducting Materials for Energy-Related Applications," in *Materials Science in Energy Technology*, G. G. Libowitz and M. S. Whittingham (editors), Academic Press, Orlando (1979), pages 491–550 and especially pages 520–533.

101. M. N. Wilson, *Superconducting Magnets*, Oxford University Press, Oxford (1983), Chapter 12.

102. R. M. White and T. H. Geballe, Reference 77, page 234.

103. P. B. Allen and B. Mitrovic, Reference 92, Table I, page 3.

104. See, for example, T. H. Geballe, "The Science of Useful Superconductors—and Beyond," *IEEE Trans. Magnetics*, **MAG-19**, 1300–1307 (May 1983), especially page 1304.

105. J. Muller, Reference 98, pages 661–663.

106. R. M. White and T. H. Geballe, Reference 77, pages 234–243.

107. C. K. N. Patel and R. C. Dynes, "Toward Room Temperature Superconductivity?," *Proc. Natl. Acad. Sci. USA*, **85**, 4945-4952 (July 1988).

108. M. C. Tinkham and C. J. Lobb. "Physical Properties of the New Superconductors," in *Solid State Physics*, D. Turnbull, and H. Ehrenreich (editors), **42**, 91–134, Academic Press, San Diego (1989).

109. K. Fitzgerald, "Superconductivity: Fact vs. Fancy," *IEEE Spectrum*, **25**, 30–41 (May 1988).

110. P. Marsh *et al.*, "Anharmonic Thermal Motion in the 93 K Superconductor $YBa_2Cu_3O_7$ using Multiple Wavelength X-ray Diffraction," *Phys. Rev. B*, **38**, 874–877 (1988).

111. See, for example, K. C. Hass, "Electronic Structure of Copper Oxide Superconductors," *Solid State Physics*, D. Turnbull and H. Ehrenreich (editors), **42**, 213–270, Academic Press, San Diego (1989), page 221.

112. R. Meservey and B. B. Schwartz, Reference 72, page 126, Table II.

113. S. Hoen *et al.*, "Oxygen Isotope Study of $YBa_2Cu_3O_7$," *Phys. Rev. B*, **39**, 2269–2278 (1989).

114. See, for example, *Novel Superconductivity*, S. A. Wolff and V. Z. Kresin (editors), Plenum Press, New York (1987), for the situation as of June 1987.

115. J. H. Miller, Jr., "The Physics of High Temperature Superconductivity," in *High Tempera-ture Superconducting Materials*, W. E. Hatfield and J. H. Miller, Jr. (editors), Marcel Dekker, New York (1988), pages 79–97.

116. M. Tinkham and C. J. Lobb, Reference 108, pages 125, 128, 131.

116a. W. A. Little, "Experimental Constraints on Theories of High Transition Temperature Superconductors," *Science*, **242**, 1390–1395 (9 December 1988).

116b. M. Lee *et al.*, "Electron Tunneling and the Energy Gap in $Bi_2Sr_2CaCu_2O_x$," *Phys. Rev. B*, **39**, 801–803 (1989), Table 1.

117. S. S. P. Parkin *et. al.*, "Model Family of High-Temperature Superconductors," *Phys. Rev. B*, **38**, 6531–6537 (1988).

118. C. Kittel, Reference 1, page 322.

119. N. W. Ashcroft and N. D. Mermin, Reference 2, pages 732–734.

120. A. C. Rose-Innes and E. H. Rhoderick, Reference 4, pages 40–53.

121. M. Tinkham, Reference 3, Chapter 1.

122. D. Larbalestier *et al.*, "High Field Superconductivity," *Physics Today*, **39**, 24–33 (March 1986).

123. M. N. Wilson, Reference 101, Chapter 12.

124. C. Kittel, Reference 1, pages 322–326, 344–348.

125. A. C. Rose-Innes and E. H. Rhoderick, Reference 4, Chapter 12.

126. In the CGS system, the magnetic induction **B** is related to the magnetic intensity **H** by the equation $\mathbf{B} = \mathbf{H} + 4\pi\mathbf{M}$, where **M** is the magnetic dipole moment per unit volume (the magnetization). In vacuum, $\mathbf{M} = 0$ and $\mathbf{B} = \mathbf{H}$. Following Kittel (Reference 1, page 317), we will denote the applied magnetic field by $\mathbf{B}_a$. The critical magnetic field would then be $\mathbf{B}_{ac}$, but it is conventional in superconductivity to use the symbol $\mathbf{H}_c$ for that quantity. We will therefore follow the usage that $\mathbf{H}_c = \mathbf{B}_{ac}$.

127. A. C. Rose-Innes and E. H. Rhoderick, Reference 4, page 22; see also J. R. Reitz, F. J. Milford, and R. W. Christy, *Foundations of Electromagnetic Theory*, Third Edition, Addison-Wesley, Reading, Mass. (1979), pages 321–325.

128. A. C. Rose-Innes and E. H. Rhoderick, Reference 4, pages 64–67.

129. C. Kittel and H. Kroemer, *Thermal Physics*, Second Edition, W. H. Freeman, San Francisco (1980), pages 252–256.

130. C. Kittel, Reference 1, pages 331–333.

131. C. Kittel, *Thermal Physics*, First Edition, John Wiley, New York (1969), Chapter 23, pages 370–374.

132. F. Reif, *Fundamentals of Statistical and Thermal Physics*, McGraw-Hill, New York (1965), pages 439–444, 455–459.

133. C. Kittel, Reference 1, page 417, Figure 11.

134. See, however, C. Kittel, Reference 1, page 333, Footnote 6, for a comment on the validity of this assumption for Type II superconductors in high fields.

135. J. Callaway, *Quantum Theory of the Solid State*, Part B, Academic Press, New York (1974), page 688.

136. M. Tinkham, Reference 3, page 2.

137. C. Kittel, *Introduction to Solid State Physics*, Fifth Edition, John Wiley, New York (1976), page 371.

138. J. K. Hulm and B. T. Matthias, "High-Field High-Current Superconductors," *Science*, **208**, 881–887 (May 23, 1980).

139. J. K. Hulm, J. E. Kunzler, and B. T. Matthias, "The Road to Superconducting Materials," *Physics Today*, **34**, 34–43 (January 1981).

140. C. Kittel, Reference 1, pages 333–336; 340–342.

141. N. W. Ashcroft and N. D. Mermin, Reference 2, pages 737–739.

142. A. C. Rose-Innes and E. H. Rhoderick, Reference 4, Chapter 3.

143. C. Kittel, Reference 1, page 337, Table 5.

144. M. Tinkham, Reference 3, page 65, equation (2-114).

145. R. M. White and T. H. Geballe, Reference 77, page 63.

146. N. W. Ashcroft and N. D. Mermin, Reference 2, page 742.

147. M. Tinkham, Reference 3, page 19.

148. T. H. Geballe and M. Beasley, Reference 100, page 500.

149. R. Meservey and B. B. Schwartz, Reference 72, page 174, Table VI.

150. This coherence length should be kept distinct from the electromagnetic or Pippard coherence length ξ_P which characterizes the range of nonlocal electromagnetic effects in superconductors. See Reference 145 and also M. Tinkham, Reference 3, page 112.

151. The wave function ψ is also the order parameter in the Ginzburg–Landau theory. See N. W. Ashcroft and N. D. Mermin, Reference 2, pages 747–749, for a brief introduction, and M. Tinkham, Reference 3, Chapter 4, for a complete discussion.

152. M. Tinkham, Reference 3, page 111.

153. A. L. Fetter and P. C. Hohenberg, "Theory of Type II Superconductors," in *Superconductivity*, R. D. Parks (editor), Marcel Dekker, New York (1969), page 821.

154. M. Tinkham, Reference 3, pages 112–113.

155. M. Tinkham, Reference 3, page 7 and pages 67–68.

156. R. M. White and T. H. Geballe, Reference 77, page 58.

157. M. Tinkham, Reference 3, page 113, equations (4-26a) and (4-26b).

158. T. H. Geballe and M. R. Beasley, Reference 100, page 502, equations (11).

159. R. M. White and T. H. Geballe, Reference 77, page 62.

160. To be really correct, we should use the Gibbs free energy to calculate the surface energy. See, for example, M. Tinkham, Reference 3, pages 114–116, and A. L. Fetter and P. C. Hohenberg, Reference 153, pages 824–827. However, since our purpose is merely to obtain a qualitative and plausible picture of the connection between ξ, λ, and the surface energy, the use of the Helmholtz free energy will be taken as sufficient.

161. M. Tinkham, Reference 3, page 93.

162. A. C. Rose-Innes and E. H. Rhoderick, Reference 4, page 80.

163. For simplicity, and because our main interest is in Type II materials, we neglect the possibility of the intermediate state in Type I superconductors. For a discussion, see A. C. Rose-Innes and E. H. Rhoderick, Reference 4, Chapter 6, and M. Tinkham, Reference 3, Chapter 3.

164. M. Tinkham, Reference 3, page 11.

165. T. H. Geballe and M. R. Beasley, Reference 100, page 503.

166. A. C. Rose-Innes and E. H. Rhoderick, Reference 4, page 195.

167. M. Tinkham, Reference 3, pages 143–157.

168. A. C. Rose-Innes and E. H. Rhoderick, Reference 4, Chapter 12.

169. See P. G. de Gennes, *Superconductivity of Metals and Alloys*, W. A. Benjamin, New York (1966), pages 71–74, for a proof that the filamentary structure has a lower free energy than a structure composed of laminar sheets of normal material.

170. Figure 8.23 is adapted with permission from Figure 12.3 on page 187 of Reference 4 by A. C. Rose-Innes and E. H. Rhoderick.

171. T. H. Geballe and M. R. Beasley, Reference 100, page 499, Figure 3(b).

172. M. Tinkham, Reference 3, page 146, Figure 5.1.

173. M. Tinkham, Reference 3, page 147, equations 5-14, 5-14(a), and 5-14(b).

174. The detailed variation of the magnetic field (in the vortex) with distance is given by Tinkham, Reference 173, above.

175. This conclusion may be obtained from equation (5-14b) of Reference 173, evaluated at $r = \xi$. The resulting value of $h(\xi)$, the microscopic magnetic flux density at $r = \xi$, is then set approximately equal to $h(0)$, at the center $r = 0$ of the vortex.

176. P. D. de Gennes, Reference 169, pages 59 and 65.

177. M. Tinkham, Reference 3, pages 149–150; P. D. de Gennes, Reference 169, pages 63–65, 80–81.

178. A. L. Fetter and P. C. Hohenberg, Reference 153, pages 840 and 846–848.

179. T. H. Geballe and M. R. Beasley, Reference 100, page 499, Figure 3(a).

180. T. H. Geballe and M. R. Beasley, Reference 100, page 504.

181. M. Tinkham, Reference 3, page 129.

182. A. C. Rose-Innes and E. H. Rhoderick, Reference 4, Section 12.5.

183. J. Callaway, Reference 135, page 671, equation (7.8.36).

184. G. Rickayzen, "The Theory of Bardeen, Cooper, and Schrieffer," in *Superconductivity*, R. D. Parks (editor), Marcel Dekker, New York (1969), pages 77–78.

185. T. H. Geballe and M. R. Beasley, Reference 100, page 498, equation (3).

186. See Problem 8.5.

187. C. Kittel, Reference 1, page 320, Table 1.

188. M. Tinkham, Reference 3, page 157, equation (5-45).

189. The discussion will be from the point of view of the parameters in equation (8.195) and will therefore neglect the important effect of the metallurgical or structural state of the material. For a discussion of these important questions, see T. H. Geballe and M. R. Beasley, Reference 100, R. M. Scanlan, *Ann. Rev. Mat. Sci.*, **10**, 113 (1980), and D. C. Larbalestier, "Towards a Microstructural Description of Superconducting Properties," *IEEE Trans. Magnetics*, **MAG-21**, 257–264 (1985).

190. C. Kittel, Reference 1, page 325, Figure 5a.

191. T. H. Geballe and M. R. Beasley, Reference 100, page 519.

192. M. R. Beasley, "Improved Materials for Superconducting Electronics," in Reference 63, page 391, Table I.

193. Ø. Fischer, "High Field Superconductors," in *Bull. Eur. Phys. Soc.*, **7**, 1–4 (1976).

194. T. H. Geballe and M. R. Beasley, Reference 100, pages 514–520.

195. J. Muller, *Rep. Prog. Phys.*, **43**, 641–687 (1980), Figures 14 and 15, page 666.

196. S. Foner, "High Critical Field Superconductors," in *Superconductivity in d- and f-Band Metals*, D. H. Douglass (editor), Plenum Press, New York (1976), pages 161–172.

197. See, for example, R. C. Morris and R. V. Coleman, "Anisotropic Superconductivity in Layer Compounds," *Phys. Rev. B*, **7**, 991–1001 (1973).

198. Ø. Fischer, "Chevrel Phases: Superconducting and Normal State Properties," *Appl. Phys.*, **16**, 1–28 (1978).

199. Ø. Fischer and M. Decroux, "The High Critical Field Chevrel Phases," *J. Magnetism and Magnetic Materials (Netherlands)*, **11**, 164–168 (1979).

200. Ø. Fischer et al., "New Ternary Molybdenum Chalcogenides," in *Superconductivity in d- and f-Band Metals*, H. Suhl and M. B. Maple (editors), Academic Press, New York (1980), pages 485–494.

201. See, for example, Ø. Fischer, Reference 193, page 4, Figure 4.

202. J. A. Woolam and S. A. Alterovitz, "Indirect Measurements of Fermi Surface Parameters

of Some Chevrel Phase Materials," *J. Magnetism and Magnetic Materials (Netherlands)*, **11**, 177–181 (1979), Table 1.

203. T. H. Geballe, "*d*- and *f*-Band Superconductivity—Some Experimental Aspects," in *Superconductivity in d- and f-Band Metals*, H. Suhl and M. B. Maple (editors), Academic Press, New York (1980), pages 1–11.

204. W. J. Gallagher, *J. Appl. Phys.*, **63**, 4216–4219 (1988).

205. T. H. Geballe and J. K. Hulm, "Superconductivity: The State that Came in from the Cold," *Science*, **239**, 367–375 (1988).

206. T. Sakakibara *et al.*, "Upper Critical Field Measurements on a Single Crystal of $YBa_2Cu_3O_x$ up to 38 T," *Jap. J. Appl. Phys.*, **26**, L1892–L1894 (1987).

207. A. C. Rose-Innes and E. H. Rhoderick, Reference 4, pages 40–41, 82–85, 202–220.

208. M. Tinkham, Reference 3, pages 99–103, 116–120, 157–174.

209. T. H. Geballe and M. R. Beasley, Reference 100, pages 506–508.

210. R. M. White and T. H. Geballe, Reference 77, pages 332–335.

211. J. K. Hulm and B. T. Matthias, Reference 138.

212. A. M. Campbell and J. E. Evetts, *Critical Currents in Superconductors*, Taylor and Francis, London (1972).

213. J. K. Hulm and B. T. Matthias, Reference 138, Figure 3, page 883.

Suggested Reading

C. KITTEL, *Introducting to Solid State Physics*, Sixth Edition, John Wiley, New York (1986). The basic background references for this chapter is Kittel's Chapter 12, which includes an introduction to the Josephson effects and to superconducting quantum interference.

R. P. FEYNMAN, R. B. LEIGHTON AND M. SANDS, *The Feynman Lectures on Physics*, Addison-Wesley, Reading, Mass. (1965), Volume III, Chapter 21. A treatment of the macroscopic quantum mechanical description of superconductivity on which our discussion is based.

A. C. ROSE-INNES AND E. H. RHODERICK, *Introduction to Superconductivity*, Second Edition, Pergamon Press, Oxford (1978). This monograph is at the intermediate level and is quite physical in its approach. It covers a wide variety of topics in superconductivity.

D. R. TILLEY AND J. TILLEY, *Superfluidity and Superconductivity*, Second Edition, Adam Hilger, Bristol (1986). This monograph discusses several areas of superconductivity, including chapters on Ginzburg–Landau theory and the vortex state.

M. TINKHAM, *Introduction to Superconductivity*, McGraw-Hill, New York (1975). This monograph is at the advanced level and covers the subject in depth and with authority.

L. SOLYMAR, *Superconductive Tunneling and Applications*, John Wiley, New York (1972). A discussion, at the intermediate level, of both the basic ideas and many kinds of applications of the Josephson effects.

A. BARONE AND G. PATERNO, *Physics and Applications of the Josephson Effect*, John Wiley, New York (1982). Devoted entirely to the Josephson effect, this book is more recent than the book by Solymar listed above. It also contains more material on microscopic theory and offers extensive references through 1979.

H. B. HALHBOHM AND H. LUBBIG, editors, *SQUID '85: Superconducting Quantum Interference Devices and their Applications*, Walter de Gruyter, Berlin (1985). These proceedings of the third international conference on the subject cover all aspects of the field. Each section of related papers begins with an appropriate review article.

R. M. WHITE AND T. H. GEBALLE, *Long Range Order in Solids*, Supplement 15 to *Solid State Physics*, H. Ehrenreich, F. Seitz, and D. Turnbull (editors), Academic Press, New York (1979). This monograph, at the advanced level, offers (among other things) a discussion of the physics of superconducting materials for the reader already familiar with BCS theory.

T. H. GEBALLE AND M. R. BEASLEY, "Superconducting Materials for Energy-Related Applications," in *Materials Science in Energy Technology*, G.G. Libowitz and M.S. Wittingham (editors), Academic Press, New York (1979), Chapter 10. Written at the advanced level for the reader already familiar with the physics of Type II superconductors, this book chapter discusses the critical magnetic fields and critical currents of technologically important materials.

J. MÜLLER AND J. L. OLSEN, editors, *Proceedings of the International Conference on High Temperature Superconductors and Materials and Mechanisms of Superconductivity, Interlaken, Switzerland 1988*, published in *Physica C*, Volumes 153–155, pages 1–1802 (June 1988). This conference (formerly known as the conference on superconductivity in d- and f-band metals) is held every few years and its papers give a view of the current status of research on superconducting materials. The conference was next held in 1989 at Stanford, California.

J. F. SCHOOLEY et al., editors, *Proceedings of the 1986 Applied Superconductivity Conference*, IEEE Transactions on Magnetics, Volume **MAG-23**, Number 2 (March 1987). The papers of this biannual conference (held in even-numbered years) cover many applications of superconductivity, including Josephson devices and superconducting materials for applications.

J. C. PHILLIPS, *Physics of High-T_c Superconductors*, Academic Press, San Diego (1989); C. P. POOLE, JR., T. DATTA, AND H. A. FARACH, *Copper Oxide Superconductors*, John Wiley and Sons, New York (1988). These recent monographs discuss the new high-temperature superconductors.

9

Physics and Applications of the Nonlinear Optical Properties of Solids

Introduction

The aim of this chapter is a discussion of some of the physics and applications of the nonlinear optical properties of solids. It begins with a review of electromagnetic wave propagation in solids and a derivation of the familiar linear relation between dielectric polarization and electric field. A more realistic anharmonic oscillator model is next introduced and this is shown to yield a polarization that is a nonlinear (specifically, a quadratic) function of the electric field. A physical picture of the nonlinear polarization and some solid state physics factors affecting the magnitude of the nonlinear susceptibility are also given. The central topic of the chapter is the propagation and interaction of three electromagnetic waves in a nonlinear medium. This results in a set of equations describing the spatial variation of the electric fields of these waves as they move through the crystal. Finally, these equations are used to discuss several applications of nonlinear solids. These are optical second harmonic generation, frequency mixing and up-conversion, and parametric amplification of optical signals.

Review of Electromagnetic Wave Propagation in Solids

We begin by writing Maxwell's equations[1]

$$\text{curl } \mathbf{H} = \frac{4\pi}{c}\,\mathbf{J} + \frac{1}{c}\,\frac{\partial \mathbf{D}}{\partial t} \tag{9.1}$$

$$\text{curl } \mathbf{E} = -\frac{1}{c}\frac{\partial \mathbf{B}}{\partial t} \tag{9.2}$$

$$\text{div } \mathbf{D} = 4\pi\varrho \tag{9.3}$$

$$\text{div } \mathbf{B} = 0 \tag{9.4}$$

using Gaussian units and in macroscopic form. The various quantities in equations (9.1)–(9.4) are macroscopic values[2] obtained by an averaging process which smooths out the microscopic details due to the atomic nature of the matter in the solid being considered. The quantity $\mathbf{E}$ is the electric field, $\mathbf{H}$ is the magnetic intensity, $\mathbf{D}$ is the electric displacement, $\mathbf{B}$ is the magnetic induction, ϱ is the free (or external[3]) electric charge density, and $\mathbf{J}$ is the free electric current density. We introduce also the relations

$$\mathbf{D} = \mathbf{E} + 4\pi\mathbf{P} \tag{9.5}$$

$$\mathbf{H} = \mathbf{B} - 4\pi\mathbf{M} \tag{9.6}$$

where $\mathbf{P}$ is the electric polarization (the electric dipole moment per unit volume) and $\mathbf{M}$ is the magnetization (the magnetic dipole moment per unit volume). Equations (9.5) and (9.6) may be substituted into Maxwell's equations (9.1)–(9.4) to give

$$\text{curl } \mathbf{H} - \frac{1}{c}\frac{\partial \mathbf{E}}{\partial t} = \frac{4\pi}{c}\left(\mathbf{J} + \frac{\partial \mathbf{P}}{\partial t}\right) \tag{9.7}$$

$$\text{curl } \mathbf{E} + \frac{1}{c}\frac{\partial \mathbf{H}}{\partial t} = -\frac{4\pi}{c}\frac{\partial \mathbf{M}}{\partial t} \tag{9.8}$$

$$\text{div } \mathbf{E} = 4\pi(\varrho - \text{div } \mathbf{P}) \tag{9.9}$$

$$\text{div } \mathbf{H} = -4\pi \text{ div } \mathbf{M} \tag{9.10}$$

Equations (9.7)–(9.10) connect the electromagnetic wave specified by the fields $\mathbf{E}$ and $\mathbf{H}$ with the properties $\mathbf{J}$, ϱ, $\mathbf{P}$, and $\mathbf{M}$ of the medium in which the wave is propagating. We will, of course, be interested in the situation when that medium is a solid. In order to solve these equations for the fields $\mathbf{E}$ and $\mathbf{H}$ of the wave, we have to know the source terms involving $\mathbf{J}$, $\mathbf{P}$, $\mathbf{M}$, and ϱ on the right-hand side of the equations.

In our discussion, we will consider only nonmagnetic solids. For these, $\mathbf{M} = 0$, and equation (9.6) gives the result that $\mathbf{H} = \mathbf{B}$. Further, we restrict ourselves to electrically neutral solids, for which there is no externally added charge, so $\varrho = 0$. Substituting these conditions into equations (9.7)–(9.10) gives

$$\text{curl } \mathbf{H} - \frac{1}{c}\frac{\partial \mathbf{E}}{\partial t} = \frac{4\pi}{c}\left(\mathbf{J} + \frac{\partial \mathbf{P}}{\partial t}\right) \tag{9.11}$$

$$\text{curl } \mathbf{E} + \frac{1}{c} \frac{\partial \mathbf{H}}{\partial t} = 0 \tag{9.12}$$

$$\text{div } \mathbf{E} = -4\pi \text{ div } \mathbf{P} \tag{9.13}$$

$$\text{div } \mathbf{H} = 0 \tag{9.14}$$

as the equations describing electromagnetic wave propagation in our non-magnetic electrically neutral solid medium.

We may obtain the wave equations for $\mathbf{E}$ and $\mathbf{H}$ from equations (9.11)–(9.14) by taking the curls of equations (9.11) and (9.12). For the electric field, we obtain from (9.12)

$$\text{curl curl } \mathbf{E} = -\frac{1}{c} \frac{\partial}{\partial t} (\text{curl } \mathbf{H}) \tag{9.15}$$

leading to

$$\text{curl curl } \mathbf{E} = -\frac{1}{c} \frac{\partial}{\partial t} \left(\frac{1}{c} \frac{\partial \mathbf{E}}{\partial t} + \frac{4\pi}{c} \mathbf{J} + \frac{4\pi}{c} \frac{\partial \mathbf{P}}{\partial t} \right) \tag{9.16}$$

on substituting equation (9.11) into equation (9.15). Using a standard vector identity for curl curl $\mathbf{E}$ in equation (9.16) and rearranging terms gives

$$\text{grad div } \mathbf{E} - \nabla^2 \mathbf{E} + \frac{1}{c^2} \frac{\partial^2 \mathbf{E}}{\partial t^2} = -\frac{4\pi}{c^2} \left(\frac{\partial \mathbf{J}}{\partial t} + \frac{\partial^2 \mathbf{P}}{\partial t^2} \right) \tag{9.17}$$

as the wave equation governing the electric field $\mathbf{E}$ in the solid.

We see that the wave equation (9.17) contains two source terms on its right-hand side. The term in the current density $\mathbf{J}$ stems from free conduction charges in the solid and the term in the polarization $\mathbf{P}$ has its origin in the bound polarization charges. The solutions of the wave equation (9.17) for the medium of interest will describe the propagation of electromagnetic waves in that medium. In metals, the source term in $\mathbf{J}$ is the more important, and the resulting solutions of the wave equation describe the familiar results of the optics of metals. In a nonconducting solid the term in the electric polarization $\mathbf{P}$ will be more important and leads to a description of dispersion, absorption, etc., in dielectric solids.

For a nonconducting dielectric material, the current density $\mathbf{J}$ will be zero, so the wave equation (9.17) becomes

$$\text{grad div } \mathbf{E} - \nabla^2 \mathbf{E} + \frac{1}{c^2} \frac{\partial^2 \mathbf{E}}{\partial t^2} = -\frac{4\pi}{c^2} \frac{\partial^2 \mathbf{P}}{\partial t^2} \tag{9.18}$$

The inhomogeneous wave equation (9.18) describes the electric field of an electromagnetic wave propagating in a nonmagnetic dielectric solid. The source term $(\partial^2 \mathbf{P}/\partial t^2)$ involves the polarization, so, in order to solve (9.18), we must set up a model of a dielectric solid from which we can calculate $\mathbf{P}$ and its time derivatives.

Before proceeding with a discussion of models of a dielectric solid, it is useful to draw some general conclusions concerning electromagnetic wave propagation in such a solid. We assume from now on that our solid is nonmagnetic (so $\mathbf{H} = \mathbf{B}$), electrically neutral (so $\varrho = 0$), and a dielectric (so $\mathbf{J} = 0$). We return to Maxwell's equations (9.1)–(9.4), which become

$$\operatorname{curl} \mathbf{H} = \frac{1}{c} \frac{\partial \mathbf{D}}{\partial t} \tag{9.19a}$$

$$\operatorname{curl} \mathbf{E} = -\frac{1}{c} \frac{\partial \mathbf{H}}{\partial t} \tag{9.19b}$$

$$\operatorname{div} \mathbf{D} = 0 \tag{9.19c}$$

$$\operatorname{div} \mathbf{H} = 0 \tag{9.19d}$$

for our solid. Since we have made no assumptions about the structure of the solid, equations (9.19) describe the electric and magnetic fields $\mathbf{E}$ and $\mathbf{H}$ in an arbitrary nonmagnetic dielectric solid.

We now consider the application of equations (9.19) to an anisotropic dielectric medium. Anisotropy[4] is the variation of some physical property with the direction in the solid along which the property is measured. For example, some electrically conducting crystals are anisotropic with respect to electrical resistivity. Isotropy[5] is the lack of dependence of a physical property on direction is a solid. An example is the optical isotropy of cubic III–V semiconductors; the refractive index does not depend on the direction of propagation of the light. If one considers the propagation of a plane monochromatic electromagnetic wave in a nonmagnetic anisotropic solid medium, the following results[6] are obtained from the Maxwell equations (9.19). If $\mathbf{k}$ is the propagation vector of the wave, then $\mathbf{k}$ and the fields $\mathbf{D}$ and $\mathbf{H}$ are all mutually perpendicular. The electric field $\mathbf{E}$ and the vectors $\mathbf{k}$ and $\mathbf{D}$ are all coplanar. This means that $\mathbf{E}$ and $\mathbf{D}$ are not parallel, and that $\mathbf{E}$ is not perpendicular to $\mathbf{k}$. In this wave, then, $\mathbf{D}$ and $\mathbf{H}$ are transverse to the direction of propagation $\mathbf{k}$, but $\mathbf{E}$ is not. These results are illustrated in Figure 9.1. We can see also from Figure 9.1 that there will be components of $\mathbf{E}$ both parallel and perpendicular to the propagation vector $\mathbf{k}$. We conclude that, for an electromagnetic wave propagating in an anisotropic dielectric, there will exist both longitudinal and transverse components of the electric field. This fact will be important when we consider wave propagation in an anisotropic nonlinear crystal in a later section.

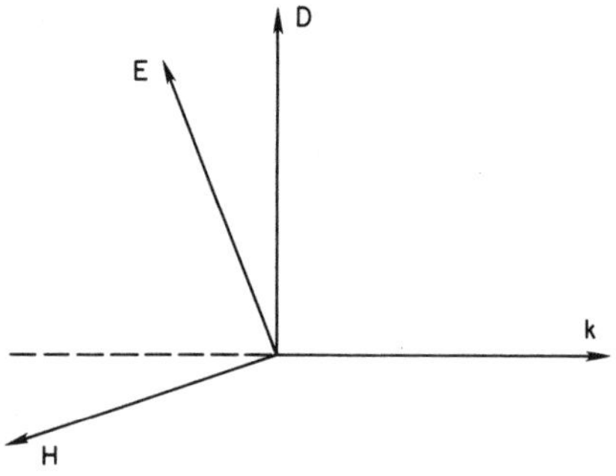

Figure 9.1. Vectors for a plane electro-magnetic wave in an anisotropic (non-magnetic) medium. The vectors **k**, **D**, and **H** are all mutually perpendicular. The vectors **E**, **D**, and **k** are coplanar and the electric field **E** is not normal to **k**.

Electric Polarization in a Dielectric Solid

Since we are interested in the propagation of electromagnetic waves, we are concerned with the polarization at optical frequencies, and hence with the electronic polarizability.[7] Because of the large mass of atomic nuclei, the ionic and dipolar polarizability cannot respond to the high-frequency electric field of the light wave. We will therefore consider only the polarization of the electron clouds of the atoms in the solid.

To consider the interaction of the incident electromagnetic wave with the electrons, we construct a model of the solid as follows. Each bound electron in an atom behaves as an isotropic damped harmonic oscillator[8] and is bound to the nucleus by fictitious "springs." For the isotropic oscillator, each of these "springs" has the same "spring constant" k for displacements in the three mutually perpendicular directions $x, y,$ and z. The restoring force on the electron is then $-k\mathbf{r}$, where $\mathbf{r}$ is the vector displacement of the electron from its quilibrium position. The model includes also a damping force $-m\gamma\dot{\mathbf{r}}$ proportional to the electron velocity $\dot{\mathbf{r}}$, where m is the electron mass and γ is the damping constant. The force due to the electric field $\mathbf{E}(t)$ of an incident electromagnetic wave is $-e\mathbf{E}(t)$, where

$$\mathbf{E}(t) = \mathbf{E}_0 \exp(-j\omega t)$$

is the time-dependent electric field of the light wave of circular frequency ω and amplitude $\mathbf{E}_0$, taken as of constant magnitude because we assume the electron's displacement is small compared to the wavelength of the wave. Writing the spring constant k as $m\omega_0^2$, where ω_0 is a constant characteristic frequency, the equation of motion of the electron is

$$m\ddot{\mathbf{r}} = -m\omega_0^2\mathbf{r} - m\gamma\dot{\mathbf{r}} - e\mathbf{E}(t)$$

If we consider an electromagnetic wave whose electric field vector $\mathbf{E}_x$ is in the x-direction, the displacement of the electron $x(t)$ is in the x-direction, and the equation of motion becomes

$$m\ddot{x} = - m\gamma\dot{x} - m\omega_0^2 x - eE_x \qquad (9.20)$$

The solution of equation (9.20) is, with $j \equiv \sqrt{-1}$,

$$x(t) = \frac{-e/m}{\omega_0^2 - \omega^2 - j\gamma\omega} E_x \qquad (9.21)$$

In obtaining equations (9.20) and (9.21), the force on the electron due to the magnetic field of the light wave has been omitted as negligible compared to the force due to the electric field.

The electric polarization **P** is the electric dipole moment per unit volume, and the electric field of the incident light wave induces an electronic dipole moment because the bound electron is displaced from its equilibrium position. Because the electric field of the light wave is in the x-direction, the induced dipole moment is in the x-direction, and has a magnitude $- ex$, where x is given by (9.21). The total electric dipole moment induced is

$$P_x = - Nex \qquad (9.22)$$

where N is the number of electrons per unit volume. Combining equations (9.21) and (9.22) gives

$$P_x = \frac{Ne^2/m}{\omega_0^2 - \omega^2 - j\gamma\omega} E_x \qquad (9.23)$$

showing that, for this model, the electronic polarization P_x is a linear function of the electric field E_x of the incident electromagnetic wave. Equation (9.23) shows also that the polarization has the same time dependence $\exp(-j\omega t)$ as the incident electric field, and so has the same frequency ω as the light wave. The isotropic dielectric medium represented by this model is thus linear.

Strictly speaking, equation (9.23) is not correct because we have neglected the electric field produced by the polarization of the solid. We should rewrite equation (9.23) to read

$$\mathbf{P} = \frac{Ne^2/m}{\omega_0^2 - \omega^2 - j\gamma\omega} \mathbf{E}_{\text{loc}} \qquad (9.24)$$

where $\mathbf{E}_{\text{loc}}$ is the local electric field at the atom whose electronic polarization we are considering. The local electric field $\mathbf{E}_{\text{loc}}$ is made up[9,10] of the sum of the applied electric field **E** of the light wave plus the total field at the atom due to the electric dipole moments (i.e., the polarization) of all the other atoms. We will consider the especially simple case of a cubic solid in a shape for which the depolarization field is zero,[3] and for which then the only contributions to E_{loc} are the Lorentz field $4\pi P/3$ and the applied electric field E of the light wave. In this case, we have

$$E_{loc} = E + 4\pi P/3 \qquad (9.25)$$

so equation (9.24) becomes

$$P = \frac{Ne^2/m}{\omega_0^2 - \omega^2 - j\gamma\omega} \left(E + \frac{4\pi}{3} P \right) \qquad (9.26)$$

which, on solving for P, yields

$$P = \frac{Ne^2/m}{(\omega_0^2 - 4\pi Ne^2/3m) - \omega^2 - j\gamma\omega} E \qquad (9.27)$$

for the polarization P as a function of the electric field E of the light wave in the solid.

Equation (9.27) gives the linear relation between the polarization and the electric field for our model of a solid. The model of a solid to which we apply equation (9.27) is an isotropic medium whose properties are independent of direction in the solid. For our isotropic model then, equation (9.27) is of the form

$$P = \chi(\omega)E \qquad (9.28)$$

where

$$\chi(\omega) \equiv \frac{Ne^2/m}{(\omega_0^2 - 4\pi Ne^2/3m) - \omega^2 - j\gamma\omega} \qquad (9.29)$$

is the frequency-dependent linear electric susceptibility. Note that, for our model of a solid in which the restoring force on the electron is linear, the susceptibility given by equation (9.29) is independent of the applied electric field. Because our model is isotropic, the electric susceptibility $\chi(\omega)$ in (9.28) is a scalar.

For a model of an anisotropic solid, we may consider a model in which the bound electron is treated as an anisotropic harmonic oscillator. The "springs" exerting restoring forces on the electron will now have different spring constants for displacements in the three different directions. For such an anisotropic oscillator, one would expect an electric field polarized in a given direction to produce an electron displacement, and hence a polarization, which is *not* parallel to the electric field direction. Thus, for an electric field E_x, one would expect the polarization P to be in a general direction, so P would have x-, y-, and z-components. Looked at in the opposite way, we might expect that, for the proper choice of electric field direction, we would obtain a polarization in the x-direction, as in

$$\mathbf{P}_x = \chi_{11}\mathbf{E}_x + \chi_{12}\mathbf{E}_y + \chi_{13}\mathbf{E}_z \tag{9.30a}$$

where the χ_{ij} are constants characteristic of the medium. Similarly we would have

$$\mathbf{P}_y = \chi_{21}\mathbf{E}_x + \chi_{22}\mathbf{E}_y + \chi_{23}\mathbf{E}_z \tag{9.30b}$$

$$\mathbf{P}_z = \chi_{31}\mathbf{E}_x + \chi_{32}\mathbf{E}_y + \chi_{33}\mathbf{E}_z \tag{9.30c}$$

Equations (9.30a–c) can be written in matrix form as

$$\begin{pmatrix} \mathbf{P}_x \\ \mathbf{P}_y \\ \mathbf{P}_z \end{pmatrix} = \begin{pmatrix} \chi_{11} & \chi_{12} & \chi_{13} \\ \chi_{21} & \chi_{22} & \chi_{23} \\ \chi_{31} & \chi_{32} & \chi_{33} \end{pmatrix} \begin{pmatrix} \mathbf{E}_x \\ \mathbf{E}_y \\ \mathbf{E}_z \end{pmatrix} \tag{9.30d}$$

where[11] the χ_{ij} are the nine components of the second-rank linear electric susceptibility tensor $\mathbf{\chi}$. We can write also (9.30d) as

$$\mathbf{P} = \mathbf{\chi}\mathbf{E} \tag{9.31}$$

an equation which is equivalent to equations (9.30a–d). These conclusions can also be reached by considering the relation between $\mathbf{P}$ and $\mathbf{E}$ in equation (9.5) as

$$4\pi\mathbf{P} = \mathbf{D} - \mathbf{E}$$

In an anisotropic crystal, as seen from Figure 9.1, the vectors $\mathbf{D}$ and $\mathbf{E}$ are not in general parallel, so $\mathbf{P}$ is not parallel to $\mathbf{E}$.

We now return to consideration of the wave equation (9.18) for our isotropic model of a dielectric solid and discuss the source term $(\partial^2\mathbf{P}/\partial t^2)$. We use the relation (9.28) between $\mathbf{P}$ and $\mathbf{E}$ to calculate[11a]

$$\frac{\partial^2\mathbf{P}}{\partial t^2} = \chi(\omega)\,\frac{\partial^2\mathbf{E}}{\partial t^2} \tag{9.32}$$

substitution of which into (9.18) gives

$$\text{grad div }\mathbf{E} - \nabla^2\mathbf{E} + \frac{1}{c^2}\,\frac{\partial^2\mathbf{E}}{\partial t^2} = -\frac{4\pi}{c^2}\,\chi(\omega)\frac{\partial^2\mathbf{E}}{\partial t^2} \tag{9.33}$$

as the wave equation. Equation (9.33) may be simplified as follows. Since the external charge density $\varrho = 0$, the Maxwell equation (9.9) becomes

$$\text{div}(\mathbf{E} + 4\pi\mathbf{P}) = 0$$

Since our dielectric solid is linear and isotropic, equation (9.28) holds, so $\mathbf{P}$ is parallel to $\mathbf{E}$, and we have

$$\text{div}[\mathbf{E} + 4\pi\chi(\omega)\mathbf{E}] = 0$$

Then

$$\text{div}([1 + 4\pi\chi(\omega)]\mathbf{E}) = \text{div}[\varepsilon(\omega)\mathbf{E}] = 0 \tag{9.34}$$

where

$$\varepsilon(\omega) \equiv 1 + 4\pi\chi(\omega) \tag{9.35}$$

defines the scalar frequency-dependent dielectric constant $\varepsilon(\omega)$. Because we are also assuming that the dielectric solid is uniform, the dielectric constant $\varepsilon(\omega)$ does not vary with position in the solid, so the divergence of ε vanishes. Equation (9.34) then becomes

$$\varepsilon(\omega) \, \text{div } \mathbf{E} = 0$$

an equation whose solution (assuming ε is not zero) is

$$\text{div } \mathbf{E} = 0 \tag{9.36}$$

Equation (9.36) states that the divergence of the electric field vanishes in a charge-free isotropic uniform linear dielectric solid. The result (9.36) stems directly from our assumptions concerning the nature of the solid.

When equation (9.36) is substituted into the wave equation (9.33), we obtain the familiar result

$$\nabla^2\mathbf{E} = \frac{1}{c^2} [1 + 4\pi\chi(\omega)] \frac{\partial^2\mathbf{E}}{\partial t^2} \tag{9.37}$$

as the wave equation for the electric field of an electromagnetic wave propagating in a uniform linear isotropic dielectric of scalar electric suscep-- tibility $\chi(\omega)$. The traveling wave solutions of (9.37) are the familiar plane waves in an isotropic solid, for which the electric field, the magnetic field, and the propagation direction are all mutually perpendicular. Such a wave is therefore transverse in an isotropic medium.

Physically, we may regard the propagation of the light wave governed by equation (9.37) in the linear dielectric in the following way. The electric field of the incident light drives the electrons in the solid. These electrons act as driven, damped harmonic oscillators whose displacement $x(t)$ is given by equation (9.21). This oscillation of electric charge at the frequency ω of the driving electric field constitutes the polarization at ω expressed by equation (9.23) and produces radiation at frequency ω. In this way, the electrons in the solid reradiate the incident light wave, which thus propagates through the solid with frequency ω.

We may also describe the process in terms of the source term in $\partial^2 P/\partial t^2$ in the wave equation (9.18). The linear relation (9.23) between P and E

means that the polarization has the same time dependence, and thus the same frequency, as the incident electric field. The source term $\partial^2 P/\partial t^2$ will also be at the frequency ω of the incident light wave, meaning that the light propagates through the solid with an unchanged frequency ω. This fact is a direct result of the linear relation (9.23) between the polarization and the electric field. Later, we will examine the results stemming from a nonlinear relation between P and E, and we go on to consider a picture of a solid in which such a nonlinear relation is the case.

Nonlinear Polarization and Nonlinear Susceptibility

In the preceding section, we examined a particular model of a dielectric solid, namely, a classical model in which the restoring force on the electrons was linear in the coordinate x. One of the properties of this linear model was expressed in equation (9.30),

$$\mathbf{P} = \chi(\omega)\mathbf{E} \qquad (9.30)$$

stating that the polarization is a linear function of the electric field. In the limit of weak electric fields, the linear relation between $\mathbf{P}$ and $\mathbf{E}$ is a good approximation for real dielectric solids. However, in general[12] it is true that

$$\mathbf{P} = f(\mathbf{E}) \qquad (9.38)$$

where $f(\mathbf{E})$ is a *nonlinear* function of the electric field $\mathbf{E}$. Often the polarization can be written as a power series in the electric field,

$$\mathbf{P} = \chi^{(1)} \cdot \mathbf{E} + \chi^{(2)} : \mathbf{EE} + \cdots \qquad (9.39)$$

where $\chi^{(1)}$ is the first-order (linear) electric susceptibility tensor used in equation (9.31) and $\chi^{(2)}$ is the second-order nonlinear susceptibility tensor, etc. The tensor $\chi^{(2)}$ is of third rank and has 27 components. We will not consider nonlinear effects higher[12–14] than the second order. Finally, we rewrite equation (9.39) in the approximate and scalar form

$$P = \chi^{(1)}E + \chi^{(2)}E^2 \qquad (9.40)$$

in order to focus attention on the effects of the second-order nonlinearity. The scalar form neglects the anisotropy of real crystals, which is discussed under the wave equation for the nonlinear crystal.

Before considering a classical model of a dielectric leading to a relation of the type (9.40) between the polarization and the electric field, it is appropriate to point out *why* we are interested in the nonlinearity expressed by the term $\chi^{(2)}E^2$. Suppose an electromagnetic wave of the form

$$E = E_0 \sin \omega t \qquad (9.41)$$

with circular frequency ω is incident on a solid whose polarization–electric-field relation contains a term in E^2, as in equation (9.40). The polarization in the solid will then be of the form

$$P = \chi^{(1)}E_0 \sin \omega t + \chi^{(2)}E_0^2 \sin^2 \omega t \qquad (9.42)$$

which may be rewritten as

$$P = \chi^{(1)}E_0 \sin \omega t - \tfrac{1}{2}\chi^{(2)}E_0^2 \cos(2\omega t) + \tfrac{1}{2}\chi^{(2)}E_0^2 \qquad (9.43)$$

using the trigonometric identity $\cos 2\theta = 1 - 2\sin^2 \theta$. From equation (9.43), we can see that, in this case, the source term $\partial^2 P/\partial t^2$ in the wave equation (9.18) will be

$$\frac{\partial^2 P}{\partial t^2} = -\omega^2 \chi^{(1)}E_0 \sin \omega t + 2\omega^2 \chi^{(2)}E_0^2 \cos(2\omega t) \qquad (9.44)$$

The key result in equations (9.43) and (9.44) is that the factor E^2 in equation (9.40) leads to the presence of the second harmonic frequency 2ω in the polarization P, and thus in the source term $\partial^2 P/\partial t^2$ in the wave equation. This result is equivalent to the existence of a term oscillating at 2ω in the displacement $x(t)$ of the electrons in our model of a solid. We would expect this oscillating electron to radiate at frequency 2ω. We conclude that a light wave of frequency ω, incident on a nonlinear crystal whose polarization is described by equation (9.40), will generate radiation at the second harmonic 2ω as well as at the fundamental frequency ω.

We may also describe this process by saying that the nonlinear term E^2 in the polarization "mixes" the incident electric field, at frequency ω, with itself, leading to a component of the polarization at 2ω as well as at ω. In the same way, a nonlinear optical medium can "mix" incident waves of frequencies ω_1 and ω_2, generating radiation at the sum and difference frequencies $\omega_1 \pm \omega_2$. It is essentially the mixing of electromagnetic waves by the term E^2 in the polarization of a nonlinear medium that is the physical basis of the applications we will consider.

Anharmonic Oscillator Model of a Nonlinear Solid

In order to discuss the physical origin of the nonlinear susceptibility, we examine a model[14-17] of a solid which is more realistic than the linear model considered earlier. In the linear model, the restoring force on the electron was the Hooke's law force $-m\omega_0^2 x$ characteristic of a harmonic oscillator. We now consider a restoring force containing terms in powers of x higher than the first and which has the form

$$-m(\lambda x^2 + \delta x^3 + \cdots)$$ (9.45)

so we will be considering the anharmonic oscillator.[18] We will limit ourselves to the first term in the series in equation (9.45). The anharmonic
term in the restoring force on the electron will thus have the form

$$-m\lambda x^2$$ (9.46)

where λ is a positive constant and x is the displacement of the electron from
its equilibrium position. The equation of motion, which was given by equation (9.20) for the linear or harmonic restoring force, becomes, on adding
the term (9.46),

$$m\ddot{x} = -m\gamma\dot{x} - m\omega_0^2 x - m\lambda x^2 - eE$$ (9.47)

where the quantities γ, ω_0, m, e, and E were defined in connection with
equation (9.20). Equation (9.47) is the equation of motion of a damped
anharmonic oscillator driven by the electric field E of the light wave. We
choose the form

$$E = E_0 \cos \omega t$$ (9.48)

for the electric field, so the equation of motion (9.47) becomes

$$\ddot{x} + \gamma\dot{x} + \omega_0^2 x + \lambda x^2 = (-eE_0/m) \cos \omega t$$ (9.49)

Equation (9.49) is a nonlinear differential equation because of the presence
of the term in x^2.

We will not solve equation (9.49) directly.[19–21] Instead, we will use
the solution[22,23] to the simpler nonlinear differential equation

$$\ddot{x} + \omega_0^2 x - \lambda x^2 = 0$$ (9.50)

which describes the undriven, undamped nonlinear oscillator, to suggest
the key features of the solution of the equation of motion (9.49). We
will assume that the constant λ is a small quantity so the solution to equation (9.50) will be close to that for a simple (linear) harmonic oscillator.
The solution of equation (9.50) may be shown to be[23]

$$x(t) = A \cos \omega_0 t - (\lambda A^2/6\omega_0^2) \cos(2\omega_0 t) + (\lambda A^2/2\omega_0^2)$$
  (9.51)

where A is a constant. Equation (9.51) gives an approximate solution to the
differential equation (9.50). The solution (9.51), which neglects terms in
λ^2, etc., is therefore correct to first order in λ.

The key physical idea in equation (9.51) is that the presence of the
nonlinear term in x^2 in the equation of motion introduces a term at the

second harmonic frequency[†] $2\omega_0$ into the expression for $x(t)$. We therefore expect that the first-order solution $x(t)$ to the equation of motion (9.49) for the driven, damped, anharmonic oscillator, will contain terms at the frequency ω of the driving electric field *and* at the second harmonic frequency 2ω. Further, by analogy with the driven, damped *harmonic* oscillator,[(24)] we expect that the solution to equation (9.49) will contain a resonant denominator in powers of $\omega_0 - \omega$, and will include the damping constant γ.

To obtain a solution of the equation of motion (9.49), we will, using the ideas outlined above and following Yariv,[(15,16)] assume a solution

$$x(t) = \tfrac{1}{2}(x_1 e^{j\omega t} + x_1{}^* e^{-j\omega t} + x_2 e^{j(2\omega)t} + x_2{}^* e^{-j(2\omega)t}) \qquad (9.52)$$

which contains oscillations at both the driving frequency ω and the second harmonic frequency 2ω. In equation (9.52), x_1 and x_2 are the amplitudes of the electron motion at the frequencies ω and 2ω, respectively, and the asterisk indicates a complex conjugate. In equation (9.52), we have ignored[‡] the nonoscillatory term contained in equation (9.51). We rewrite equation (9.49) in the form

$$\ddot{x} + \gamma\dot{x} + \omega_0{}^2 x + \lambda x^2 = (-eE_0/2m)(e^{j\omega t} + e^{-j\omega t}) \qquad (9.53)$$

by using the complex form for $\cos \omega t$. We do not use the usual complex form $E_0 \exp(j\omega t)$ for the driving electric field because[(10)] of the presence of the term in x^2 in the differential equation.

If the assumed solution (9.52) is substituted into the differential equation (9.53), one obtains[(15,16)]

$$
\begin{aligned}
[-\tfrac{1}{2}\omega^2 &(x_1 e^{j\omega t} + x_1{}^* e^{-j\omega t}) - 2\omega^2(x_2 e^{2j\omega t} + x_2{}^* e^{-2j\omega t}) \\
&+ \tfrac{1}{2}j\omega\gamma[x_1 e^{j\omega t} - x_1{}^* e^{-j\omega t}] + j\omega\gamma(x_2 e^{2j\omega t} - x_2{}^* e^{-2j\omega t}) \\
&+ \tfrac{1}{2}\omega_0{}^2(x_1 e^{j\omega t} + x_1{}^* e^{-j\omega t} + x_2 e^{2j\omega t} + x_2{}^* e^{-2j\omega t}) \\
&+ \tfrac{1}{4}\lambda(x_1{}^2 e^{2j\omega t} + x_2{}^2 e^{4j\omega t} + 2x_1 x_1{}^* + 2x_1 x_2 e^{3j\omega t} \\
&+ 2x_1{}^* x_2 e^{j\omega t} + 2x_2 x_2{}^* + x_1{}^{*2} e^{-2j\omega t} \\
&+ x_2{}^{*2} e^{-4j\omega t} + 2x_1{}^* x_2 e^{-3j\omega t} + 2x_1 x_2{}^* e^{-j\omega t})] \\
&= (-eE_0/2m)(e^{j\omega t} + e^{-j\omega t}) \qquad (9.54)
\end{aligned}
$$

Equation (9.54) must hold in order that the assumed equation (9.52) be a solution of the equation of motion (9.53). For equation (9.54) to be true, the coefficients of the term in each power of $\exp(j\omega t)$ must be the same on both sides of the equation. Considering first the term in $\exp(j\omega t)$, the coefficient of $\exp(j\omega t)$ on the left-hand side of equation (9.54) must equal the coefficient on the right-hand side, so we have

---

[†] If terms in higher powers of λ had been retained in the solution (9.51), then terms at higher harmonic frequencies would have been present.

[‡] The nonoscillatory or DC term represents a nonzero average displacement of the electron and leads to the effect called optical rectification. See, for example, the book by Baldwin.[(25)]

$$x_1(-\omega^2 + j\gamma\omega + \omega_0^2) + x_1{}^*x_2\lambda = -eE_0/m \qquad (9.55)$$

We now make the assumption that the magnitude of the term $x_1{}^*x_2\lambda$ on the left-hand side of equation (9.55) may be neglected compared to the magnitude of the term $x_1(\omega_0^2 - \omega^2 + j\gamma\omega)$. This approximation is expressed[9] as

$$|\lambda x_2| \ll [(\omega_0^2 - \omega^2)^2 + \gamma^2\omega^2]^{1/2} \qquad (9.56)$$

This assumption is physically reasonable[†] because we are, in equation (9.47), tacitly assuming that the anharmonic contribution $m\lambda x^2$ to the restoring force is small compared to the harmonic term $m\omega_0^2 x$. We would then expect that the magnitude x_2 of the electron displacement at the frequency 2ω will be relatively small, so the product λx_2 will also be small. With the assumption (9.56), equation (9.55) becomes

$$x_1 = \frac{-eE_0/m}{\omega_0^2 - \omega^2 + j\gamma\omega} \qquad (9.57)$$

which is the same result as equation (9.21) obtained for the linear model discussed earlier. This is, of course, a consequence of neglecting the anharmonic term in (9.55) when obtaining equation (9.57).

We may now find the expression for the amplitude x_2 of the electron motion at the second harmonic frequency 2ω. Equating the coefficients of the terms in $\exp[j(2\omega)t]$ on both sides of equation (9.54) gives

$$x_2(-2\omega^2 + j\gamma\omega + \tfrac{1}{2}\omega_0^2) + \tfrac{1}{4}\lambda x_1^2 = 0 \qquad (9.58)$$

On substituting equation (9.57) for the amplitude x_1 into equation (9.58), we obtain

$$x_2(\omega_0^2 - 4\omega^2 + 2j\gamma\omega) + \frac{1}{2}\lambda\frac{(e^2/m^2)E_0^2}{(\omega_0^2 - \omega^2 + j\gamma\omega)^2} = 0 \qquad (9.59)$$

which may be rewritten as

$$x_2 = \frac{-\lambda(e^2/m^2)E_0^2}{2(\omega_0^2 - \omega^2 + j\gamma\omega)^2(\omega_0^2 - 4\omega^2 + 2j\gamma\omega)} \qquad (9.60)$$

[†] We may make an estimate of the validity of the approximation (9.56) as follows. Considering visible frequencies, ω will be of the order of $2\pi \times 10^{15}$ sec^{-1}, so the right-hand side of equation (9.56) will of the order of 10^{31} sec^{-2}. The term in $\gamma^2\omega^2$ is negligible because[26] γ is of the order of 10^8 sec^{-1}. Then, using Garrett's value[27] of $\lambda \approx 10^{39}$ cm^{-1} sec^{-2} (where Garrett's v is our λ), we may estimate $|\lambda x_2|$. We take the amplitude x_2 as a small fraction, say 1%, of the amplitude x_1, which in turn is estimated to be of the order of 10^{-8} cm, so we take $x_2 \approx 10^{-10}$ cm. This gives a value of $|\lambda x_2| \approx 10^{29}$ sec^{-2}, which is small compared to the estimated magnitude of 10^{31} sec^{-2} of the right-hand side of equation (9.56). We are therefore justified in using the approximation.

Equation (9.60) gives the amplitude x_2 of the electron motion at the second harmonic frequency 2ω.

Having found the electron displacements x_1 and x_2 at the frequencies ω and 2ω, respectively, we may now find the total polarization P as

$$P = N(-e)[x(t)] = \tfrac{1}{2}N(-e)(x_1 e^{j\omega t} + x_2 e^{j(2\omega)t} + \text{c.c.}) \qquad (9.61)$$

where we have used equation (9.52) for $x(t)$, and have written "c.c." for the complex conjugate terms. Substituting equations (9.57) and (9.60) for x_1 and x_2 into equation (9.61) gives

$$P = \frac{Ne^2 E_0 e^{j\omega t}}{2m(\omega_0^2 - \omega^2 + j\gamma\omega)} + \frac{Ne^3 \lambda (E_0 e^{j\omega t})^2}{4m^2(\omega_0^2 - \omega^2 + j\gamma\omega)^2(\omega_0^2 - 4\omega^2 + 2j\gamma\omega)}$$
$$(9.62)$$

where we have written $[E_0 \exp(j\omega t)]^2$ in place of $E_0^2 \exp(2j\omega t)$, and have omitted the complex conjugates for simplicity.

Equation (9.62) gives the total time- and frequency-dependent polarization P as a function of the electric field $E_0 \exp(j\omega t)$ of the light wave incident on the dielectric solid. Recalling from equation (9.19) that

$$E = E_0 \exp(j\omega t) \qquad (9.19)$$

we can write equation (9.62) in the form

$$P = P^{(\omega)} + P^{(2\omega)} = \chi_L^{(\omega)}E + d^{(2\omega)}E^2 \qquad (9.63)$$

where

$$\chi_L^{(\omega)} \equiv \frac{Ne^2/m}{\omega_0^2 - \omega^2 + j\gamma\omega} \qquad (9.64)$$

and

$$d^{(2\omega)} \equiv \frac{\lambda(Ne^3/m^2)}{2(\omega_0^2 - \omega^2 + j\gamma\omega)^2(\omega_0^2 - 4\omega^2 + 2j\gamma\omega)} \qquad (9.65)$$

and the presence of the complex conjugates in equation (9.62) provides a factor of 2 in the expressions for $\chi_L^{(\omega)}$ and $d^{(2\omega)}$. The quantities $P^{(\omega)}$ and $P^{(2\omega)}$ in equation (9.63) are, respectively, the components of the polarization at the frequencies ω and 2ω. Since $\chi_L^{(\omega)}$ is the ratio of the polarization $P^{(\omega)}$ to the first power of the electric field E, $\chi_L^{(\omega)}$ is the linear electric susceptibility. The expression (9.64) is the same as that given in equation (9.29) if the local field correction is omitted in the latter equation.

The polarization $P^{(2\omega)}$ at the second harmonic frequency 2ω is often called the nonlinear polarization because it is proportional to E^2. The quantity $d^{(2\omega)}$ in equation (9.65) is the ratio of the nonlinear polarization

to the square of the electric field and is called the nonlinear electric susceptibility or the nonlinear optical coefficient. We note from equation (9.65)
that the nonlinear susceptibility $d^{(2\omega)}$ arises from the presence of the anharmonic force term $m\lambda x^2$ because the expression for $d^{(2\omega)}$ contains the
anharmonic force constant λ. We have therefore obtained from the anharmonic oscillator model the nonlinear dependence of the polarization
on the electric field expressed by equation (9.63), which is of the same form
as equation (9.39) when only second-order nonlinearity is considered. Note
also that our assumed solution (9.52) of the nonlinear equation of motion
(9.53) is correct only to first order in the anharmonic force constant λ
because no terms in $\exp(3j\omega t)$, etc. were used. A perturbation solution
correct to higher order in λ would have produced a solution $x(t)$ containing
harmonic frequencies higher[17] than 2ω and would have led to an expression
for the polarization containing terms in E^3, etc.

Summary of the Physical Picture of Nonlinear Polarization

We may summarize our physical picture of nonlinear polarization as
follows. We considered a classical[28] model of a solid containing an anharmonic term $m\lambda x^2$ in the restoring force acting on an electron. The
presence of this nonlinear term in the equation of motion led to the electron
displacement $x(t)$ having Fourier components at the frequency ω of the
incident light wave and at the second harmonic 2ω as well. This, in turn,
led to a time-dependent electric polarization with components[29] $P^{(\omega)}$ and
$P^{(2\omega)}$ at the frequencies ω and 2ω, respectively. The component $P^{(\omega)}$ of the
induced polarization is directly proportional to the incident electric field E
and the linear electric susceptibility $\chi_L^{(\omega)}$ is just that found for the harmonic
oscillator model of the solid. The component $P^{(2\omega)}$ is proportional to E^2,
and the nonlinear electric susceptibility $d^{(2\omega)}$ depends directly on the
anharmonic force constant λ.

We now have a simple classical model leading to the second-order
nonlinear polarization and nonlinear susceptibility expressed in equation
(9.63). With that expression giving the polarization as a function of the
electric field, we can next calculate the source term $\partial^2 P/\partial t^2$ in the wave
equation (9.18) governing electromagnetic wave propagation in a solid
with a nonlinear electric susceptibility. Before doing that, however, we
discuss the nonlinear susceptibility $d^{(2\omega)}$ itself.

Tensor Nature of the Nonlinear Susceptibility

In the preceding sections, the results were derived on the basis of a *scalar* model of a dielectric which ignored the anisotropy of real crystals. This viewpoint, while useful for discussing the basic physical ideas involved, is oversimplified. In a vector treatment, we can see from equation (9.39) that equation (9.63) would become

$$\mathbf{P} = \mathbf{P}^{(\omega)} + \mathbf{P}^{(2\omega)} = \boldsymbol{\chi}_L{}^{(\omega)} \cdot \mathbf{E} + \mathbf{d}^{(2\omega)} : \mathbf{EE} \qquad (9.66)$$

where $\boldsymbol{\chi}_L{}^{(\omega)}$ is the second-rank first-order (linear) susceptibility tensor and $\mathbf{d}^{(2\omega)}$ is the third-rank second-order nonlinear susceptibility tensor. The tensor[30] $\mathbf{d}^{(2\omega)}$ has 27 components $d_{ijk}^{(2\omega)}$, so the components $P_i^{(2\omega)}$ of the nonlinear polarization vector would be expressed[31] as

$$P_i^{(2\omega)} = \sum_j \sum_k d_{ijk}^{(2\omega)} E_j E_k \qquad (9.67)$$

where the indices i, j, and k run over the three dimensions x, y, and z. (A contracted notation is often used in which d_{ijk} is replaced by d_{il}, where $i = 1, 2, 3$ represents x, y, z, and $l = 1, 2, 3, 4, 5, 6$ represents xx, yy, zz, yz, xz, xy.) In some crystals,[32] the symmetry of the lattice reduces the number of independent nonzero values of $d_{ijk}^{(2\omega)}$, and, in certain cases (to be discussed later) symmetry causes the tensor to vanish.

For simplicity, however, in our treatment of the applications of the nonlinear susceptibility, we will ignore its tensor nature and use the scalar equations developed so far.

Solid State Physics Factors Affecting the Nonlinear Susceptibility

We now discuss some of the properties of solids that affect the magnitude[33] of the second-order nonlinear susceptibility $d^{(2\omega)}$.

We note from equation (9.65) that the nonlinear susceptibility $d^{(2\omega)}$ is directly proportional to the anharmonic force constant λ. Consider the effect on the magnitude of λ of the crystal in question having a center of symmetry. The anharmonic restoring force $-m\lambda x^2$ corresponds to a term

$$U(x) = m\lambda x^3/3 \qquad (9.68)$$

in the potential energy U of an electron in the crystal. If a center of symmetry is present, the potential energy must be invariant if the position vector $\mathbf{r}$ of a point is replaced by $-\mathbf{r}$. In our one-dimensional discussion, this corresponds to replacing x by $-x$, so a center of symmetry requires that

$$U(x) = U(-x) \qquad (9.69)$$

for the potential energy to be invariant under the transformation. For the potential energy given by equation (9.68), the requirement (9.69) becomes

$$m\lambda x^3 = -m\lambda x^3 \qquad (9.70)$$

an equation which must be true for all values of x. Equation (9.70) requires that the anharmonic force constant λ be equal to zero. The physical consequence of inversion invariance is thus that a crystal with a center of symmetry cannot have a term proportional to x^2 in the restoring force on the electron. If $\lambda = 0$, then equation (9.65) tells us that $d^{(2\omega)} = 0$ in a crystal with a center of symmetry, and such a crystal has no second-order nonlinear terms in its optical polarization. Crystals lacking a center of symmetry may have nonzero values of the components $d_{ijk}^{(2\omega)}$ of the second-order nonlinear susceptibility tensor $d^{(2\omega)}$. An example of a centrosymmetric crystal is NaCl; examples of noncentrosymmetric crystals that have nonzero values[34,35] of some of the $d_{ijk}^{(2\omega)}$ are GaAs, KH_2PO_4 (potassium dihydrogen phosphate, known as KDP), $LiNbO_3$, and SiO_2.

To discuss factors affecting the magnitude of the nonlinear susceptibility $d^{(2\omega)}$, we recall from equation (9.65) that

$$d^{(2\omega)} \equiv \frac{\lambda(Ne^3/m^2)}{2(\omega_0^2 - \omega^2 + j\gamma\omega)^2(\omega_0^2 - 4\omega^2 + 2j\gamma\omega)} \qquad (9.65)$$

and that the linear susceptibility $\chi_L^{(\omega)}$ at frequency ω is

$$\chi_L^{(\omega)} \equiv \frac{Ne^2/m}{\omega_0^2 - \omega^2 + j\gamma\omega} \qquad (9.64)$$

From equation (9.64), we can see that $\chi_L^{(2\omega)}$, the *linear* susceptibility at frequency 2ω, is given by

$$\chi_L^{(2\omega)} \equiv \frac{Ne^2/m}{\omega_0^2 - 4\omega^2 + 2j\gamma\omega} \qquad (9.71)$$

On substituting expressions (9.64) and (9.71) into equation (9.65), one obtains

$$d^{(2\omega)} = (m\lambda/2N^2e^3)(\chi_L^{(\omega)})^2(\chi_L^{(2\omega)}) \tag{9.72}$$

Equation (9.72) is an expression for the nonlinear susceptibility $d^{(2\omega)}$ as a function of the linear susceptibilities $\chi_L^{(\omega)}$ and $\chi_L^{(2\omega)}$ at frequencies ω and 2ω, and of the solid state material parameters N and λ. Defining a quantity $\delta^{(2\omega)}$ by the relation

$$\delta^{(2\omega)} \equiv m\lambda/2N^2e^3 \tag{9.73}$$

equation (9.72) can be rewritten in the form

$$\delta^{(2\omega)} = \frac{d^{(2\omega)}}{(\chi_L^{(\omega)})^2(\chi_L^{(2\omega)})} \tag{9.74}$$

It has been found experimentally[27,36] that $\delta^{(2\omega)}$ (or more precisely, its three dimensional analog[36] δ_{ijk}) is remarkably constant for solids whose values of the nonlinear susceptibilities vary by four orders of magnitude. This result is often called Miller's phenomenological rule concerning nonlinear susceptibilities.

The fact that $\delta^{(2\omega)}$ is roughly constant for different materials suggests that the ratio λ/N^2 of the anharmonic force constant to the square of the electron density is approximately constant[17] for different solids. We next rewrite equation (9.74) as

$$d^{(2\omega)} = \delta^{(2\omega)}(\chi_L^{(\omega)})^2(\chi_L^{(2\omega)}) \tag{9.75}$$

Since $\delta^{(2\omega)}$ is approximately constant for different solids, equation (9.75) suggests that the large differences observed in the nonlinear susceptibility $d^{(2\omega)}$ for different materials are due to differing values of the linear susceptibilities $\chi_L^{(\omega)}$ and $\chi_L^{(2\omega)}$. We may carry this idea a little further by assuming that the frequency dependence of $\chi_L^{(\omega)}$ is small between the frequencies ω and 2ω, justifying the approximation of $\chi_L^{(2\omega)}$ by $\chi_L^{(\omega)}$. With this assumption, equation (9.75) becomes

$$d^{(2\omega)} = \delta^{(2\omega)}(\chi_L^{(\omega)})^3 \tag{9.76}$$

stating that the nonlinear susceptibility $d^{(2\omega)}$ is, to a good approximation, directly proportional[37] to the cube of the linear susceptibility $\chi_L^{(\omega)}$ for a given crystal. We would therefore expect large values of the nonlinear susceptibility in solids with large values of the *linear* electric susceptibility at optical frequencies. The linear susceptibility is related to the dielectric constant $\varepsilon(\omega)$ at the frequency ω by the relation

$$\varepsilon(\omega) = 1 + 4\pi\chi_L{}^{(\omega)} \tag{9.77}$$

and the dielectric constant is related to the refractive index $n(\omega)$ at frequency ω by

$$\varepsilon(\omega) = (n(\omega))^2 \tag{9.78}$$

for frequencies not near a resonant frequency ω_0 and if the damping, and hence the absorption, is small. It is therefore expected that crystals with large values of the refractive index (or, alternatively, of the optical dielectric constant) will have large values of the linear susceptibility, leading, through equation (9.76), to large values of the nonlinear susceptibility.

Table 9.1 illustrates these ideas by showing values of the nonlinear susceptibility[38] d_{xyz} (units are discussed below) for 10.6-μm radiation for the noncentrosymmetric (but optically isotropic) zinc blende crystals GaP, GaAs, and GaSb. Also shown are the index of refraction[39] n at a wavelenght of 10 μm and the calculated optical dielectric constant $\varepsilon = n^2$. From Table 9.1, we see that the magnitude of the nonlinear susceptibility increases as the optical dielectric constant increases from GaP to GaAs to GaSb. Another example is found in tellurium, which exhibits the very high nonlinear susceptibility[40] of 1600×10^{-9} esu. This figure correlates well with its high values[41] of 6.2 and 4.85 for the refractive index for 8-μm infrared radiation polarized, respectively, parallel and perpendicular to the c axis of the tellurium crystal. It may be noted that this value also correlates well with the high polarizability[42] of the tellurium ion, since materials with high electronic polarizabilities should have high values of the linear susceptibility.

To summarize, the solid state factors favoring relatively large values of the second-order nonlinear susceptibility include a lack of (inversion) symmetry and a high *linear* susceptibility, as reflected by a large value of the optical dielectric constant and refractive index. More detailed models of crystals may be used to calculate and explain values of the nonlinear

Table 9.1. Nonlinear Susceptibility d_{xyz} of GaP, GaAs, and GaSb

Crystal	d_{xyz} (esu)	n	$\varepsilon = n^2$
GaSb	650×10^{-9}	3.84	14.8
GaAs	215×10^{-9}	3.20	10.2
GaP	99×10^{-9}	2.90	8.4

susceptibility. For a discussion of these models, the reader is referred to the literature.[43]

Magnitude of the Nonlinear Susceptibility

We can now make an estimate of the magnitude of the nonlinear susceptibility $d^{(2\omega)}$ using equation (9.72), employing Gaussian units[44] because they are commonly used in the literature. From equation (9.72), we have

$$d^{(2\omega)} = (m\lambda/2N^2e^3)(\chi_L^{(\omega)})^2(\chi_L^{(2\omega)}) \qquad (9.72)$$

so we need an estimate of the linear susceptibilities $\chi_L^{(\omega)}$ and $\chi_L^{(2\omega)}$. From equation (9.77) for the dielectric constant ε,

$$\chi_L^{(\omega)} = (1/4\pi)(\varepsilon(\omega) - 1) \qquad (9.79)$$

Using GaAs as an example, $\varepsilon = 10.2$ from Table 9.1, so $\chi_L^{(\omega)} = 0.732$, and we approximate the value of $\chi_L^{(2\omega)}$ by $\chi_L^{(\omega)}$ on assuming little dispersion. We use Garrett's value[27] of $\lambda \approx 10^{39}$ cm^{-1} sec^{-2} for the anharmonic force constant, take the electron density $N \approx 10^{23}$ cm^{-3}, $m = 9.1 \times 10^{-28}$ g and the charge $e = 1.6 \times 10^{-19}$ C $= 4.8 \times 10^{-10}$ esu (or $g^{1/2}$ cm$^{3/2}$ sec^{-1}). Then, from equation (9.72), one obtains

$$d^{(2\omega)} = \frac{(9.1 \times 10^{-28})(10^{39})(0.732)^3}{2(10^{46})(4.8 \times 10^{10})^3} \quad \frac{\text{g cm}^{-1}\text{ sec}^{-2}}{\text{cm}^{-6}\text{ g}^{3/2}\text{ cm}^{9/2}\text{ sec}^{-3}}$$

leading to the value

$$d^{(2\omega)} = 161 \times 10^{-9}\text{ g}^{-1/2}\text{ cm}^{1/2}\text{ sec} \qquad (9.80)$$

which is also written as $d^{(2\omega)} = 161 \times 10^{-9}$ esu. The magnitude of this value of $d^{(2\omega)}$ is in agreement with that of the experimental value for GaAs quoted in Table 9.1. The experimental values[35] of nonlinear susceptibilities[†] range from about 1×10^{-9} esu for quartz to 1600×10^{-9} esu for tellurium.

It is instructive to obtain an idea of the magnitude of the induced polarization $P^{(2\omega)}$ at the second harmonic frequency 2ω relative to the

---

[†] It is interesting to note here that the *third-order* nonlinear susceptibility is of the order[45] of 10^{-12} esu, several orders of magnitude smaller than the values of the second-order susceptibility quoted above.

polarization $P^{(\omega)}$ at the fundamental frequency ω. From equation (9.63), the ratio

$$\frac{P^{(2\omega)}}{P^{(\omega)}} = \frac{d^{(2\omega)}E^2}{\chi_L^{(\omega)}E} = \frac{d^{(2\omega)}}{\chi_L^{(\omega)}}E \qquad (9.81)$$

is found to depend on the magnitude E of the electric field of the light wave, as well as on the ratio $d^{(2\omega)}/\chi_L^{(\omega)}$ of the nonlinear and linear susceptibilities. Using GaAs as our example again, taking $d^{(2\omega)} = 215 \times 10^{-9}$ esu and $\chi_L^{(\omega)} = 0.732$, one finds that

$$P^{(2\omega)}/P^{(\omega)} = (2.94 \times 10^{-7})E \qquad (9.82)$$

where the ratio will be dimensionless if E is expressed in dynes (esu)$^{-1}$ or statvolts cm^{-1} or g$^{1/2}$ cm$^{-1/2}$ sec^{-1}. From equation (9.82), we can see that the relative magnitude of $P^{(2\omega)}$ will be very small unless the electric field E is quite large. For this reason, nonlinear optics began to develop only with the invention of the laser, which produces intense light beams in which the peak value of the electric field may often be greater than, say, 3×10^9 V m^{-1},[45] which is equal to 10^5 statvolts cm^{-1}. For electric fields of this magnitude, we can see from equation (9.82) that $P^{(2\omega)}$ will be of the order of a few percent of the linear polarization $P^{(\omega)}$.

Wave Equation for the Nonlinear Crystal

We now return to the general wave equation obtained earlier,

$$\text{grad div } \mathbf{E} - \nabla^2\mathbf{E} + \frac{1}{c^2}\frac{\partial^2\mathbf{E}}{\partial t^2} = -\frac{4\pi}{c^2}\frac{\partial^2\mathbf{P}}{\partial t^2} \qquad (9.18)$$

which describes the electric field $\mathbf{E}$ of an electromagnetic wave in a crystal whose time-dependent polarization is $\mathbf{P}(t)$. We consider the wave equation for a crystal in which $\mathbf{P}$ is a nonlinear function of the electric field $\mathbf{E}$, as in

$$\mathbf{P} = \boldsymbol{\chi}_L^{(\omega)} \cdot \mathbf{E} + \mathbf{d}^{(2\omega)} : \mathbf{EE} \qquad (9.83)$$

We will therefore be considering only second-order nonlinear effects; these are due to the term $\mathbf{d}^{(2\omega)} : \mathbf{EE}$ in the polarization. In equation (9.83), $\boldsymbol{\chi}_L^{(\omega)}$ is the linear second-rank electric susceptibility tensor and $\mathbf{d}^{(2\omega)}$ is the second-order nonlinear third-rank electric susceptibility tensor.

As mentioned earlier, a crystal possessing a center of inversion symmetry cannot have a second-order nonlinearity in its relation $\mathbf{P} = f(\mathbf{E})$

between the polarization and the electric field. For such a crystal, an example of which is NaCl, $\mathbf{d}^{(2\omega)} = 0$. In general, we must consider anisotropic (and noncentrosymmetric) crystals. An example is tellurium, which has a relatively large value of $\mathbf{d}^{(2\omega)}$ and is optically anisotropic. It is well known[46] that the divergence of the electric field does not necessarily vanish in a charge-free ($\varrho = 0$) anisotropic dielectric solid. This is a manifestation of the fact, pointed out earlier, that there will in general exist longitudinal (as well as transverse) components of the electric field of an electromagnetic wave in an anisotropic medium. In order to discuss second-order nonlinear optical effects in anisotropic crystals, the proper procedure[47] is to consider the longitudinal component $\mathbf{E}_l$ and the transverse component $\mathbf{E}_t$ of the electric field $\mathbf{E}$, and the corresponding longitudinal and transverse components $\mathbf{P}_l^{NL}$ and $\mathbf{P}_t^{NL}$ of the nonlinear part $\mathbf{P}^{NL}$ of the polarization. The result of this approach, using the general wave equation (9.18), is two wave equations,[48] one for $\mathbf{E}_t$ and one for $\mathbf{E}_l$. The wave equation for the transverse electric field $\mathbf{E}_t$ in the nonlinear medium is of the form

$$\nabla^2\mathbf{E}_t - \frac{1}{c^2}\,(1 + 4\pi\boldsymbol{\chi}_L{}^{(\omega)}) \cdot \frac{\partial^2\mathbf{E}_t}{\partial t^2} = \frac{4\pi}{c^2}\,\frac{\partial^2\mathbf{P}_t{}^{NL}}{\partial t^2} \qquad (9.84)$$

where $\boldsymbol{\chi}_L{}^{(\omega)}$ in (9.84) is the linear susceptibility tensor in expression (9.83). Equation (9.84) has no term in div $\mathbf{E}_t$ because it describes only the transverse component $\mathbf{E}_t$ of the electric field.

Since the treatment in this book is designed to be introductory and simplified, we will not discuss the wave equation for the longitudinal components of the electric field and nonlinear polarization. We will treat the electric field in the nonlinear dielectric as if only the transverse component were present.[49] With this proviso, we drop the "t" indicating "transverse component" in equation (9.84), so that the equation becomes

$$\nabla^2\mathbf{E} - \frac{1}{c^2}\,[1 + 4\pi\boldsymbol{\chi}_L{}^{(\omega)}] \cdot \frac{\partial^2\mathbf{E}}{\partial t^2} = \frac{4\pi}{c^2}\,\frac{\partial^2}{\partial t^2}\,\mathbf{P}^{NL} \qquad (9.85)$$

Further, using the notation of equations (9.66) and (9.83), we have

$$\mathbf{P}^{NL} = \mathbf{d}^{(2\omega)} : \mathbf{EE} \qquad (9.86)$$

Substituting (9.86) into (9.85), we obtain the wave equation

$$\nabla^2\mathbf{E} - \frac{1}{c^2}\,[1 + 4\pi\boldsymbol{\chi}_L{}^{(\omega)}] \cdot \frac{\partial^2\mathbf{E}}{\partial t^2} = \frac{4\pi}{c^2}\,\mathbf{d}^{(2\omega)} : \frac{\partial^2}{\partial t^2}\,(\mathbf{EE}) \qquad (9.87)$$

for the transverse electric field $\mathbf{E}$ in a crystal exhibiting a second-order nonlinearity.

In order to discuss second-order nonlinear optical effects, we can use equation (9.87). However, in the interest of pedagogic simplicity and, hopefully, physical clarity, we will make the following oversimplification. We will treat all of the quantities in equation (9.87), vector, tensor, and dyad alike, as if they were scalars. By doing this, we are tacitly considering a model of a solid in which the polarization and electric field are parallel, so the model is, in that sense, nonanisotropic. (Such a hypothetical model of an optical medium would probably be expected to have a center of symmetry and so not actually exhibit a second-order nonlinearity.) However, this somewhat unrealistic model of a nonlinear solid (perhaps reminiscent of "jellium" in that it does not really exist) will allow a mathematically simple treatment appropriate for an introductory discussion and will also still present the essential physical features of the second-order nonlinear processes. The reader interested in realistic and complete discussions is referred to the literature.[13,47]

In this spirit, we rewrite the wave equation (9.87) in a scalar form as

$$\nabla^2 E - \frac{1}{c^2}(1 + 4\pi\chi_L{}^{(\omega)})\frac{\partial^2 E}{\partial t^2} - \frac{4\pi}{c^2}d^{(2\omega)}\frac{\partial^2}{\partial t^2}(E^2) = 0 \quad (9.88)$$

where $d^{(2\omega)}$ is the magnitude of the second-order nonlinear susceptibility and E is the magnitude of the electric field. Substituting the frequency-dependent dielectric constant $\varepsilon(\omega)$ defined by

$$\varepsilon(\omega) \equiv 1 + 4\pi\chi_L{}^{(\omega)} \quad (9.89)$$

into equation (9.88), we obtain

$$\nabla^2 E - \frac{1}{c^2}\varepsilon(\omega)\frac{\partial^2 E}{\partial t^2} - \frac{4\pi}{c^2}d^{(2\omega)}\frac{\partial^2}{\partial t^2}(E^2) = 0 \quad (9.90)$$

as the scalar wave equation for the electric field magnitude E in our dielectric medium with a second-order nonlinearity described by the term in $d^{(2\omega)}$. We will use equation (9.90) to discuss the propagation and interaction, via the nonlinear term in E^2, of electromagnetic waves in a nonlinear crystal.

Wave Propagation and Interaction in a Nonlinear Crystal

To discuss the physics of a number of nonlinear optical devices, it is necessary to consider the propagation and interaction of several electromagnetic waves in a nonlinear crystal. The propagation of these waves is governed by the wave equation (9.90). We consider the waves

$$E_1(\omega_1, z) = \tfrac{1}{2}\{E_{10}(z)\exp[j(\omega_1 t - k_1 z)] + \text{c.c.}\} \tag{9.91}$$

$$E_2(\omega_2, z) = \tfrac{1}{2}\{E_{20}(z)\exp[j(\omega_2 t - k_2 z)] + \text{c.c.}\} \tag{9.92}$$

$$E_3(\omega_3, z) = \tfrac{1}{2}\{E_{30}(z)\exp[j(\omega_3 t - k_3 z)] + \text{c.c.}\} \tag{9.93}$$

with frequencies $\omega_1, \omega_2, \omega_3$. In equations (9.91)–(9.93), k_1, k_2, and k_3 are the wave numbers, c.c. means complex conjugate, and we indicate an explicit z dependence of the amplitudes $E_{10}(z)$, $E_{20}(z)$, and $E_{30}(z)$ of these plane waves moving in the z direction. This is done because, as we will see, the nonlinear interaction between the various waves can change the amplitudes. The aim is to find expressions giving the z dependence of the amplitudes E_{10}, E_{20}, and E_{30}.

As mentioned above, we will not proceed in the most general manner[50] but instead will ignore the anisotropy of real solids and consider the three waves above, all of whose wave vectors are in the same direction. Following Yariv,[51] the electric field E at any point z is given by

$$E = E_1(\omega_1, z) + E_2(\omega_2, z) + E_3(\omega_3, z) \tag{9.94}$$

We want to consider the wave equation (9.90), which we rewrite as

$$\nabla^2 E - \frac{1}{c^2}\,\varepsilon(\omega)\,\frac{\partial^2 E}{\partial t^2} = \frac{4\pi}{c^2}\,d^{(2\omega)}\,\frac{\partial^2}{\partial t^2}\,(E^2) \tag{9.95}$$

where E is given by equation (9.94), E^2 is given by

$$E^2 = EE^* \tag{9.96}$$

and the asterisk indicates the complex conjugate. The next step is to substitute expressions (9.94) and (9.96) for E and E^2 into the wave equation (9.95); but, since this leads to an extremely long result, we shall merely describe what happens to the various terms in (9.95).

First, since (9.94) states that E is not a function of x or y, the term in $\nabla^2 E$ becomes

$$\nabla^2 E = \frac{\partial^2}{\partial z^2}\,E_1(\omega_1, z) + \frac{\partial^2}{\partial z^2}\,E_2(\omega_2, z) + \frac{\partial^2}{\partial z^2}\,E_3(\omega_3, z) \tag{9.97}$$

Second, the term in $\partial^2 E/\partial t^2$ is given by

$$\frac{\partial^2 E}{\partial t^2} = \frac{\partial^2}{\partial t^2}\,E_1(\omega_1, z) + \frac{\partial^2}{\partial t^2}\,E_2(\omega_2, z) + \frac{\partial^2}{\partial t^2}\,E_3(\omega_3, z) \tag{9.98}$$

The result of equations (9.97) and (9.98) is that the left-hand side of the wave equation (9.95) is a sum of terms, each of which is a function of one of the frequencies ω_1 or ω_2 or ω_3.

If we consider the right-hand side

$$\frac{4\pi}{c^2} d^{(2\omega)} \frac{\partial^2}{\partial t^2} (E^2) \qquad (9.99)$$

of the wave equation, we can see that E^2 will contain all of the sums and differences of the three frequencies ω_1, ω_2, and ω_3. Thus, E^2 will contain terms oscillating at the frequencies zero, $2\omega_1$, etc., $\omega_1 + \omega_2$, etc., and $\omega_1 - \omega_2$, etc. The conclusion is that the driving term (9.99) on the right-hand side of the wave equation (9.95) will *not* in general contain terms oscillating at the frequencies ω_1, ω_2, and ω_3 of the left-hand side of the equation. In general then, there will not be traveling wave solutions of (9.95) because the driving term (9.99) is not synchronous (i.e., does not contain the same frequencies) with the left-hand side of the wave equation.

However, if the driving term *is* synchronous with the right-hand side of the wave equation, then there will be traveling wave solutions of (9.95). We may consider one such situation by choosing the case[52] for which

$$\omega_1 + \omega_2 = \omega_3 \qquad (9.100)$$

A result of equation (9.100) is that the driving term (9.99), which includes a factor

$$-\tfrac{1}{2}(\omega_1 + \omega_2)^2 E_{10} E_{20} \exp[j(\omega_1 + \omega_2)t - j(k_1 + k_2)z] \qquad (9.101)$$

(and its complex conjugate) in the sum frequency $\omega_1 + \omega_2$, will, because of the condition expressed by (9.100), contain a term

$$-\tfrac{1}{2}\omega_3^2 E_{10} E_{20} \exp[j\omega_3 t - j(k_1 + k_2)z] \qquad (9.102)$$

Since the requirement (9.100) has made the expression (9.101) oscillate at the frequency ω_3, as in (9.102), this term can "drive" the terms at ω_3 on the left-hand side of the wave equation. In the same way, the terms oscillating at $\omega_3 - \omega_2$ and $\omega_3 - \omega_1$ in the source term (9.99) can drive oscillations at frequencies ω_1 and ω_2, respectively.

We may consider the requirement (9.100), rewritten

$$\hbar\omega_1 + \hbar\omega_2 = \hbar\omega_3 \qquad (9.103)$$

as expressing the conservation of energy in a process in which photons of energies $\hbar\omega_1$ and $\hbar\omega_2$ are destroyed and a photon of energy $\hbar\omega_3$ is created. Equation (9.103) can also be thought of as describing the inverse process, in which a photon $\hbar\omega_3$ is destroyed with the creation of photons of energies $\hbar\omega_1$ and $\hbar\omega_2$. Equations (9.100) and (9.103) thus describe processes in which energy flows from electromagnetic fields of frequency ω_3 to fields at frequencies ω_1 and ω_2, or vice versa. It is this interchange of energy between electromagnetic fields of different frequencies, due to the nonlinear susceptibility of the solid, that is the physical basis of the various applications we will discuss.

We return to consideration of the wave equation subject to the condition that $\omega_1 + \omega_2 = \omega_3$. There will be terms $\partial^2 E_3/\partial z^2$ and $\partial^2 E_3/\partial t^2$, oscillating at frequency ω_3, on the left-hand side of the wave equation. The wave equation for E_3 at frequency ω_3 is

$$\frac{\partial^2 E_3}{\partial z^2} - \frac{1}{c^2}\varepsilon(\omega_3)\frac{\partial^2 E_3}{\partial t^2} = \frac{4\pi}{c^2}d^{(2\omega)}\left(-\frac{1}{2}\omega_3^2 E_{10}E_{20}e^{j[\omega_3 t-(k_1+k_2)z]} + \text{c.c.}\right)$$

(9.104)

where $\varepsilon(\omega_3)$ is the dielectric constant at frequency ω_3, E_3 is a function of ω_3 given by (9.93), and c.c. means complex conjugate.

We now consider the various terms in equation (9.104) in order to obtain an expression for the variation in space (i.e., with z coordinate) of the electric field $E_3(\omega_3, z)$. From equation (9.93), we have

$$\frac{\partial^2 E_3}{\partial z^2} = \frac{1}{2}\left[(-k_3^2)E_{30} - 2jk_3\frac{dE_{30}}{dz} + \frac{d^2E_{30}}{dz^2}\right]e^{j(\omega_3 t-k_3 z)} + \text{c.c.} \qquad (9.105)$$

and that

$$\frac{\partial^2 E_3}{\partial t^2} = \frac{1}{2}(-\omega_3^2 E_{30}e^{j(\omega_3 t-k_3 z)} + \text{c.c.}) \qquad (9.106)$$

Results (9.105) and (9.106) are to be substituted into the form (9.104) of the wave equation describing the electromagnetic field oscillating at frequency ω_3.

We next make the assumption[53] that the relative change in the amplitude E_{30} per wavelength is small because the nonlinear susceptibility is small compared to the linear susceptibility. Since dE_{30}/dz is the change in E_{30} per unit length, $\lambda_3(dE_{30}/dz)$ is the change in E_{30} per wavelength λ_3, and the assumption states that this is small compared to the amplitude E_{30} itself. This condition[54] is expressed as

$$E_{30} \gg \lambda_3\left|\frac{dE_{30}}{dz}\right| \qquad (9.107)$$

which can be rewritten as

$$k_3 \left| \frac{dE_{30}}{dz} \right| \gg \left| \frac{d^2 E_{30}}{dz^2} \right| \tag{9.108}$$

The assumption contained in equation (9.108) allows us to neglect the term in $d^2 E_{30}/dz^2$ in equation (9.105), which then becomes

$$\frac{\partial^2 E_3}{\partial z^2} = \frac{1}{2}\left(-k_3{}^2 E_{30} - 2jk_3 \frac{dE_{30}}{dz} \right) e^{j(\omega_3 t - k_3 z)} + \text{c.c.} \tag{9.109}$$

On substituting the expressions (9.109) and (9.106) for $\partial^2 E_3/\partial z^2$ and $\partial^2 E_3/\partial t^2$ into the wave equation (9.104), one obtains, on cancelling the factors of $\frac{1}{2}$,

$$\left(-k_3{}^2 E_{30} - 2jk_3 \frac{dE_{30}}{dz} \right) e^{j(\omega_3 t - k_3 z)} + \left(-k_3{}^2 E_{30}^* + 2jk_3 \frac{dE_{30}^*}{dz} \right) e^{-j(\omega_3 t - k_3 z)}$$

$$+ \frac{\varepsilon(\omega_3)}{c^2} \omega_3{}^2 (E_{30} e^{j(\omega_3 t - k_3 z)} + E_{30}^* e^{-j(\omega_3 t - k_3 z)})$$

$$= -\frac{4\pi}{c^2} d^{(2\omega)} \omega_3{}^2 (E_{10} E_{20} e^{j[\omega_3 t - (k_1 + k_2)z]} + E_{10}^* E_{20}^* e^{-j[\omega_3 t - (k_1 + k_2)z]}) \tag{9.110}$$

which can be rewritten as

$$\left[\left(-k_3{}^2 E_{30} - 2jk_3 \frac{dE_{30}}{dz} + \frac{\omega_3{}^2 \varepsilon(\omega_3)}{c^2} E_{30} \right) e^{j(\omega_3 t - k_3 z)} \right.$$

$$\left. + \left(-k_3{}^2 E_{30}^* + 2jk_3 \frac{dE_{30}^*}{dz} + \frac{\omega_3{}^2 \varepsilon(\omega_3)}{c^2} E_{30}^* \right) e^{-j(\omega_3 t - k_3 z)} \right]$$

$$= \frac{-4\pi}{c^2} d^{(2\omega)} \omega_3{}^2 (E_{10} E_{20} e^{j[\omega_3 t - (k_1 + k_2)z]} + E_{10}^* E_{20}^* e^{-j[\omega_3 t - (k_1 + k_2)z]}) \tag{9.111}$$

We note next from the wave equation (9.95) that the phase velocity v of a wave is given by

$$v^2 = c^2/\varepsilon \tag{9.112}$$

Since it is also true that

$$v = \omega/k \tag{9.113}$$

we have the result that

$$k^2 = \varepsilon \omega^2/c^2 \tag{9.114}$$

Applying this result to the wave E_3 we are considering,

$$k_3{}^2 = \varepsilon(\omega_3)\omega_3{}^2/c^2 \tag{9.115}$$

which when substituted into equation (9.111), gives

$$\left(-2jk_3\frac{dE_{30}}{dz}\right)\exp[j(\omega_3t-k_3z)]+\left(2jk_3\frac{dE_{30}^*}{dz}\right)\exp[-j(\omega_3t-k_3z)]$$

$$=-\frac{4\pi}{c^2}d^{(2\omega)}\omega_3{}^2(E_{10}E_{20}e^{j[\omega_3t-(k_1+k_2)z]}+E_{10}^*E_{20}^*e^{-j[\omega_3t-(k_1+k_2)z]})\ (9.116)$$

Equating coefficients of $\exp(j\omega_3t)$ on both sides of equation (9.116) leads to

$$e^{-jk_3z}\frac{dE_{30}}{dz}=\frac{4\pi}{2jc^2}\frac{\omega_3{}^2}{k_3}d^{(2\omega)}E_{10}E_{20}\exp[-j(k_1+k_2)z]\qquad(9.117)$$

which can be written as

$$\frac{dE_{30}}{dz}=-j\frac{2\pi}{c^2}\frac{\omega_3{}^2}{k_3}d^{(2\omega)}E_{10}E_{20}\exp[-j(k_1+k_2-k_3)z]\qquad(9.118)$$

Equating the coefficients of $\exp(-j\omega_3t)$ in equation (9.116) leads to the complex conjugate of equation (9.118). In view of equation (9.115), one finds that

$$\omega_3{}^2/c^2k_3=\omega_3/c\varepsilon_3{}^{1/2}\qquad(9.119)$$

where the notation $\varepsilon_3\equiv\varepsilon(\omega_3)$ has been introduced.

Using result (9.119), the expression (9.118) can be rewritten as

$$\frac{dE_{30}}{dz}=-j\frac{2\pi\omega_3}{c\varepsilon_3{}^{1/2}}d^{(2\omega)}E_{10}E_{20}\exp[-j(k_1+k_2-k_3)z]\qquad(9.120)$$

an equation that describes the variation in space of the amplitude $E_{30}(z)$ of the electric field $E_3(\omega_3,z)$ of frequency ω_3. Using the units of $d^{(2\omega)}$ discussed in an earlier section, a dimensional analysis of equation (9.120) shows that the right-hand side has the same units (i.e., electric field per unit length) as does the left-hand side.

In the same manner as that in which equation (9.120) was obtained, expressions for the spatial variation of the amplitudes $E_{10}(z)$ and $E_{20}(z)$ may be found, giving the full set of equations below. These are

$$\frac{dE_{10}^*}{dz}=j\frac{2\pi\omega_1}{c\varepsilon_1{}^{1/2}}d^{(2\omega)}E_{20}E_{30}^*\exp[-j(\Delta k)z]\qquad(9.121)$$

$$\frac{dE_{20}^*}{dz}=j\frac{2\pi\omega_2}{c\varepsilon_2{}^{1/2}}d^{(2\omega)}E_{10}E_{30}^*\exp[-j(\Delta k)z]\qquad(9.122)$$

$$\frac{dE_{30}}{dz}=-j\frac{2\pi\omega_3}{c\varepsilon_3{}^{1/2}}d^{(2\omega)}E_{10}E_{20}\exp[-j(\Delta k)z]\qquad(9.123)$$

where $\varepsilon_1 \equiv \varepsilon(\omega_1)$, etc. are the dielectric constants of the nonlinear crystal at the frequencies ω_1, ω_2, and ω_3, and where the notation

$$\Delta k \equiv k_1 + k_2 - k_3 \qquad (9.124)$$

has been introduced. In the derivation of (9.121)–(9.123), it has been tacitly assumed that the nonlinear susceptibility $d^{(2\omega)}$ is independent of frequency. This is a good approximation[55] when all of the frequencies involved are far from resonances.

Equations (9.121)–(9.123) describe the spatial variation of the amplitudes $E_{10}(z)$, $E_{20}(z)$, and $E_{30}(z)$ of the three interacting waves $E_1(\omega_1, z)$, $E_2(\omega_2, z)$, and $E_3(\omega_3, z)$ in a dielectric,[56] subject to the condition (9.100) that $\omega_1 + \omega_2 = \omega_3$. The interaction is via the nonlinear term $d^{(2\omega)}E^2$ in the polarization given by equation (9.63); this is reflected in the presence of the nonlinear susceptibility $d^{(2\omega)}$ in each of these equations. The fact that the three waves are interacting is seen in the result that equations (9.121)–(9.123) are coupled differential equations because the spatial variation of any one amplitude is a function of the other amplitudes. Finally, as we shall discuss in detail later, the spatial variation of each amplitude is a function of the quantity Δk defined by equation (9.124).

Equations (9.21)–(9.123) are the basic equations governing the interaction of three electromagnetic waves (of frequencies ω_1, ω_2, ω_3) in the nonlinear solid, subject to the requirement $\omega_1 + \omega_2 = \omega_3$. We now apply these equations to the discussion of several processes of device interest in nonlinear solids.

Optical Second Harmonic Generation

The first application of the nonlinear optical properties of solids that we shall consider is the generation of the second harmonic frequency 2ω of a light wave of frequency ω. The physical process considered is one in which light of frequency ω is incident on a crystal with a nonlinear susceptibility $d^{(2\omega)}$. The nonlinearity produces a component of the electric polarization at frequency 2ω, resulting in the generation of radiation at the second harmonic frequency 2ω. One can also speak of the incident radiation of frequency ω being mixed with itself to produce radiation at the sum frequency $\omega + \omega = 2\omega$. Experimentally, a laser is necessary as the source of incident radiation in order that the incident electric field be large. The original experiment[57] involved a ruby laser beam, with a wavelength of 6943 Å, incident on a quartz crystal. The radiation emerging from the crystal was found to contain radiation at one-half the incident wavelength, i.e., at the second harmonic frequency.

We will use equations (9.121)–(9.123) to discuss the process of optical second harmonic generation. Consider a nonlinear crystal of length L, upon whose face at $z = 0$ is incident an input wave

$$E_1(\omega, z) = E_{10} \exp[j(\omega t - k_1 z)] \qquad (9.125)$$

of frequency ω. Equation (9.125) is obtained from equation (9.91) by setting $\omega_1 = \omega$. This input wave is then mixed with itself in the nonlinear crystal to produce the output wave at the second harmonic frequency 2ω. Using the condition

$$\omega_3 = \omega_1 + \omega_2 \qquad (9.100)$$

and the fact that $\omega_2 = \omega$ since identical frequencies are being mixed, we obtain $\omega_3 = 2\omega$, leading to the choice of

$$E_3(2\omega, z) = E_{30} \exp[j(2\omega t - k_3 z)] \qquad (9.126)$$

as the output wave. Equation (9.126) is obtained from (9.93). The complex conjugate term has been ignored for simplicity, and we keep in mind that the amplitudes E_{10} and E_{30} of the input and output waves are functions of the space coordinate z.

With $\omega_3 = 2\omega$, equation (9.123) becomes

$$\frac{dE_{30}}{dz} = \frac{-4\pi j\omega\, d^{(2\omega)}}{c\varepsilon_3^{1/2}} E_{10}^2 \exp[-j(2k_1 - k_3)z] \qquad (9.127)$$

since $E_{20} = E_{10}$ because two identical waves are mixed in this process. In addition, because $\omega_1 = \omega_2$, the wave numbers k_1 and k_2 are equal, so, in this case,

$$\Delta k \equiv k_1 + k_2 - k_3 = 2k_1 - k_3 \qquad (9.128)$$

a result that has also been included in equation (9.127). The nonlinear susceptibility $d^{(2\omega)}$ is given by (9.65) or (9.72). Equation (9.127) describes the spatial variation of the amplitude E_{30} of the second harmonic wave in the crystal. Our aim is to integrate this equation in order to obtain the amplitude $E_{30}(z)$ as a function of space coordinate z in the crystal.

In general, as seen from equation (9.91), the amplitude E_{10} of the input wave at frequency ω will be a function of z. In order to integrate (9.127) easily, we consider the simple case in which the fractional decrease in the energy of the input wave E_1, due to conversion to the second harmonic, is small. This case[58] is a physically reasonable one to discuss because the amplitude E_{10} of the incident wave (from, say, a laser) is large, and the magnitude of the nonlinear interaction is small. With this assumption, we may treat the amplitude E_{10} in equation (9.127) as a constant. If we integrate that equation over the length of the crystal from $z = 0$ to $z = L$, we obtain

$$E_{30}(L) - E_{30}(0) = \frac{-4\pi j\omega\, d^{(2\omega)} E_{10}^2}{c\varepsilon_3^{1/2}} \int_0^L \exp[-j(\varDelta k)z]\, dz \qquad (9.129)$$

In (9.129), $E_{30}(z)$ is the magnitude of the amplitude E_{30} of the second harmonic wave at the point z of the crystal, and $\varDelta k = 2k_1 - k_3$ from equation (9.128).

We next choose the boundary condition that there be no input wave at the second harmonic frequency at $z = 0$, so we have

$$E_{30}(0) = 0 \qquad (9.130)$$

The amplitude $E_{30}(L)$ of the second harmonic wave at the output $z = L$ of the crystal is found by integrating (9.129), giving

$$E_{30}(L) = \frac{-4\pi j\omega\, d^{(2\omega)} E_{10}^2}{c\varepsilon_3^{1/2}} \frac{e^{-j(\varDelta k)L} - 1}{-j(\varDelta k)} \qquad (9.131)$$

Multiplying (9.131) by its complex conjugate and multiplying and dividing by L^2 gives

$$E_{30}(L)E_{30}^*(L) = \frac{16\pi^2\omega^2[d^{(2\omega)}]^2 E_{10}^4 L^2}{c^2\varepsilon_3} \frac{(e^{-j(\varDelta k)L} - 1)(e^{j(\varDelta k)L} - 1)}{(\varDelta k)^2 L^2} \qquad (9.132)$$

which, using the trigonometric identity $1 - \cos x = 2\sin^2(x/2)$, becomes

$$E_{30}(L)E_{30}^*(L) = \frac{16\pi^2\omega^2[d^{(2\omega)}]^2 E_{10}^4 L^2}{c^2\varepsilon_3} \left(\frac{\sin[\tfrac{1}{2}(\varDelta k)L]}{\tfrac{1}{2}(\varDelta k)L} \right)^2 \qquad (9.133)$$

We now relate the square $E_{30}E_{30}^*$ of the electric field amplitude of the second harmonic wave to the intensity. Recalling[59] that the intensity[†] I of a plane electromagnetic wave is the time-averaged energy per unit time per unit area transmitted by the wave, we have the result that the intensity $I_{30}(L)$ of the second harmonic wave at $z = L$ is given by[59]

$$I_{30}(L) = (n_3 c/8\pi)E_{30}(L)E_{30}^*(L) \qquad (9.134)$$

In equation (9.134), n_3 is the index of refraction of the crystal at the second

---

[†] Strictly speaking, the term irradiance[59] should be used for the (average) energy per unit time per unit area transmitted by the wave. However, intensity seens to be a more common usage.

harmonic frequency $\omega_3 = 2\omega$. Substituting (9.134) into (9.133), and using the fact that $n_3^2 = \varepsilon_3$, we find

$$I_{30}(L) = \frac{2\pi\omega^2[d^{(2\omega)}]^2 E_{10}^4 L^2}{c\varepsilon_3^{1/2}} \left(\frac{\sin[\frac{1}{2}(\Delta k)L]}{\frac{1}{2}(\Delta k)L} \right)^2 \qquad (9.135)$$

This equation gives the intensity $I_{30}(L)$ of the wave of frequency 2ω at the output end $z = L$ of the crystal.

Equation (9.135) shows that the intensity $I_{30}(L)$ is proportional to the factor

$$\left(\frac{\sin[\frac{1}{2}(\Delta k)L]}{\frac{1}{2}(\Delta k)L} \right)^2 \qquad (9.136)$$

indicating by comparison with results[60] of optics that an interference effect is taking place. We consider first the variation of $I_{30}(L)$ as a function of the length L of the crystal for a given *nonzero* value of Δk. From equation (9.135), we have

$$I_{30}(L) = \frac{2\pi\omega^2[d^{(2\omega)}]^2 E_{10}^4}{c\varepsilon_3^{1/2}} \left(\frac{2}{\Delta k} \right)^2 \sin^2\left[\frac{1}{2}(\Delta k)L \right] \qquad (9.137)$$

showing that $I_{30}(L)$ is proportional to $\sin^2[\frac{1}{2}(\Delta k)L]$, a function that has zeros at $(\Delta k)L = 0, 2\pi, 4\pi, \ldots$ and has maxima at $(\Delta k)L = \pi, 3\pi, 5\pi, \ldots$ etc. Figure 9.2 shows $I_{30}(L)$ from equation (9.137) plotted (in arbitrary units) as a function of L [which is expressed in units of $\pi/(\Delta k)$], for an

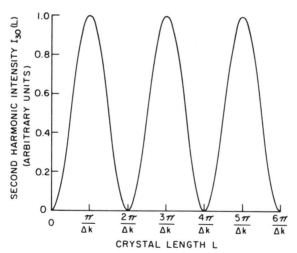

Figure 9.2. Second harmonic intensity $I_{30}(L)$ from equation (9.137) plotted (in arbitrary units as a function of crystal length L [in units of $\pi/(\Delta k)$) for an arbitrary nonzero value of $\Delta k = 2k_1 - k_3$.

arbitrary nonzero value of $\Delta k = 2k_1 - k_3$. From this figure, it can be seen that the intensity of the second harmonic wave varies periodically with the length L of the crystal, reaching a maximum at $L = \pi/\Delta k$, $3\pi/\Delta k$, ... etc. (We continue to keep in mind that we are considering the case for which $\Delta k \neq 0$; the significance of the case for which $\Delta k = 0$ will be discussed later.) The length

$$L_c \equiv \pi/\Delta k \qquad (9.138)$$

at which a maximum in the second harmonic intensity occurs is called the coherence length. The periodicity shown in Figure 9.2 is due to the term of the form (9.136) in the expression (9.135) for the second harmonic intensity $I_{30}(L)$, and this periodicity is due to the presence of an interference effect. It can be shown[61,62] that, for $\Delta k \neq 0$, interference takes place between the second harmonic wave generated at one point $z = z_1$ of the crystal and the second harmonic wave generated at a second point $z = z_2$. The phase differences between these waves are such that the following takes place. Starting from $z = 0$ (the input face of the crystal), the input wave at frequency ω will generate the second harmonic at 2ω over the first coherence length $0 \leq z \leq L_c$ of the crystal. The second harmonic intensity $I_{30}(L)$ reaches a maximum for a crystal length L equal to L_c. However, if L is greater than the coherence length, the second harmonic wave will beat with the input wave in the second coherence length $L_c \leq z \leq 2L_c$. The result of this interference is the generation of the difference frequency $2\omega - \omega = \omega$. Energy is thus transferred from the second harmonic to the fundamental and the intensity of the second harmonic wave decreases, reaching zero at a crystal length $L = 2L_c$. In this way, for $\Delta k \neq 0$, energy is interchanged between the fundamental and second harmonic waves. Since $I_{30}(L)$ has its maximum value for a crystal length L equal to the coherence length L_c, the coherence length is the maximum length of the crystal useful in producing the second harmonic.

We may discuss the magnitude of the coherence length, using equation (9.138), by calculating $\Delta k = 2k_1 - k_3$. Since $k_1 = \omega n_1/c$ and $k_3 = 2\omega n_3/c$, where n_1 and n_3 are the refractive indices of the crystal at the frequencies ω and 2ω, respectively, we have

$$\Delta k = (2\omega/c)(n_1 - n_3) \qquad (9.139)$$

for second harmonic generation. Because of dispersion, the refractive indices n_1 and n_3 will not usually be equal. In general, then, Δk will not be zero and the process of second harmonic generation will not be efficient because of the interference effects which result in energy transfer back and forth between the fundamental and second harmonic waves. This inefficiency is reflected in the typical magnitude of the coherence length L_c, which, from (9.138) and (9.139), can be written as

$$L_c = \frac{\pi c}{2\omega(n_1 - n_3)} = \frac{\lambda_0}{4(n_1 - n_3)} \tag{9.140}$$

where λ_0 is the wavelength in free space of the fundamental wave of frequency ω. From equation (9.140), we can see that the coherence length will be of the order of $25\,\lambda_0$ on using an estimate of 0.01 for the quantity $n_1 - n_3$. This leads to a magnitude of perhaps 0.001 cm for the coherence length if λ_0 is a visible wavelength. We conclude that L_c is generally small unless the difference Δk in wave number can be very small, or preferably, equal to zero.

Next, we consider the intensity $I_{30}(L)$ when $\Delta k = 0$. From equation (9.135), we see that the second harmonic intensity $I_{30}(L)$ is proportional to

$$\left(\frac{\sin[\frac{1}{2}(\Delta k)L]}{[\frac{1}{2}(\Delta k)L]} \right)^2 \tag{9.141}$$

Figure 9.3 shows $I_{30}(L)$ from equation (9.135) plotted (in arbitrary units) as a function of Δk, which is expressed in units of π/L where L is the fixed value of the crystal length. We recall[60] that the function (9.141) has its maximum value of unity for $\frac{1}{2}(\Delta k)L = 0$, and has the minimum value of zero for $\frac{1}{2}(\Delta k)L = \pm\pi, \pm2\pi, \ldots$ etc. This means that the intensity $I_{30}(L)$ has its maximum for $\Delta k = 0$ and has zeros at $\Delta k = 2\pi/L, 4\pi/L$, etc., as shown in the figure. It is clear that, for a given crystal length L, the second harmonic intensity $I_{30}(L)$ is largest when $\Delta k = 0$, and falls rapidly with

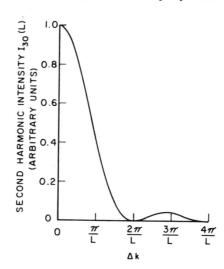

Figure 9.3. Second harmonic intensity $I_{30}(L)$ from equation (9.135) plotted (in arbitrary units) as a function of $\Delta k = 2k_1 - k_3$ for a fixed value of the crystal length L. The quantity Δk is plotted in units of π/L.

increasing nonzero values of Δk. It is therefore desirable to arrange things such that $\Delta k = 0$, a procedure called phase matching or index matching, and which is discussed briefly later. Using equation (9.138), the same conclusions may be restated in terms of the coherence length. The second harmonic intensity is a maximum when the coherence length is infinite (for $\Delta k = 0$), and falls sharply as the coherence length decreases (for $\Delta k \neq 0$).

Having examined the effect of Δk (or, alternatively, of the coherence length) on the second harmonic intensity $I_{30}(L)$, we may now examine the influence of other factors on the intensity. From the definition of intensity introduced in equation (9.134), the intensity $I_{10}(0)$ of the input fundamental wave of frequency ω at $z = 0$ is given by

$$I_{10}(0) = (n_1 c/8\pi)E_{10}(0)E_{10}^*(0) \tag{9.142}$$

where n_1 is the refractive index at frequency ω. Recalling our assumption that the amplitude E_{10} is approximately constant, equation (9.142) becomes

$$I_{10}(0) = (n_1 c/8\pi)E_{10}^2 \equiv I_{10} \tag{9.143}$$

Using equation (9.143), the expression (9.135) for $I_{30}(L)$ may be written as

$$I_{30}(L) = \frac{128\pi^3\omega^2(d^{(2\omega)})^2 L^2}{c^3\varepsilon_1\varepsilon_3^{1/2}} I_{10}^2\left(\frac{\sin x}{x}\right)^2 \tag{9.144}$$

where $x \equiv (\Delta k)L/2$ and $\varepsilon_1 = n_1^2$ is the dielectric constant at frequency ω. From equation (9.144), we see that the second harmonic intensity is proportional to the *square* of the input intensity at the fundamental frequency. Further, if we assume that the situation in which $\Delta k = 0$ has been achieved, then $[(\sin x)/x]^2 = 1$, and equation (9.144) gives us

$$[I_{30}(L)]_{max} = \frac{128\pi^3\omega^2(d^{(2\omega)})^2 L^2 I_{10}^2}{c^3\varepsilon_1\varepsilon_3^{1/2}} \tag{9.145}$$

as the expression for the *maximum* second harmonic intensity. For the case for which $\Delta k = 0$ and interference effects do not decrease the output, equation (9.145) shows that the second harmonic intensity $I_{30}(L)$ is proportional to the square of the crystal length L. Finally, we see from the expressions above that, in any case, $I_{30}(L)$ is proportional to the square of the nonlinear susceptibility $d^{(2\omega)}$ and thus will depend on the crystal used in the experiment.

We may calculate the efficiency e of second harmonic generation, defined by

$$e \equiv I_{30}(L)/I_{10}(0) = I_{30}(L)/I_{10} \tag{9.146}$$

From equation (9.144), we find that

$$e = \frac{128\pi^3\omega^2(d^{(2\omega)})^2L^2I_{10}}{c^3\varepsilon_1\varepsilon_3^{1/2}} \left(\frac{\sin x}{x}\right)^2 \tag{9.147}$$

showing that the *efficiency* of second harmonic generation is proportional to the intensity I_{10} of the input wave at the fundamental frequency. If the condition $\Delta k = 0$ is achieved, then the factor $[(\sin x)/x]^2 = 1$ and the maximum efficiency is given by

$$e_{max} = \frac{128\pi^3\omega^2(d^{(2\omega)})^2L^2I_{10}}{c^3\varepsilon_1\varepsilon_3^{1/2}} \tag{9.148}$$

From equations (9.147) and (9.148), the efficiency e is seen to depend on the input intensity I_{10}, so one way to increase e is to increase I_{10}. A method[63] of achieving this is to place the nonlinear crystal inside a laser resonator, which then supplies the input wave. In such a situation, the input intensity can be very large, thereby increasing the efficiency e.

It might appear, from equations (9.147) and (9.148), that the conversion efficiency e could become larger than unity. This difficulty is only apparent, however, and stems from the assumption that the input wave is of constant amplitude, used in deriving these results. Actually, calculation[58] of the conversion efficiency for the case of a depleted input wave shows that the efficiency approaches unity asymptotically as the input intensity becomes very large. The assumption of constant input amplitude is a good approximation[58] for conversion efficiencies e less than about 0.1.

To summarize, we can see from the results developed in this section that efficient second harmonic generation is favored by the following factors. First, use of a crystal with a high value of the nonlinear susceptibility. Second, a high value of the intensity of the input wave at the fundamental frequency. Third, achievement as closely as possible of the condition $\Delta k = 0$, in order that the second harmonic intensity $I_{30}(L)$ and the efficiency e be maximized.

The aim of this section has been the use of the coupled-amplitude equations (9.121)–(9.123) to discuss, in an introductory and approximate way, some aspects of the physics of harmonic generation. For more complete discussions, including more realistic treatments, the reader is referred to the books by Yariv[15,16] and the review article by Akhmanov et al.[64] The latter paper includes a discussion of nonlinear materials for second harmonic generation applications.

Phase Matching (Index Matching) in Second Harmonic Generation

In our discussion of second harmonic generation, we considered the effect on the second harmonic intensity $I_{30}(L)$ of the quantity Δk defined by equation (9.124) and given by

$$\Delta k = 2k_1 - k_3 \qquad (9.128)$$

for the case of second harmonic generation. It was shown that $I_{30}(L)$ and the efficiency e were both maximized for $\Delta k = 0$. The aim of this section is a brief discussion of the meaning of the condition $\Delta k = 0$ in the process of second harmonic generation.

If $\Delta k = 0$, then equation (9.128) states that

$$2k_1 = k_3 \qquad (9.149)$$

where k_1 is the wave vector of the fundamental of frequency ω and k_3 is that of the second harmonic of frequency 2ω. Since the phase velocity v is given by

$$v = \omega/k \qquad (9.150)$$

for a wave of frequency ω, the condition (9.149) is equivalent to

$$v(\omega) = v(2\omega) \qquad (9.151)$$

where $v(\omega)$ and $v(2\omega)$ are the phase velocities of the waves of frequencies ω and 2ω, respectively. The condition (9.149) that $\Delta k = 0$ in second harmonic generation therefore is equivalent to requiring that the phase velocities $v(\omega)$ and $v(2\omega)$ of the fundamental and second harmonic be equal. Since the phase velocity $v = c/n$, where n is the index of refraction, equation (9.151) is equivalent to

$$n(\omega) = n(2\omega) \qquad (9.152)$$

where $n(\omega)$ and $n(2\omega)$ are the refractive indices of the nonlinear crystal at the frequencies ω and 2ω, respectively. The condition that $\Delta k = 0$ is thus equivalent to requiring that the refractive indices at the fundamental and second harmonic frequencies be equal. For this reason, the process of making $\Delta k = 0$ is called index matching. Since, as described in the previous section, having $\Delta k \neq 0$ results in a phase difference and interference between the fundamental and the second harmonic, the attainment of $\Delta k = 0$ is also referred to as phase matching.

The condition for phase matching in the second harmonic generation process is found, from (9.152), to be the equality of the refractive index for the fundamental and second harmonic waves. Since the nonlinear crystal will in general exhibit dispersion, $n(\omega)$ and $n(2\omega)$ will not generally be equal. For this reason, phase matching will not be achieved unless special steps are taken to satisfy the condition contained in equation (9.152). A variety of methods have been used experimentally to obtain phase matching. A discussion of these techniques, while containing much interesting physics,[66-68] is outside the scope of this introduction, and the interested reader is referred to the literature.[62,64,65]

From the point of view of applications, phase matching is important in order to maximize the intensity and efficiency of second harmonic generation by making $\Delta k = 0$ so the factor $[(\sin x)/x]^2$ in (9.144) and (9.147) is close to unity. The same phase-matching requirement exists for high efficiency in other nonlinear mixing processes, including parametric amplification and frequency up-conversion.

Frequency Mixing and Up-Conversion

We consider in this section the mixing of two different frequencies ω_1 and ω_2 to give a third frequency ω_3, again subject to the condition (9.100) that

$$\omega_3 = \omega_1 + \omega_2 \qquad (9.100)$$

This process converts a lower frequency ω_1 to a higher frequency ω_3 and is often called frequency up-conversion.[69,70] It is the mixing process of which second harmonic generation is the special case for which $\omega_1 = \omega_2$ and $\omega_3 = 2\omega_1$. We will refer to ω_1 as the input frequency which is mixed with the pump frequency ω_2 to produce the output frequency $\omega_3 = \omega_1 + \omega_2$. An example is the mixing of infrared photons (ω_1) with a strong laser source (ω_2) to produce visible photons (ω_3) for which sensitive detectors are available.

We return to the coupled-amplitude equations (9.121)–(9.123) to discuss this process and assume that a negligible fraction of the pump frequency (ω_2) photons is converted. The pump signal amplitude E_{20} may thus be considered as approximately constant, so we have

$$dE_{20}/dz = 0 \tag{9.153}$$

It is assumed also that phase matching has been achieved,[71] so we have

$$\Delta k \equiv k_1 + k_2 - k_3 = 0 \tag{9.154}$$

from the definition (9.124) of Δk. Following Yariv,[69] the coupled-amplitude equations (9.121) and (9.123) then become

$$\frac{dE_{10}^*}{dz} = j \frac{2\pi\omega_1}{c\varepsilon_1^{1/2}} d^{(2\omega)} E_{20} E_{30}^* \tag{9.155}$$

$$\frac{dE_{30}}{dz} = -j \frac{2\pi\omega_3}{c\varepsilon_3^{1/2}} d^{(2\omega)} E_{10} E_{20} \tag{9.156}$$

where $E_{10}(z)$ and $E_{30}(z)$ are, respectively, the amplitudes of the input and output waves. Defining a quantity β_1 by the relation

$$\beta_1 \equiv 2\pi\omega_1 d^{(2\omega)} E_{20}/c\varepsilon_1^{1/2} \tag{9.157}$$

where E_{20} is the constant electric field of the pump signal, and β_3 by the analogous relation

$$\beta_3 \equiv 2\pi\omega_3 d^{(2\omega)} E_{20}/c\varepsilon_3^{1/2} \tag{9.158}$$

the differential equations (9.155) and (9.156) become

$$\frac{dE_{10}}{dz} = -j\beta_1 E_{30} \tag{9.159}$$

$$\frac{dE_{30}}{dz} = -j\beta_3 E_{10} \tag{9.160}$$

In (9.159), the complex conjugate of (9.155) was taken, and both E_{10} and E_{30} are functions of the space coordinate z along the nonlinear crystal.
Equations (9.159) and (9.160) are a pair of coupled first-order differential equations for the electric fields $E_{10}(z)$ and $E_{30}(z)$ of the input and output waves. The solutions are

$$E_{10}(z) = A \cos \lambda z + B \sin \lambda z \tag{9.161}$$

$$E_{30}(z) = C \cos \lambda z + D \sin \lambda z \tag{9.162}$$

where $A, B, C,$ and D are constants to be determined by the boundary conditions on the problem, and the quantity λ is defined by

$$\lambda^2 \equiv \beta_1\beta_3 = 4\pi^2\omega_1\omega_3[d^{(2\omega)}]^2E_{20}^2/c^2\varepsilon_1^{1/2}\varepsilon_3^{1/2} \tag{9.163}$$

Using the symbols $E_{10}(0)$ and $E_{30}(0)$ for the values of $E_{10}(z)$ and $E_{30}(z)$ at the input face $z = 0$ of the crystal, satisfaction of either of the original differential equations (9.159) or (9.160) determines A, B, C, and D, leading to

$$E_{10}(z) = E_{10}(0) \cos \lambda z - j(\beta_1/\beta_3)^{1/2}E_{30}(0) \sin \lambda z \tag{9.164}$$

$$E_{30}(z) = E_{30}(0) \cos \lambda z - j(\beta_3/\beta_1)^{1/2}E_{10}(0) \sin \lambda z \tag{9.165}$$

We choose next the boundary condition that there shall be no photons of frequency ω_3 at the input face $z = 0$ of the crystal, so

$$E_{30}(0) = 0 \tag{9.166}$$

and the solutions (9.164) and (9.165) then become

$$E_{10}(z) = E_{10}(0) \cos \lambda z \tag{9.167}$$

$$E_{30}(z) = -j(\beta_3/\beta_1)^{1/2}E_{10}(0) \sin \lambda z \tag{9.168}$$

These equations give the input electric field $E_{10}(z)$ and the output electric field $E_{30}(z)$ as functions of distance z in the nonlinear crystal. From equation (9.163), these solutions are functions of the frequencies ω_1 and ω_3, the nonlinear susceptibility $d^{(2\omega)}$, the pump electric field E_{20}, and the dielectric constants ε_1 and ε_3, through the dependence on β_1, β_3, and λ in equations (9.167) and (9.168).

We may calculate the intensities of the input (ω_1) and output (ω_3) waves as functions of z, obtaining

$$I_{10}(z) = (n_1c/8\pi)E_{10}(z)E_{10}^*(z) = (n_1c/8\pi)[E_{10}(0)]^2 \cos^2 \lambda z \tag{9.169}$$

$$I_{30}(z) = (n_3c/8\pi)E_{30}(z)E_{30}^*(z) = (n_3c/8\pi)(\beta_3/\beta_1)[E_{10}(0)]^2 \sin^2 \lambda z \tag{9.170}$$

where $n_1 = \varepsilon_1^{1/2}$ and $n_3 = \varepsilon_3^{1/2}$ are the refractive indices of the nonlinear crystal at the frequencies ω_1 and ω_3. From equations (9.157) and (9.158), one finds that

$$\beta_3/\beta_1 = \omega_3\varepsilon_1^{1/2}/\omega_1\varepsilon_3^{1/2} = \omega_3n_1/\omega_1n_3 \tag{9.171}$$

Further, the input intensity $I_{10}(0)$ at $z = 0$ is given by

$$I_{10}(0) = (n_1c/8\pi)[E_{10}(0)]^2 \tag{9.172}$$

Substituting (9.171) and (9.172) into (9.169) and (9.170) gives

$$I_{10}(z) = I_{10}(0) \cos^2 \lambda z \tag{9.173}$$

$$I_{30}(z) = (\hbar\omega_3/\hbar\omega_1)I_{10}(0) \sin^2 \lambda z \qquad (9.174)$$

for the input intensity $I_{10}(z)$ and the output intensity $I_{30}(z)$ as functions of distance z in the crystal. In obtaining (9.174), numerator and denominator were multiplied by $\hbar$ in order to convert the frequencies into photon energies.

We can discuss the magnitude of the quantity λ defined by equation (9.163) by rewriting that expression as

$$\lambda^2 = 32\pi^3\omega_1\omega_3(d^{(2\omega)})^2I_{20}/c^3n_1n_2n_3 \qquad (9.175)$$

on using the relation $I_{20} = (cn_2/8\pi)E_{20}^2$ for the intensity I_{20} of the pump wave of frequency ω_2, and the fact that $n_1{}^2 = \varepsilon_1$ and $n_3{}^2 = \varepsilon_3$. A dimensional analysis of equation (9.175) shows that λ has units of cm^{-1}, as it should, because λz in equation (9.173) and (9.174) must be dimensionless.

We may calculate λ for a typical example[72] of up-conversion in which the 10.6 μm wavelength output of a CO_2 laser is to be mixed with the 1.06 μm output of a neodymium doped YAG laser to produce a near-infrared signal which can be readily detected with a photomultiplier tube. In this example, then, the input (CO_2 laser) frequency $\omega_1 = 1.78 \times 10^{14}$ sec^{-1}, the pump (YAG laser) frequency $\omega_2 = 1.78 \times 10^{15}$ sec^{-1}, and the output frequency $\omega_3 = \omega_2 + \omega_1 = 1.96 \times 10^{15}$ sec^{-1}, corresponding to a wavelength of 0.964 μm in the near infrared. The nonlinear crystal used is Ag_3AsS_3, known as "proustite," which is quite transparent[73] between 0.6 and 13 μm and has a value[74] of $d^{(2\omega)}$ of approximately 2.5×10^{-22} (MKS) $\cong 7 \times 10^{-8}$ esu. Taking the refractive index values[73] $n_1 \cong n_2 \cong n_3 \cong 2.6$, and using a value of 10^{11} ergs sec^{-1} cm^{-2} for the pump intensity I_{20} of the YAG laser, we calculate a value of $\lambda = 1.8 \times 10^{-2}$ cm^{-1} from equation (9.175).

Since, from equation (9.173),

$$I_{10}(1/\lambda) = I_{10}(0)[\cos 1]^2 = 0.292I_{10}(0) \qquad (9.176)$$

we can see the physical meaning of λ. The quantity $1/\lambda$ is the distance in the crystal over which the input intensity $I_{10}(z)$ at ω_1 decreases to 0.292 of its initial value $I_{10}(0)$. For this example, the distance $1/\lambda = 56$ cm. Using the value of $\hbar\omega_3/\hbar\omega_1 = 11.0$ for this example, the output intensity at frequency ω_3 is given by

$$I_{30}(z) = (11.0)\{\sin^2[(1.80 \times 10^{-2})z]\}I_{10}(0) \qquad (9.177)$$

a function that is plotted in Figure 9.4. This figure shows $I_{30}(z)$ in units of the initial input intensity $I_{10}(0)$, as a function of crystal length z in centimeters. For a crystal length of 1 cm, the curve shows that I_{30} is equal to

$0.00356I_{10}(0)$; for a crystal length of 5 cm, $I_{30} = 0.0889I_{10}(0)$.

Further, the conversion efficiency e may be defined as

$$e \equiv I_{30}(z)/I_{10}(0) = (\omega_3/\omega_1) \sin^2 \lambda z \qquad (9.178)$$

on using equation (9.174). For the example, given above, $e = 0.0036$ for a length of 1 cm, and $e = 0.089$ for a length of 5 cm, showing that useful percentages of the input intensity $I_{10}(0)$ may be up-converted in this situation. Since the maximum value of $\sin^2 \lambda z$ is unity, the largest value of the conversion efficiency is $e_{\max} = \omega_3/\omega_1$, which we may write as

$$e_{\max} = \hbar\omega_3/\hbar\omega_1 \qquad (9.179)$$

This equation shows that the maximum value of e is the ratio of the energies of the output and input photons and hence is larger than unity. This value of $e_{\max} = \hbar\omega_3/\hbar\omega_1$ corresponds to the situation in which *all* of the input photons of energy $\hbar\omega_1$ are converted to output photons of energy $\hbar\omega_3$. The energy necessary to do this comes, of course, from the pump photons of energy $\hbar\omega_2$, which are destroyed whenever a photon of energy $\hbar\omega_3$ is created and a photon of energy $\hbar\omega_1$ is destroyed.

Further, if λz is small compared to unity, then we may approximate

$$\sin \lambda z \simeq \lambda z \qquad (9.180)$$

and equations (9.174) and (9.178) become

$$I_{30}(z) \simeq (\hbar\omega_3/\hbar\omega_1)\lambda^2 I_{10}(0)z^2 \qquad (9.181)$$

$$e \simeq (\omega_3/\omega_1)\lambda^2 z^2 \qquad (9.182)$$

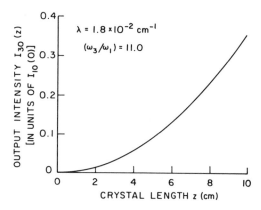

Figure 9.4. Output intensity $I_{30}(z)$, at frequency ω_3, from equation (9.177) [in units of the initial input intensity $I_{10}(0)$] plotted as a function of crystal length z (in cm) for the case of $\lambda = 1.8 \times 10^{-2}$ cm^{-1} and $\omega_3/\omega_1 = 11.0$ used in the up-conversion example in the text.

for the output intensity at frequency ω_3 and for the conversion efficiency. For the example plotted in Figure 9.4, λz is less than 0.18 and equation (9.181) for $I_{30}(z)$ is a good approximation to equation (9.177) for the values of z shown in the figure.

Finally, it should be emphasized that results obtained in this section for the output intensity and for the conversion efficiency refer to the largest possible values of these quantities. This is because we assumed phase matching ($\Delta k = 0$) in setting up the equations (9.155) and (9.156) which were solved to describe the up-conversion process. If $\Delta k \neq 0$, then the values of the output intensity and the conversion efficiency will be reduced[75] below the largest possible values.

For a discussion of experimental details of up-conversion processes and various applications, the reader is referred to the references, particularly those by Zernike and Midwinter[69] and by Warner.[70]

Parametric Amplification

In this section, we consider a signal wave of frequency ω_1 and an intense pump wave of frequency ω_3 both incident on a nonlinear crystal. Assuming that ω_3 is greater than ω_1, parametric amplification is the process by which energy is transferred from the pump wave to the signal wave, thereby amplifying the latter. At the same time, an idler wave of frequency

$$\omega_2 = \omega_3 - \omega_1 \tag{9.183}$$

is produced. In terms of photons, a pump photon $\hbar\omega_3$ is destroyed with the simultaneous creation of a signal photon $\hbar\omega_1$ and an idler photon $\hbar\omega_2$.

Since equation (9.183) is equivalent to the condition that $\omega_1 + \omega_2 = \omega_3$, we may again use the coupled amplitude equations (9.121)–(9.123) to discuss parametric amplification.[76,77] We will assume that the pump wave is sufficiently intense that it is undepleted, so the amplitude $E_{30}(z)$ is a constant, and

$$dE_{30}/dz = 0 \tag{9.184}$$

For simplicity, we assume also that phase matching[78] has been achieved, so

$$\Delta k = k_3 - k_1 - k_2 = 0 \tag{9.185}$$

With the conditions (9.184) and (9.185), the coupled amplitude equations

(9.121) and (9.122) become

$$\frac{dE_{10}^*}{dz} = j\frac{2\pi\omega_1}{c\varepsilon_1^{1/2}} d^{(2\omega)} E_{20} E_{30}^* \tag{9.186}$$

$$\frac{dE_{20}^*}{dz} = j\frac{2\pi\omega_2}{c\varepsilon_2^{1/2}} d^{(2\omega)} E_{10} E_{30}^* \tag{9.187}$$

These equations describe the variation in space of the amplitudes $E_{10}(z)$ and $E_{20}(z)$ of, respectively, the electric fields of the signal and idler waves. As were equations (9.155) and (9.156) for the up-conversion process, they are a pair of coupled differential equations.

Introducing the symbols α_1 and α_2 by

$$\alpha_1 = 2\pi\omega_1 d^{(2\omega)} E_{30}^*/c\varepsilon_1^{1/2} \tag{9.188}$$

$$\alpha_2 = 2\pi\omega_2 d^{(2\omega)} E_{30}^*/c\varepsilon_2^{1/2} \tag{9.189}$$

we may rewrite equations (9.186) and (9.187) as

$$\frac{dE_{10}^*}{dz} = j\alpha_1 E_{20} \tag{9.190}$$

$$\frac{dE_{20}^*}{dz} = j\alpha_2 E_{10} \tag{9.191}$$

It is useful to note the differences (signs and complex conjugates) between these equations and the analogous ones [(9.159) and (9.160)] for the up-conversion process. These differences lead, of course, to very different solutions.

The coupled differential equations (9.190) and (9.191) have the solution

$$E_{10}(z) = Ae^{\delta z} + Be^{-\delta z} \tag{9.192}$$

$$E_{20}(z) = Ce^{\delta z} + De^{-\delta z} \tag{9.193}$$

where A, B, C, and D are constants, and the quantity δ is defined by

$$\delta^2 \equiv \alpha_1\alpha_2 = 4\pi^2\omega_1\omega_2(d^{(2\omega)})^2 E_{30}^2/c^2\varepsilon_1^{1/2}\varepsilon_2^{1/2} \tag{9.194}$$

Next, we prescribe the boundary condition that there is no idler wave input to the crystal at the input face $z = 0$. This means that

$$E_{20}(0) = 0 \tag{9.195}$$

leading to the result that $D = -C$, so equation (9.193) for $E_{20}(z)$ becomes

$$E_{20}(z) = C(e^{\delta z} - e^{-\delta z}) = 2C \sinh(\delta z) \tag{9.196}$$

Substituting equation (9.196) for $E_{20}(z)$ into the differential equation (9.191) for (dE_{20}^*/dz) leads to the results

$$A = B \tag{9.197}$$

$$C = (j\alpha_2 A/\delta) = j(\alpha_2/\alpha_1)^{1/2}A \tag{9.198}$$

which, when substituted into equations (9.192) and (9.196), give

$$E_{10}(z) = 2A \cosh(\delta z) \tag{9.199}$$

$$E_{20}(z) = 2jA(\alpha_2/\alpha_1)^{1/2} \sinh(\delta z) \tag{9.200}$$

From (9.199), we see that the value $E_{10}(0)$ of $E_{10}(z)$ at $z = 0$ is equal to the constant $2A$, and we finally obtain the solutions

$$E_{10}(z) = E_{10}(0) \cosh(\delta z) \tag{9.201}$$

$$E_{20}(z) = j(\alpha_2/\alpha_1)^{1/2}E_{10}(0) \sinh(\delta z) \tag{9.202}$$

These equations describe the spatial variation of the electric fields $E_{10}(z)$ and $E_{20}(z)$ of the signal and idler waves for the case in which there is no idler (ω_2) wave input at $z = 0$.

We next use the definition of the intensity I as

$$I(z) = (cn/8\pi)EE^* \tag{9.203}$$

where n is the refractive index, to obtain the intensities $I_{10}(z)$ and $I_{20}(z)$ of the signal and idler waves as functions of position z in the crystal. We find

$$I_{10}(z) = (cn_1/8\pi)E_{10}E_{10}^* = (cn_1/8\pi)[E_{10}(0)]^2 \cosh^2(\delta z) \tag{9.204}$$

$$I_{20}(z) = (cn_2/8\pi)E_{20}E_{20}^* = (cn_2/8\pi)(\alpha_2/\alpha_1)[E_{10}(0)]^2 \sinh^2(\delta z) \tag{9.205}$$

where $n_1 = \varepsilon_1^{1/2}$ and $n_2 = \varepsilon_2^{1/2}$ are the refractive indices of the crystal at the signal and idler frequencies ω_1 and ω_2. The signal intensity at the input $z = 0$ is, from (9.204),

$$I_{10}(0) = (cn_1/8\pi)[E_{10}(0)]^2 \tag{9.206}$$

so the expressions for the intensities become

$$I_{10}(z) = I_{10}(0) \cosh^2(\delta z) \tag{9.207}$$

$$I_{20}(z) = (n_2\alpha_2/n_1\alpha_1)I_{10}(0) \sinh^2(\delta z) \tag{9.208}$$

From the behavior of the functions $\cosh(\delta z)$ and $\sinh(\delta z)$ as z increases from zero, we see that the intensities $I_{10}(z)$ and $I_{20}(z)$ increase as the waves move along the crystal (i.e., as z increases). Since $\cosh^2(\delta z)$ is greater than unity for z greater than zero, the signal intensity increases from its initial value $I_{10}(0)$ as the signal wave moves through the crystal. In this manner, the signal wave is amplified, the energy to do so coming, of course, from the pump wave. Similarly, since $\sinh^2(\delta z)$ is greater than zero for positive values of z, the idler wave intensity $I_{20}(z)$ increases from its input value of zero as the wave moves through the crystal. From equation (9.207), the signal wave undergoes, in a distance z, an amplification by a factor a, where

$$a \equiv I_{10}(z)/I_{10}(0) = \cosh^2(\delta z) \tag{9.209}$$

Let us make some numerical estimates concerning parametric amplification in a representative case. From equation (9.194),

$$\delta^2 = 4\pi^2 \omega_1 \omega_2 (d^{(2\omega)})^2 E_{30}^2/c^2 n_1 n_2 \tag{9.210}$$

which, on substituting the expression

$$I_{30} = (cn_3/8\pi)E_{30}^2 \tag{9.211}$$

for the constant intensity of the pump wave, becomes

$$\delta^2 = 32\pi^3 \omega_1 \omega_2 (d^{(2\omega)})^2 I_{30}/c^3 n_1 n_2 n_3 \tag{9.212}$$

We calculate the value of δ for an amplification experiment[79] in which a signal wavelength of 0.633 μm was amplified using a pump wavelength of 0.348 μm. The nonlinear crystal was ammonium dihydrogen phosphate (ADP), with a nonlinear susceptibility value[80] $d^{(2\omega)} = 1.50 \times 10^{-9}$ esu, and a refractive index[71] in the visible of approximately 1.5. The pump power was 2×10^{13} ergs sec^{-1} cm^{-2}. Using equation (9.212), we obtain $\delta = 0.06$ cm^{-1}. From equation (9.209), we can see that the physical significance of δ is that $1/\delta$ is the distance z over which the signal intensity $I_{10}(z)$ is amplified by a factor of $(\cosh^2 1) = 2.38$ over its input value $I_{10}(0)$. For the experiment described here, the ADP crystal was 8 cm long, so the factor $\delta z = 0.48$, and the amplification factor a is equal to $\cosh^2 0.48 = 1.25$. This is only a modest degree of amplification. From (9.209), we see that the amplification factor a depends on δ, which itself depends on, among other things, the pump intensity I_{30}. For this reason,[81] optical parametric amplification does not usually lead to large values of gain unless the pump intensity is extremely high. Because of this, one of the main reasons for interest in parametric amplification is its connection with optical parametric oscillation. In this process, the nonlinear crystal is placed in an appropriate optical resonant cavity, generating oscillations at the signal and idler frequencies. For a discussion of the parametric oscillator, the reader is referred to the references.[82-84]

Summary

How does the physics of optically nonlinear solids result in useful applications?

The key physical point is that the nonlinear term in the relation between polarization P and electric field E mixes waves of different frequencies in the crystal. The term in P that is proportional to E^2 gives rise to a component of the polarization at frequency $\omega_1 \pm \omega_2$ when the frequencies ω_1 and ω_2 are present. Since the wave equation is "driven" by a term $\partial^2 P/\partial t^2$, the result is an oscillating electric field at frequency (for example), $\omega_1 + \omega_2$, as well as at ω_1 and ω_2. This mixing process is exploited to generate second harmonic $(\omega_1 = \omega_2 = \omega; \ \omega_1 + \omega_2 = 2\omega)$ and sum $(\omega_1 + \omega_2 = \omega_3)$ frequencies. For high conversion efficiencies it is necessary to use large (i.e., laser) input electric fields since the nonlinear susceptibilities are small.

Problems

9.1. *Nonlinear Polarization in Tellurium.* (a) Estimate the value of the electric field for which the magnitude of the second harmonic polarization $P^{(2\omega)}$ in Te is equal to 0.1% of the linear polarization $P^{(\omega)}$. Use the values $d^{(2\omega)} = 1600 \times 10^{-9}$ esu and 6.2 for the refractive index n. (b) Calculate the value of $P^{(2\omega)}$ for this value of the applied electric field. Compare this value with some value of the polarization observed in a solid.

9.2. *Dimensional Analysis.* Using the units for electric field and nonlinear susceptibility introduced in the text, show that the right-hand side of equation (9.120) has units appropriate for dE_{30}/dz, i.e., has units of electric field per unit length.

9.3. *Derivation of Coupled Amplitude Equations.* Using the derivation of equation (9.120) for dE_{30}/dz as a guide, derive equation (9.121) for dE_{10}^*/dz, or the equivalent complex conjugate equation for dE_{10}/dz. It will be necessary to calculate the driving term $d^2(E^2)/dt^2$, where E is given by (9.94), in the wave equation (9.95). This tedious task is required because it is necessary to find the driving terms at frequency ω_1 on the right-hand side of the equation analogous to (9.104) which describes the electric field E_1.

9.4. *Efficiency of Second Harmonic Generation in Quartz.* Calculate the maximum efficiency of second harmonic generation in quartz, using as an input the $\lambda = 6940$ Å line from a ruby laser for which the electric field amplitude is 10^5 V cm^{-1}. (Find the input intensity in ergs sec^{-1} cm^{-2} by converting this electric field to statvolts cm^{-1}). Take the index of refraction of quartz as 1.5, the crystal length as 1 cm, and the nonlinear susceptibility as 10^{-9} esu.

9.5. *Photon Fluxes in Up-Conversion.* For the up-conversion of photons of frequency ω_1 to frequency ω_3, show that, at any point z of the crystal, the flux of ω_1 photons plus the flux of ω_3 photons is equal to the flux of ω_1 photons at the point $z = 0$.

9.6. *Manley–Rowe Relations.* Consider the coupled amplitude equations (9.121)– (9.123) and multiply each derivative by the complex conjugate of the electric field, thus obtaining the quantities $E_{10}(dE_{10}^{*}/dz)$, etc. Relate the three quantities so obtained to the intensities $I_{10} = (cn_1/8\pi)E_{10}E_{10}^{*}$, etc., and thereby to the fluxes N_1, N_2, N_3 of photons of frequency ω_1, ω_2, ω_3. Show that

$$\frac{dN_1}{dz} = \frac{dN_2}{dz} = -\frac{dN_3}{dz}$$

These equations are one form of the Manley–Rowe relations. (These useful results are discussed, for example, in the book by Zernike and Midwinter.[52])

References and Comments

1. See, for example, J. D. Jackson, *Classical Electrodynamics*, Second Edition, John Wiley (1975), page 226, equation (6.70).
2. See, for example, J. D. Jackson, Reference 1, Section 6.7.
3. See, for example, J. R. Reitz, F. J. Milford, and R. W. Christy, *Foundations of Electromagnetic Theory*, Third Edition, Addison-Wesley, Reading, Mass. (1979), page 85; J. D. Jackson, Reference 1, page 144, equation (4.29).
4. See, for example, D. Turnbull in *The McGraw-Hill Encyclopedia of Science and Technology*, McGraw-Hill, New York (1982), Volume 1, page 545; A. J. Dekker, *Solid State Physics*, Prentice-Hall, New York (1957), pages 27–29.
5. See, for example, D. Turnbull, Reference 4, Volume 7, page 411.
6. L. D. Landau and E. M. Lifschitz, *Electrodynamics of Continuous Media*, Addison-Wesley, Reading, Mass. (1960), Section 77, pages 315–316.
7. C. Kittel, *Introduction to Solid State Physics*, Sixth Edition, John Wiley, New York (1986), page 370.
8. See, for example, K. R. Symon, *Mechanics*, Second Edition, Addison-Wesley, Reading, Mass. (1960), pages 104–108.
9. C. Kittel, Reference 7, pages 360–368.
10. N. W. Ashcroft and N. D. Mermin, *Solid State Physics*, Holt, Rinehart, and Winston, New York (1976), pages 539–542.
11. J. F. Nye, *Physical Properties of Crystals*, Oxford University Press (1957), Chapter 4.
11a. Simply differentiating equation (9.28) is not really correct. See R. Becker and F.

Sauter, *Electromagnetic Fields and Interactions*, Blaisdell Publishing Co., New York (1964), Volume I, Section 57, especially pages 235 and 243, for the proper approach using the Fourier integral. Further discussion may be found in J. D. Jackson, Reference 1, Section 7.10, and in L. D. Landau and E. M. Lifshitz, Reference 4, Section 58.

12. Y. R. Shen, "Recent Advances in Nonlinear Optics," *Reviews of Modern Physics*, **48**, 1–32 (1976).

12a. Y. R. Shen, *The Principles of Nonlinear Optics*, John Wiley, New York (1984), pages 4–5. We follow Shen's notation of boldface symbols for tensors; see also his pages 38–40.

13. J. Ducuing, in *Quantum Optics*, R. J. Glauber (editor), Proceedings of the International School of Physics "Enrico Fermi," Course XLII, Academic Press, New York (1969), page 448.

14. N. Bloembergen, *Nonlinear Optics*, W. A. Benjamin, New York (1965).

15. A. Yariv, *Quantum Electronics*, Second Edition, John Wiley, New York (1975), pages 413–418.

16. A. Yariv, in *Topics in Solid State and Quantum Electronics*, W. D. Hershberger (editor), John Wiley, New York (1972), pages 280–290.

17. C. G. B. Garrett, "Nonlinear Optics, Anharmonic Oscillators, and Pyroelectricity," *IEEE Journal of Quantum Electronics*, **QE-4**, 70–84 (1968).

18. See, for example, C. Kittel, W. D. Knight, M. A. Ruderman, A. C. Helmholz, and B. J. Moyer, *Mechanics* (Berkeley Physics Course, Volume 1), Second Edition, McGraw-Hill, New York (1973), pages 224–226.

19. J. J. Stoker, *Nonlinear Vibrations*, Interscience Publishers, New York (1950), Chapter 4.

20. L. A. Pipes and L. R. Harvill, *Applied Mathematics for Engineers and Physicists*, Third Edition, McGraw-Hill, New York (1970), Chapter 15.

21. G. C. Baldwin, *An Introduction to Nonlinear Optics*, Plenum Press, New York (1969), page 141.

22. F. W. Constant, *Theoretical Physics*, Addison-Wesley, Reading, Mass. (1954), Section 6-6, pages 93–97.

23. J. B. Marion, *Classical Dynamics of Particles and Systems*, Second Edition, Academic Press, New York (1970), Section 5.5, pages 165–167.

24. See, for example, C. Kittel, W. D. Knight, M. A. Ruderman, A. C. Helmholz and B. J. Moyer, Reference 18, page 226.

25. G. C. Baldwin, Reference 21, pages 98–99.

26. F. Seitz, *Modern Theory of Solids*, McGraw-Hill, New York (1940), page 637.

27. C. G. B. Garrett and F. N. H. Robinson, "Miller's Phenomenological Rule for Computing Nonlinear Susceptibilities," *IEEE Journal of Quantum Electronics*, **QE-2**, 328–329 (1966).

28. A. Yariv, Reference 15 Appendix 4, and N. Bloembergen, Reference 14 Chapter 2, give quantum mechanical discussions.

29. A. Yariv, Reference 16, pages 283–285, gives figures exhibiting the polarization components at the fundamental and second harmonic frequencies.

30. J. F. Nye, Reference 11, Chapter 7, pages 110–115.

31. A. Yariv, Reference 16, page 287.

32. A. Yariv, Reference 15, pages 410–411, gives a table of the form of the nonlinear susceptibility tensor for various crystal classes.

33. J. Ducuing, Reference 13, Sections 3.2.1–3.2.3, pages 435–439.

34. A. Yariv, Reference 15, page 416, gives a table of values of $d_{ijk}^{(2\omega)}$ for several crystals using the contracted d_{il} notation mentioned earlier. Yariv explains this notation on page 409. J. F. Nye, Reference 11, page 113, also discusses it, as do T. S. Moss, G. J. Burrell, and B. Ellis, *Semiconductor Opto-Electronics*, John Wiley, New York (1973), page 249.

35. B. F. Levine, "Bond-Charge Calculation of Nonlinear Optical Susceptibilities for Various Crystal Structures," *Physical Review B*, **7**, 2600–2626 (1973), gives values of nonlinear susceptibilities (with references) for a large number of crystals. See also the paper by B. F. Levine and C. G. Bethea, *Applied Physics Letters*, **20**, 272–274 (1972). The book by Yariv, Reference 15, page 416, gives a table of values of $d^{(2\omega)}$ in units of $\frac{1}{9} \times 10^{-22}$ MKS. (To convert $d^{(2\omega)}$ in these units to $d^{(2\omega)}$ in esu, divide $d^{(2\omega)}$ (MKS) by 3.68×10^{-15}, as described by F. Zernike and J. E. Midwinter, *Applied Nonlinear Optics*, John Wiley, New York (1973), pages 52–53.) Finally, mention should be made of the article by S. K. Kurtz, "Measurement of Nonlinear Optical Susceptibilities," in *Quantum Electronics: A Treatise*, H. Rabin and C. L. Tang (editors), Academic Press, New York (1975), Volume 1 (Nonlinear Optics, Part A), Section 3, pages 209–281. This article describes methods of experimental measurement and gives values of $d^{(2\omega)}$ for a number of crystals important in nonlinear applications.

36. R. C. Miller, "Optical Second Harmonic Generation in Piezoelectric Crystals," *Applied Physics Letters*, **5**, 17–19 (1964).

37. J. A. Giordmaine, "Nonlinear Optics," *Physics Today*, **21**, 39–44 (January 1969).

38. B. F. Levine, Reference 35, page 2608, Table III.

39. B. O. Seraphin and H. E. Bennett, in *Semiconductors and Semimetals*, R. K. Willardson and A. C. Beer (editors), Volume 3, Academic Press, New York (1967), Chapter 12, pages 520, 526, and 511. The value of n quoted for GaAs is the average of the two values given in this reference.

40. B. F. Levine, Reference 35, page 2621, Table XVI.

41. R. A. Smith, *Semiconductors*, Cambridge University Press (1959), page 386.

42. C. Kittel, Reference 7, pages 368–372.

43. Y. R. Shen, Reference 12, Section II, pages 2–4.

44. See, for example, C. Kittel, W. D. Knight, M. A. Ruderman, A. C. Helmholz, and B. J. Moyer, Reference 18, pages 67–70. A useful table of conversion factors is given by G. Joos, *Theoretical Physics*, Second Edition, Hafner Publishing Co., New York (1950), page 836.

45. N. Bloembergen, "The Stimulated Raman Effect," *American Journal of Physics*, **35**, 989–1023 (1967), page 996.

46. J. F. Nye, Reference 11, page 76.

47. Y. R. Shen, Reference 12a, Sections 3.1, 3.3.

48. Y. R. Shen, Reference 12a, equations (3.21a) and (3.21b).

49. The amplitudes of the longitudinal and transverse components of the electric field in a nonlinear anisotropic crystal can be calculated using equations (6.16) of Reference 12a.

50. See, for example, J. Ducuing, Reference 13, Sections 2.3 and 3.3 pages 433–435 and 439–448. This paper also gives references to the original work, including that of J. A. Armstrong, N. Bloembergen, J. Ducuing, and P. S. Pershan, "Interactions between Light Waves in a Nonlinear Dielectric," *Physical Review*, **127**, 1918–1939 (1962). This paper is reproduced in the book by N. Bloembergen, Reference 14.

51. A. Yariv, Reference 16, pages 290–292.

52. See N. Bloembergen, Reference 45, page 995, for a treatment of the interaction of four waves (with frequencies ω_1, ω_2, ω_3, ω_4) subject to the condition that $\omega_1 + \omega_4 = \omega_2 + \omega_3$.

53. J. A. Armstrong, N. Bloembergen, J. Ducuing, and P. S. Pershan, "Interactions between Light Waves in a Nonlinear Dielectric," *Physical Review*, **127**, 1918–1939 (1962), page 1928.

54. J. Ducuing, Reference 13, page 434, equation (31).

55. A. Yariv, Reference 15, page 409; Y. R. Shen, Reference 47, page 26.

56. See, for example, A. Yariv, Reference 15, page 421, for a treatment of the case of a solid with a nonzero value of the electrical conductivity.

57. P. A. Franken, A. E. Hill, C. W. Peters, and G. Weinreich, "Generation of Optical Harmonics," *Physical Review Letters*, **7**, 118–119 (1961).

58. See, for example, A. Yariv, Reference 15, Section 16.6, for a discussion of second harmonic generation for the case of a depleted input wave.

59. See, for example, G. R. Fowles, *Introduction to Modern Optics*, Second Edition, Holt, Rinehart, and Winston, New York (1975), page 25.

60. See, for example, G. R. Fowles, Reference 59, Section 5.4.

61. See, for example, G. C. Baldwin, Reference 21, Section 4.3; J. Ducuing, Reference 13, page 441.

62. F. Zernike and J. E. Midwinter, *Applied Nonlinear Optics*, John Wiley, New York (1973), Chapter 3.

63. See A. Yariv, Reference 16, pages 298–303, for a discussion.

64. S. A. Akhmanov, A. I. Kovrygin, and A. P. Sukhorukov, "Optical Harmonic Generation and Optical Frequency Multipliers," in *Quantum Electronics: A Treatise*, H. Rabin and C. L. Tang (editors), Academic Press, New York (1975), Volume 1 (Nonlinear Optics, Part B), Section 8, pages 475–586.

65. See, for example, A. Yariv, Reference 15, pages 424–428; A. Yariv, Reference 16, pages 294–298, G. C. Baldwin, Reference 21, pages 88–97.

66. See, for example, M. Born and E. Wolf, *Principles of Optics*, Fourth Edition, Pergamon Press, New York (1970), page 671.

67. See, for example, G. R. Fowles, Reference 59, Section 6.7.

68. See, for example, A. Yariv, Reference 15, Section 5.4.

69. See, for example, F. Zernike and J. E. Midwinter, Reference 62, Chapter 6; A. Yariv, Reference 15, pages 454–456; A. Yariv, Reference 16, pages 321–325.

70. J. Warner, "Difference Frequency Generation and Up-Conversion," in *Quantum Electronics: A Treatise*, H. Rabin and C. L. Tang (editors), Academic Press, New York (1975), Volume 1 (Nonlinear Optics, Part B), Section 10, pages 703–737.

71. F. Zernike and J. E. Midwinter, Reference 62, Section 6.4; J. Warner, Reference 70, pages 707–712. These references also discuss the situation when Δk given by (9.154) is not zero.

72. A. Yariv, Reference 16, page 325.

73. F. Zernike and J. E. Midwinter, Reference 62, page 99.

74. A. Yariv, Reference 15, Table 16.2, page 416.

75. J. Warner, Reference 70, page 705.

76. See, for example, A. Yariv, Reference 15, Section 17.1; A. Yariv, Reference 16, pages 304–313; F. Zernike and J. E. Midwinter, Reference 62, Sections 7.1–7.3.

77. J. A. Giordmaine in *Quantum Optics*, R. J. Glauber (editor), Proceedings of the International School of Physics "Enrico Fermi", Course XLII, Academic Press, New York (1969), pages 499–502.

78. See especially A. Yariv, Reference 16, pages 311–313, and F. Zernike and J. E. Midwinter, Reference 62, Section 7.3, for discussions of phase mismatching in parametric amplification.
79. J. A. Giordmaine, Reference 77, page 502, Table II.
80. A. Yariv, Reference 15, Table 16.2, page 416.
81. A. Yariv, Reference 16, page 311.
82. See, for example, A. Yariv, Reference 16, Sections 7.9–7.11.
83. R. L. Byer, "Optical Parametric Oscillators," in *Quantum Electronics: A Treatise*, H. Rabin and C. L. Tang (editors), Academic Press, New York (1975), Volume 1 (Nonlinear Optics, Part B). Section 9, pages 587–702.
84. J. A. Giordmaine, Reference 77, pages 502–509.

Suggested Reading

G. R. FOWLES, *Introduction to Modern Optics*, Second Edition, Holt, Rinehart and Winston, New York (1975). Chapter 6 of this senior-level optics text provides an introduction to the optics of solids; our treatment of polarization and wave propagation in a linear dielectric is similar to Fowles' sections 6.2–6.4.

A. YARIV, in *Topics in Solid State and Quantum Electronics*, W. D. Hershberger (editor), John Wiley, New York (1972), Chapter 7. This collection of articles contains a chapter by Yariv on optical second harmonic generation which discusses, among other things, nonlinear polarization in solids. Our discussion parallels the treatment of Yariv.

A. YARIV, *Quantum Electronics*, Second Edition, John Wiley, New York (1975). This advanced textbook covers many topics in quantum electronics including, in Chapters 16 and 17, a detailed discussion of nonlinear optics, second harmonic generation, and parametric amplification more extensive than that in Yariv's article quoted above.

A. YARIV, *Optical Electronics*, Third Edition, Holt, Rinehart, and Winston, New York (1985). This text covers many of the topics discussed in Yariv's *Quantum Electronics*, but at a more introductory level. Chapter 8 discusses nonlinear optics and its applications.

N. BLOEMBERGEN, *Non-Linear Optics*, W. A. Benjamin, New York (1965). This lecture note and reprint volume discusses its subject, including the quantum theory of nonlinear susceptibilities, at the advanced level. It includes reprints of several important original papers.

G. C. BALDWIN, *Non-Linear Optics*, Plenum Press, New York (1969). This short introductory book has brief but useful discussions of many of the topics we have covered.

J. J. STOKER, *Nonlinear Vibrations*, Interscience Publishers, New York (1950). This short book is a good introduction to nonlinear oscillations and is clear and readable.

R. J. GLAUBER (editor), *Quantum Optics* [Proceedings of the International School of Physics "Enrico Fermi," Course LXII] Academic Press, New York (1969). This collection of tutorial lectures at the advanced level includes several on various aspects of nonlinear optics. Especially pertinent is that by J. Ducuing on nonlinear optical processes.

N. BLOEMBERGEN, "The Stimulated Raman Effect," *Am. J. Phys.* **35**, 989–1023 (1967). A tutorial and review article discussing many topics in the physics of nonlinear optics.

H. RABIN and C. L. TANG (editors), *Quantum Electronics: A Treatise*, Academic Press, New York (1975), Volume I on *Nonlinear Optics*, Parts A and B. For those wishing a treatise on nonlinear optics at the advanced level, these books offer articles covering, among other topics, the measurement of nonlinear optical susceptibilities, optical harmonic generation, optical parametric oscillators, and frequency up-conversion.

F. ZERNIKE and J. E. MIDWINTER, *Applied Nonlinear Optics*, John Wiley, New York (1973)· This short book is rather terse in style, but covers a wide variety of topics in the physics and applications of nonlinear optics.

Y. R. SHEN, *The Principles of Nonlinear Optics*, John Wiley, New York (1984). Covering many topics, this treatise discusses the subject at the advanced level.

10

Ferromagnetic Materials

Introduction

This chapter presents a discussion of ferromagnetic materials and some of their applications. After a review of the basic terminology of magnetism, the exchange interaction is discussed in some detail, with emphasis on the Heisenberg model. Next, the concept of magnetic domains is presented (relying heavily on the classic paper by Kittel and Galt) and used to describe the magnetization curve and hysteresis. All of these topics are brought together to treat a major application of ferromagnetic materials, the permanent magnet, including some of the newer materials. Finally, a few other applications of ferromagnetic materials are discussed briefly.

Review of Magnetism

We begin reviewing some fundamental ideas about magnetism. We recall that moving electric charges are the source of magnetic fields, and the fundamental[1-3] magnetic field $\mathbf{B}$ is called the magnetic induction or the magnetic-flux density. The magnetic field $\mathbf{B}$ appears in Maxwell's equations which, in Gaussian units, are

$$\text{curl } \mathbf{E} = (-1/c)(\partial \mathbf{B}/\partial t) \tag{10.1}$$

$$\text{curl } \mathbf{H} = (1/c)(\partial \mathbf{D}/\partial t) + (4\pi/c)\mathbf{J} \tag{10.2}$$

$$\text{div } \mathbf{D} = 4\pi\varrho \tag{10.3}$$

$$\text{div } \mathbf{B} = 0 \tag{10.4}$$

In these equations, $\mathbf{E}$ is the electric field, ϱ is the free electric charge density, and $\mathbf{J}$ is the free charge current density. The auxilary vectors $\mathbf{D}$ and $\mathbf{H}$ are defined by

$$\mathbf{D} = \mathbf{E} + 4\pi\mathbf{P} \tag{10.5}$$

$$\mathbf{H} = \mathbf{B} - 4\pi\mathbf{M} \tag{10.6}$$

where $\mathbf{P}$ is the electric polarization (discussed in Chapter 9) and $\mathbf{M}$ is the magnetic dipole moment per unit volume, or magnetization, of the matter under consideration; $\mathbf{H}$ is usually called the magnetic intensity.[4] All of the quantities in (10.1)–(10.6) are macroscopic values obtained by a suitable averaging process. While dimensionally identical, the unit of B is called the gauss and that of H the oersted, both in the Gaussian CGS system of units we will employ. In this system, the unit[4] of magnetic dipole moment is the erg/gauss and the unit of magnetization is the erg (gauss-cm)$^{-3}$.

All matter exhibits a magnetization $\mathbf{M}$ when a magnetic field $\mathbf{B}$ is applied to it. This effect is sometimes described by defining the magnetic susceptibility χ_M by the equation

$$\mathbf{M} = \chi_M\mathbf{B} \tag{10.7}$$

where χ_M (a scalar for an isotropic material[5]) is called the volume magnetic susceptibility (because M is the magnetic dipole moment per unit volume of material) and is dimensionless. The more usual definition of χ_M is,[6] however, the relation

$$\mathbf{M} = \chi_M\mathbf{H} \tag{10.8}$$

which, when combined with (10.6), gives

$$\mathbf{B} = (1 + 4\pi\chi_M)\mathbf{H} = \mu\mathbf{H} \tag{10.9}$$

In equation (10.9), the quantity $\mu \equiv (1 + 4\pi\chi_M)$ is the magnetic permeability and is generally used, rather than the susceptibility, in discussing ferromagnetism. The linear relation (10.9) between $\mathbf{B}$ and $\mathbf{H}$ is true for many solids, but not for all. In particular, the relation between $\mathbf{B}$ and $\mathbf{H}$ for ferromagnetic materials is not a simple linear one.

Diamagnetism[7–9] occurs in all atoms and is the induction of a magnetic dipole moment $\mathbf{m}$ by an external applied magnetic field $\mathbf{B}$. The direction of the induced magnetic moment $\mathbf{m}$ is opposite to the direction of $\mathbf{B}$, so $\mathbf{m}$

and **B** are antiparallel, leading to a negative value of the susceptibility χ_M. Diamagnetism is a small effect, with χ_M for purely diamagnetic substances having magnitudes[10] in the neighborhood of 10^{-6}. While diamagnetism occurs in all cases, it is only readily observed in atoms (such as rare gases) with no permanent magnetic dipole moment of their own to mask the diamagnetic effect.

If an atom does have a permanent magnetic dipole moment **m** (spin and/or orbital), then the potential energy U of the dipole in an applied magnetic field **B** is

$$U = -\mathbf{m} \cdot \mathbf{B} \tag{10.10}$$

Since the potential energy U is minimized if **m** is parallel to **B**, the effect of the applied magnetic field is to tend to align the magnetic dipole moments with the field, producing a magnetization **M** parallel to **B**. This leads to a positive value of the magnetic susceptibility χ_M. This effect is called paramagnetism[11] and, in addition to atoms, ions, and solids, is exhibited by the conduction electrons in metals. Since the permanent magnetic moments of atoms which lead to paramagnetism are usually larger than induced diamagnetic moments, diamagnetic behavior is masked by paramagnetism when the latter is present.

Ferromagnetism

A ferromagnet is a solid that exhibits a (spontaneous) magnetization in the absence of an applied magnetic field. The basic reason for the existence of ferromagnetism is that the magnetic moments of the individual atoms are ordered. If ↑ indicates the magnetic moment of an atom, a simple ferromagnet will have the atomic magnetic moments ordered, as shown in Figure 10.1, in which all of the magnetic moments are taken to be equal in magnitude. Since all of the atomic magnetic moments are aligned in the same direction, there exists a net nonzero magnetic moment in the ferromagnet.

It is believed[12] (at least for the transition metals) that the atomic mag-

Figure 10.1. Schematic picture of the arrangement of atomic magnetic moments in a simple ferromagnet. Each arrow represents a single atomic magnetic moment and all of the magnetic moments are parallel.

netic moments in ferromagnets are almost entirely due to electron spin, with only a small contribution from the orbital motion of the electrons, because experimental values of the gyromagnetic (or magnetomechanical) ratio are close to (e/mc), the value expected for a spin, rather than an orbital, magnetic moment. The orbital contribution is generally neglected in discussing ferromagnetic solids, and one usually refers to "spins" when describing atomic magnetic moments in these substances.

We have seen that the susceptibility χ_M was independent of the applied magnetic field for the cases of diamagnetism (χ_M negative) and paramagnetism (χ_M positive). For ferromagnetic materials, the susceptibility χ_M and permeability μ are not constants independent of the applied magnetic field. One generalizes the relation between **B** and **H** to include this by writing

$$\mathbf{B} = \mu(\mathbf{H})\mathbf{H} \qquad (10.11)$$

where the permeability $\mu(\mathbf{H})$ is generally written as a function of the magnetic intensity **H**.

A ferromagnetic material has a spontaneous nonzero value of the magnetization, even in the absence of an applied magnetic field. We can see that the magnitude M of the magnetization **M** will depend on the applied magnetic field. The maximum value of M for a given ferromagnetic material is called the saturation magnetization M_s, and corresponds to the state in which all of the atomic magnetic moments are parallel. Typical values[13] of M_s (at 300 K) are 1700 gauss for iron, and 1400 gauss for cobalt. The saturation magnetization at 0 K is denoted by $M_s(0)$. The effective magneton number n_B for a ferromagnetic material is defined by

$$M_S(0) \equiv n_B N \mu_B \qquad (10.12)$$

where the Bohr magneton μ_B is defined as $(e\hbar/2mc)$, equal to 9.27×10^{-21} erg/gauss, and N is the number of atoms (or formula units) per unit volume. At 0 K, we expect all of the atomic magnetic moments μ to be parallel, so we have

$$M_S(0) = N\mu \qquad (10.13)$$

with each atomic magnetic moment μ equal to $n_B \mu_B$. For example, for nickel, $n_B = 0.61$, so each nickel atom acts as if it had a total magnetic moment (orbital plus spin) of about 0.6 Bohr magnetons. (Nonintegral values of the magneton number n_B are common, and may be explained in terms of the band, or itinerant, model[14,15] of ferromagnetism.)

It is observed that the spontaneous magnetization of a ferromagnet disappears when it is heated above a characteristic temperature T_c called the Curie temperature. The Curie temperature separates the ordered ferromagnetic phase (at temperatures below T_c) from a disordered paramagnetic phase (at temperatures above T_c). At temperatures above T_c, thermal energy destroys the long-range order of the parallel magnetic moments of the ferromagnetic phase. However, the atomic magnetic moments may still be aligned by an applied external magnetic field, so the material is paramagnetic at temperature above T_c, but there is no spontaneous magnetic moment.

The Exchange Interaction

The question we address is the interaction which tends to align the magnetic moments of the atoms parallel to each other, as in Figure 10.1, thus producing the ferromagnetic phase. One description of this interaction is called the direct exchange model,[16–19] also called the Heisenberg model of ferromagnetism. (This model is also called the localized model in that the atomic magnetic moments are localized in space on atomic sites.)

We may approach a microscopic description of the direct exchange interaction by discussing first a simpler, but related, question. Consider the electrostatic interaction between the overlapping electron clouds of two atoms, A and B, each of which has one electron, labeled 1 and 2. We will use the notation that ψ_A is the space part of the wave function of the electron on atom A, and ψ_B is the space part of the wave function of the electron on atom B, a is the spin wave function for spin "up", (that is, the spin quantum number $m_s = -1/2$) and β is the spin wave function for spin "down" (that is, $m_s = +1/2$).

The Pauli principle then says that the total wave function Ψ (composed of the product of space and spin parts) for an electron must be antisymmetric under interchange of the coordinates of identical particles. This means that Ψ must be antisymmetric in the coordinates of the two electrons. There are two possibilities for the total wave function Ψ for the two electrons. These are

$$\Psi_s \equiv \text{(symmetric space function)} \times \text{(antisymmetric spin function)} \quad (10.14)$$

$$\Psi_t \equiv \text{(antisymmetric space function)} \times \text{(symmetric spin function)} \quad (10.15)$$

First, we can have a total wave function of the type (10.14),

$$\Psi_s = [\psi_A(1)\psi_B(2) + \psi_A(2)\psi_B(1)] [\alpha(1)\beta(2) - \alpha(2)\beta(1)] \qquad (10.16)$$

where $\psi_A(1)$ is the space wave function for electron 1 on atom A, and so on, and $\alpha(1)$ means that electron 1 is in the spin "up" or "↑" state, with $m_s = -1/2$, and so on. We can see in the wave function (10.16) that exchanging the coordinates of electrons 1 and 2 leaves the space part of Ψ_s unchanged, so the space part of the wave function given by (10.16) is symmetric. Exchanging the spin coordinates of Ψ_s multiplies Ψ_s by (-1), so the spin part of wave function (10.16) is antisymmetric. Hence, overall, the wave function Ψ_s is antisymmetric in the coordinates of electrons 1 and 2, as required by the Pauli principle.

The second possibility for the total wave function is given by equation (10.15), so we have either

$$\Psi_t = [\psi_A(1)\psi_B(2) - \psi_A(2)\psi_B(1)] [\alpha(1)\alpha(2)] \qquad (10.17)$$

or

$$\Psi_t = [\psi_A(1)\psi_B(2) - \psi_A(2)\psi_B(1)] [\beta(1)\beta(2)] \qquad (10.18)$$

or

$$\Psi_t = [\psi_A(1)\psi_B(2) - \psi_A(2)\psi_B(1)] [\alpha(1)\beta(2) + \alpha(2)\beta(1)] \qquad (10.19)$$

For each of the three possible wave functions Ψ_t for the two electrons given by (10.17), (10.18), and (10.19), the space part is antisymmetric in the coordinates of electrons 1 and 2. All of three possible spin parts are symmetric in the exchange of coordinates of electrons 1 and 2, so Ψ_t is antisymmetric overall, as required by the Pauli principle. We conclude that the wave function Ψ_s and the three wave functions Ψ_t (for the two electrons on atoms A and B) satisfy the Pauli principle.

We consider next the wave function Ψ_s, given by (10.16), which is[20,21] a singlet state in which the total spin angular momentum is equal to zero. Since there are three possible wave functions Ψ_t, given by equations (10.17)–(10.19), this state is called a triplet state. We may, approximately, think of the singlet state as one in which the electron spins are "antiparallel,", and the triplet states as having the electron spins "parallel," so the total spin angular momentum is not zero for the triplet.

Next, we consider the total energy E of the two electrons in the states Ψ_s and Ψ_t.[22] We have two atoms, A and B, and two electrons, 1 and 2. The Hamiltonian operator $\mathcal{H}$ for the problem is[23]

$$\mathcal{H} = (-\hbar^2/2m)(\nabla_1^2 + \nabla_2^2) + V(\mathbf{r}_1, \mathbf{r}_2) \qquad (10.20)$$

where $(-\hbar^2/2m)\nabla_1^2$ is the kinetic energy operator for electron 1, and so on. In the potential energy $V(\mathbf{r}_1, \mathbf{r}_2)$ due to the Coulomb interaction of electrons with electrons, electrons with ions, and ions with ions, $\mathbf{r}_1$ is the position vector of electron 1, and so forth. Assuming that the wave functions Ψ_s and Ψ_t are normalized, the expectation values of the total energy in the two-electron states Ψ_s and Ψ_t are, respectively,

$$E_s = \int \Psi_s^* \mathcal{H} \Psi_s \, dr_1 \, dr_2 \qquad (10.21a)$$

$$E_t = \int \Psi_t^* \mathcal{H} \Psi_t \, dr_1 \, dr_2 \qquad (10.21b)$$

and where the integrations are over the coordinates r_1 and r_2 of electrons 1 and 2. One may show[24] that the energy difference $(E_s - E_t)$ between the singlet and triplet states is

$$(E_s - E_t) = 2\int dr_1 \, dr_2 \, [\psi_A^*(1) \, \psi_B^*(2)] \, V(\mathbf{r}_1, \mathbf{r}_2) \, [\psi_A(2) \, \psi_B(1)] \equiv 2J \quad (10.22)$$

The integral J defined by (10.22) is called the exchange integral because it is a matrix element between the states $[\psi_A(1)\psi_B(2)]$ and $[\psi_A(2)\psi_B(1)]$ which differ in the exchange of the coordinates of electrons 1 and 2.

The key result embodied in equation (10.22) is that the energy difference $(E_s - E_t)$ between the singlet and triplet states depends on the exchange integral J. If J is positive, then E_s is larger than E_t, and the triplet state Ψ_t, in which the spins of the two electrons are parallel, is of lower energy. If the exchange integral J is negative, then E_s is smaller than E_t, and the singlet state Ψ_s, in which the spins are opposite, is energetically favored. We conclude that a positive value of the exchange integral J favors parallel electron spins. If we generalize this approach to more than two spins, we see that a positive value of J favors a parallel alignment of the electron spins and hence favors ferromagnetism. In that case, one speaks of the "exchange interaction" between electrons aligning the electron spins and producing ferromagnetism.

From expression (10.22), we can see that the sign and magnitude of J will depend on the details of the atomic electronic charge distribution represented by the wave functions ψ and on the electrostatic potential energy V in the integral. When one calculates the exchange integral J, one usually finds it to be negative,[25-27] favoring antiparallel electron spins (as in, for example, the H_2 molecule). This fact suggests that the Heisenberg model of a direct exchange interaction between spin magnetic moments, localized on particular atoms, requires reconsideration. It is also found that the localized direct exchange model cannot explain the observed nonintegral values of the effective magneton number n_B defined by equation (10.12). An alternative is the itinerant or band model,[28,29] in which the electrons contributing spin magnetic moments are not localized in space, but are in Bloch or band states which extend throughout the crystal. In spite of the shortcomings of the Heisenberg exchange model of ferromagnetism, we will use it because it offers a simple way to treat domains, the aspect of ferromagnetic material perhaps most important for applications.

While it should be emphasized that the exchange interaction is electrostatic in nature, it is usual to discuss exchange[30,31] *as if* it actually involved adjacent electron spins. We begin by briefly reviewing some ideas about the magnetic moment of the electrons in an atom. For an atom (in free space), the magnetic moment μ is given classically by

$$\mu = (-g\mu_B/\hbar)\mathbf{J} \tag{10.23}$$

where the total angular momentum vector $\mathbf{J}$ of the electrons in the atom is the sum of the total orbital angular momentum $\mathbf{L}$ and the total spin angular momentum $\mathbf{S}$ for the atom. The quantity g is the Landé g-factor[32,33] for the atom. Since the experimental evidence indicates the total magnetic moment of ferromagnetic atoms is almost entirely spin in origin, we may say $\mathbf{L} \cong 0$, so $\mathbf{J} \cong \mathbf{S}$. In that case, the g-factor is $g = 2$, and the classical magnetic moment vector μ is given by

$$\mu = (-2\mu_B/\hbar)\mathbf{S} \tag{10.24}$$

where μ_B is the Bohr magneton. Quantum mechanically, if Σ^2 is the operator for the square of the magnitude of the total spin angular momentum, then we recall[34] that

$$\Sigma^2\varphi = [S(S+1)\hbar^2]\varphi \tag{10.25}$$

where φ is the spin wave function for an eigenstate of Σ^2 and S is the total spin angular momentum quantum number. For the singlet state Ψ_s, we have $S = 0$ and (10.25) becomes

$$\Sigma^2 \Psi_s = 0 \cdot \Psi_s = 0 \qquad (10.26)$$

For the triplet state Ψ_t, we have $S = 1$, and

$$\Sigma^2 \Psi_t = [2\hbar^2] \Psi_t \qquad (10.27)$$

If we denote the spin angular momentum operator for electron 1 by σ_1, and so on, we can write

$$\Sigma^2 = (\sigma_1 + \sigma_2)^2 = \sigma_1^2 + \sigma_2^2 + 2\sigma_1 \cdot \sigma_2 \qquad (10.28)$$

For either Ψ_s or Ψ_t, we have

$$\Sigma^2 \Psi_s = \sigma_1^2 \Psi_s + \sigma_2^2 \Psi_s + (2\sigma_1 \cdot \sigma_2) \Psi_s = 0 \qquad (10.29)$$

$$\Sigma^2 \Psi_t = \sigma_1^2 \Psi_t + \sigma_2^2 \Psi_t + (2\sigma_1 \cdot \sigma_2) \Psi_t = [2\hbar^2] \Psi_t \qquad (10.30)$$

However, we know also that Ψ_s and Ψ_t are eigenfunctions of the one-electron spin operators σ_1^2 and σ_2^2, with eigenvalue $(3\hbar^2/4)$, so

$$\sigma_1^2 \Psi_s = \sigma_2^2 \Psi_s = \left[\frac{1}{2} \left(\frac{1}{2} + 1 \right) \hbar^2 \right] \Psi_s = \left(\frac{3}{4} \hbar^2 \right) \Psi_s \qquad (10.31)$$

$$\sigma_1^2 \Psi_t = \sigma_2^2 \Psi_t = \left[\frac{1}{2} \left(\frac{1}{2} + 1 \right) \hbar^2 \right] \Psi_t = \left(\frac{3}{4} \hbar^2 \right) \Psi_t \qquad (10.32)$$

Substituting the result (10.31) into (10.29) gives

$$[(3\hbar^2/2) + 2\sigma_1 \cdot \sigma_2] \Psi_s = 0 \qquad (10.33)$$

or

$$(\sigma_1 \cdot \sigma_2) \Psi_s = (-3\hbar^2/4) \Psi_s \qquad (10.34)$$

Equation (10.34) says that the singlet state wave function Ψ_s is an eigenfunction of the operator $(\sigma_1 \cdot \sigma_2)$ with eigenvalue $(-3\hbar^2/4)$. Similarly, substituting result (10.32) into (10.30) gives

$$[(3\hbar^2/2) + 2\sigma_1 \cdot \sigma_2]\Psi_t = (2\hbar^2)\Psi_t \tag{10.35}$$

so

$$(\sigma_1 \cdot \sigma_2)\Psi_t = (\hbar^2/4)\Psi_t \tag{10.36}$$

We see from (10.36) that the triplet state wave function Ψ_t is an eigenfunction of $(\sigma_1 \cdot \sigma_2)$ with eigenvalue $(\hbar^2/4)$.

Consider next the Hamiltonian $\mathcal{H}_{\text{spin}}$ defined by[35]

$$\mathcal{H}_{\text{spin}} \equiv \frac{1}{4}(E_s + 3E_t) - (1/\hbar^2)(E_s - E_t)(\sigma_1 \cdot \sigma_2) \tag{10.37}$$

which operates on the total wave function (either Ψ_s or Ψ_t) of our two-electron system. Then

$$\mathcal{H}_{\text{spin}}\Psi_s = \frac{1}{4}(E_s + 3E_t)\Psi_s - (1/\hbar^2)(E_s - E_t)(-3\hbar^2/4)\Psi_s = E_s\Psi_s \tag{10.38}$$

using the result (10.34), so Ψ_s is an eigenfunction of $\mathcal{H}_{\text{spin}}$, with eigenvalue E_s given by (10.21a). Similarly, Ψ_t is an eigenfunction of $\mathcal{H}_{\text{spin}}$ with eigenvalue E_t, given by (10.21b), since

$$\mathcal{H}_{\text{spin}}\Psi_t = \frac{1}{4}(E_s + 3E_t) - (1/\hbar^2)(E_s - E_t)(\hbar^2/4)\Psi_t = E_t\Psi_t \tag{10.39}$$

on using (10.36). Since, from (10.22), the exchange integral J is defined as $(1/2)(E_s - E_t)$, the operator $\mathcal{H}_{\text{spin}}$ in (10.37) becomes

$$\mathcal{H}_{\text{spin}} = \frac{1}{4}(E_s + 3E_t) - (2J/\hbar^2)(\sigma_1 \cdot \sigma_2) \tag{10.40}$$

where, from (10.38) and (10.39), the hamiltonian $\mathcal{H}_{\text{spin}}$ has as its eigenvalues the total energy of the two-electron system. Defining the zero of total energy so that $(1/4)(E_s + 3E_t) \equiv 0$, equation (10.40) becomes

$$\mathcal{H}_{\text{spin}} = (-2J/\hbar^2)(\sigma_1 \cdot \sigma_2) \tag{10.41}$$

Equation (10.41) is called the Heisenberg Hamiltonian[36] and the direct exchange model of ferromagnetism that it describes is also called the Heisenberg model. It should be emphasized that the exchange integral J, given by (10.22), is entirely *electrostatic* in nature, so the direct exchange interaction described by $\mathscr{H}_{spin}$ is also electrostatic in nature even though $\mathscr{H}_{spin}$ is expressed in terms of the electron spin operators σ_1 and σ_2. The interaction energy between the electrons (of spin σ_1 and σ_2) is electrostatic in nature, and spin enters the description of the interaction as a consequence of the Pauli Principle.

If we use the direct exchange interaction to describe two electrons localized on atomic sites i and j, we see that (10.41) becomes

$$\mathscr{H}_{spin} = (-2J_{ij}/\hbar^2)(\sigma_i \cdot \sigma_j) \qquad (10.42)$$

where J_{ij} is the exchange integral of the type (10.22) between electrons on the atoms i and j. For many purposes, including ours, the operators σ_i and σ_j in the Hamiltonian (10.42) can be replaced by *classical* spin angular momentum *vectors* S_i and S_j, so the exchange interaction energy U_{ij} is, from (10.42), given by[37]

$$U_{ij} = -2J_{ij}(S_i \cdot S_j) \qquad (10.43)$$

In (10.43), the spin angular momenta are in multiples of $\hbar$, so the factor of $\hbar^2$ in (10.42) has disappeared in (10.43). We should also note that, while the exchange energy U_{ij} given by (10.43) is written in terms of the electron spins on adjacent atoms, the vectors S_i and S_j really represent[38] the total magnetic moments (orbital plus spin) of atoms i and j. As mentioned earlier, the atomic magnetic moments in ferromagnetic solids are due almost entirely to spin, so we will refer to the vectors in (10.43) as spins.

From equation (10.43), we can see that the exchange energy U_{ij} is negative (a) if J_{ij} is positive and the spins S_i and S_j are parallel, or (b) if J_{ij} is negative and the spins S_i and S_j are antiparallel. We conclude that ferromagnetism (that is, parallel spins on adjacent atoms) is energetically favored if the exchange integral is positive. As mentioned earlier, the exchange integral is often not positive, so the Heisenberg exchange model is not a complete explanation of ferromagnetism. In spite of its deficiencies, we will use this model because it will suffice to discuss the aspect of ferromagnetism (that is, domains) of key importance for application to material properties.

Finally, we may obtain an estimate of the magnitude of the exchange integral J. Equation (10.22) says $2J$ is equal to the energy difference $(E_s - E_t)$ between the singlet and triplet states, so we expect J to be of the order of the electrostatic energy splitting between atomic states, which is typically[39] of the order of 0.1 eV. In fact, one can show[40] that J has the form

$$J = ak_B T_c$$

where k_B is Boltzmann's constant, T_c is the Curie temperature, and a is a constant (with magnitudes in the range 0.1–0.5) which depends on the crystal structure and atomic magnetic moment of the ferromagnetic solid. For example, for iron, $a = 0.15$, $T_c = 1040$ K, so $J = 160k_B = 2.2 \times 10^{-14}$ erg = 0.014 eV.

Magnetic Domains

If one examines a piece of ferromagnetic material, it will probably have a magnetization much less than the saturation magnetization at that temperature, or the material may even appear "unmagnetized" and without a spontaneous magnetic moment. Since it is also true that the atomic spin magnetic moments are (more-or-less) aligned parallel on an atomic scale, how can these two observations be reconciled?

Consider a ferromagnetic solid in which all of the spin magnetic moments are parallel. This situation is shown schematically in Figure 10.2(a), in which the arrows represent magnetic moments. If we think of the ferromagnet as a "bar magnet" with North (N) and South (S) "poles" on its surfaces as shown, there is an external magnetic field associated with the ferromagnetic sample. We call a region of ferromagnetic material in which

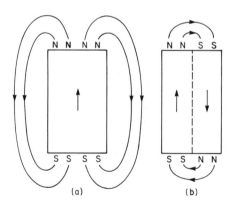

Figure 10.2. Ferromagnetic structures composed (a) of a single domain, and (b) of two domains. The two-domain structure has a smaller magnetic energy because of the smaller spatial extent of the magnetic field (after Kittel,[40a] Figure 9).

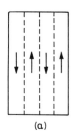
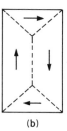
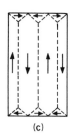

Figure 10.3. Ferromagnetic structures show multidomain arrangements with lower magnetic energy than those in Figure 10.2. (b) and (c) exhibit "domains of closure" (after Kittel,[40a] Figure 9).

(a)   (b)   (c)

all of the magnetic moments are parallel a ferromagnetic domain,[41–44] so Figure 10.2(a) shows a single domain. The magnetic energy associated with the magnetic field of the single domain in Figure 10.2(a) may be compared with the energy of the structure show in Figure 10.2(b), in which there are two domains with opposite direction of magnetization. In the two-domain structure, the energy of the magnetic field due to the surface "poles" is much less because of the smaller spatial extent of the field, so the two-domain structure has a lower energy than the single-domain structure. The magnetic energy can be lowered still further by the formation of structures with a more complex domain arrangement. Examples of multidomain structures are shown in Figure 10.3, in which the arrows indicate the direction of the magnetization in a domain. In summary, the existence of domain structure is due to the possibility of lowering the ferromagnet magnetic energy by going from a saturated single-domain structure to one like that shown in Figure 10.3(a). In Figure 10.3(a), with N domains, the magnetic energy is reduced[45] to about $(1/N)$ of the single-domain configuration energy because of the smaller spatial extent of the magnetic field of the N-domain structure. Further, in domain structures like those in Figures 10.3(b) and 10.3(c), the magnetic energy is reduced to zero[46] because there are no surfaces with a nonzero density of magnetic "poles" present and no magnetic field associated with the magnetization of the sample. In such a case, one speaks of the formation of "domains of closure." Domains are of macroscopic size; the size and shape depend[46a] on the crystal itself, as well as on strain and applied magnetic field. A representative domain size[46a] is 10^{-3} cm for a crystal of dimension of the order of 1 cm.

The existence of domains is the reason why the magnetization of a ferromagnetic specimen may be much less than the saturation value M_s of the magnetization (for example, 1700 gauss for iron at 300 K) expected for a single-domain configuration like Figure 10.2(a). However, as we will discuss later, the application of an external magnetic field can "magnetize" a multidomain structure [like Figure 10.3(c)], thereby increasing its magnetization to a value close to the saturation value.

We next want to discuss the structure and dynamics of ferromagnetic domains and will consider two important contributions to the domain energy, and also the energy of the transition region, called the Bloch wall, between domains.[47-50]

We discuss first the exchange energy, given by the Heinsenberg Hamiltonian (10.41) or (10.42). Considering two electrons on neighboring atoms (1) and (2), the exchange energy U_x between them is given by equation (10.43) as

$$U_x = -2J\mathbf{S}_1 \cdot \mathbf{S}_2 \qquad (10.44)$$

In equation (10.44), J is the exchange integral between atoms (1) and (2), and $\mathbf{S}_1$ and $\mathbf{S}_2$ are the classical spin angular momentum vectors (in units of $\hbar$) discussed earlier. For two adjacent identical atoms, each with spin angular momentum vector $\mathbf{S}$ of magnitude S, in units of $\hbar$ (or total spin[51] quantum number S for the atom), the exchange energy U_x is given by

$$U_x = -2JS^2 \cos \varphi \qquad (10.45)$$

where φ is the angle between the adjacent spin vectors. Common values of S in equation (10.45) are (as discussed below for iron) $S = (1/2)$, $S = 1$. If the angle φ is small, we may write

$$\cos \varphi \cong 1 - \frac{1}{2}\varphi^2 \qquad (10.46)$$

so the exchange energy becomes

$$U_x(\varphi) = -2JS^2 + JS^2\varphi^2 \qquad (10.47)$$

between the adjacent atoms. If we choose the zero of exchange energy to be the case of parallel spins, when $\varphi = 0$, so

$$U_x(0) = -2JS^2 \equiv 0 \qquad (10.48)$$

then equation (10.47) becoms[52]

$$U_x(\varphi) = JS^2\varphi^2 \qquad (10.49)$$

Equation (10.49) gives the difference in exchange energy $U_x(\varphi)$ between two spins (of magnitude S in units of $\hbar$) making a small angle φ with each other when the exchange integral between the adjacent atoms is J. We note particularly that, in this model, the exchange energy U_x is perfectly isotropic in that it depends only on the angle φ between the spins and is not related to any crystallographic direction in the ferromagnetic crystal. (We will see presently that this apparent isotropy is in fact not correct.)

It is useful at this point to discuss the value of S in a typical case, that of iron. Assuming, as discussed earlier, that the atomic magnetic moment μ is almost completely due to spin, we have, classically, from equation (10.24) that the magnitude of the magnetic moment is

$$\mu = (2\mu_B/\hbar)\,S \tag{10.50}$$

where the spin angular momentum magnitude S is in units of $\hbar$. Since the effective magneton number n_B is, from equations (10.12) and (10.13), related to μ by

$$\mu = n_B\mu_B \tag{10.51}$$

we obtain

$$n_B\hbar = 2S \tag{10.52}$$

For iron,[53] $n_B = 2.22$, so we obtain $S = 1.11\hbar$, or (as S is usually quoted in units of $\hbar$), $S = 1.11$. This value is frequently seen in the literature[54,55] as $S \cong 1$ for iron. We will use this value of S later in discussing the width of the transition region between domains.

Next, we discuss the anisotropy energy. As pointed out earlier, the isotropy of the exchange energy U_x contained in equation (10.49) is not really correct. The energy of interaction between spins *does* have an anisotropic dependence on absolute orientation, that is, on the angle between the spin magnetic moment and the crystal axes. This orientation-dependent energy is called the anisotropy energy.[56–59] The reason this orientation dependence is absent from the exchange energy is that the spin–orbit interaction[60] was neglected in our discussion of the exchange energy. In general, there will be a coupling between the spins and the electronic charge density of the atoms, that is, between the spins and the orbital motion of the electrons. Since the charge density (or, alternatively, the orbital motion) of the electrons reflects the crystal structure of the lattice, it is plausible that the spin–orbit interaction results in a dependence of the magnetization (for a given applied field) on direction in the crystal.

This result is observed experimentally in the existence of "easy" and "hard" directions of magnetization in many ferromagnetic materials.[61] It is observed that it requires more energy, and hence a larger applied field, to magnetize such a crystal to saturation in a hard direction than in an easy direction. An example is cobalt, for which the saturation magnetization M_s is about 1400 gauss at room temperature, and whose hexagonal close-packed crystal structure is shown schematically in Figure 10.4. Figure 10.5 shows experimental data[62] on the magnetization as a function of applied magnetic field for cobalt. The two directions shown are parallel to the c-axis of the hexagonal crystal structure and a direction (normal to the c-axis) in the basal plane shown in Figure 10.4. Is is evident from the data of Figure 10.5 that a larger applied field is required to magnetize cobalt to saturation along a hard direction (in the basal plane) than is necessary along the easy direction parallel to the c-axis. The higher applied field required reflects[63] the increase in energy (that is, the anisotropy energy) required to magnetize the cobalt crystal in the hard direction.

Experimentally, it is found[64] that the anisotropy energy density U_a can be expressed as a series of even powers of some angle (or direction cosine) specifying the direction of the magnetization relative to an appropriate crystal axis. (It is reasonable that only even powers occur owing to the equivalence of directions specified by positive and negative angles.) For hexagonal cobalt, one finds[64] that the anisotropy energy density U_a has the form

$$U_a = K_1 \sin^2\theta + K_2 \sin^4\theta \qquad (10.53a)$$

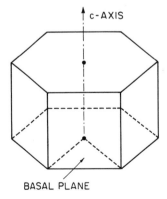

c-AXIS

BASAL PLANE

Figure 10.4. Hexagonal close-packed crystal structure (schematic) of cobalt, showing the c-axis and the basal plane normal to the c-axis.

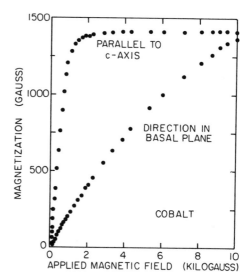

Figure 10.5. Magnetization of cobalt as a function of applied magnetic field for a direction parallel to the c-axis of the hexagonal structure and for a direction in the basal plane (after Kaya[62]).

where θ is the angle between the magnetization direction and the c-axis of the hexagonal structure. For cubic iron, the expression analogous to equation (10.53a) is[64,64a]

$$U_a = K_1 (a_1^2 a_2^2 + a_2^2 a_3^2 + a_3^2 a_1^2) + K_2 a_1^2 a_2^2 a_3^2 \tag{10.53b}$$

where a_1, a_2, and a_3 are the direction cosines of the direction of magnetization relative to the cube edges of the iron unit cell. The quantities K in (10.53a) and (10.53b) are called anisotropy constants, and are temperature-dependent constants independent of the directions specified by the angle θ or the direction cosines a. For hexagonal cobalt[65] at room temperature, $K_1 = 4.1 \times 10^6$ erg cm^{-3} and $K_2 = 1.4 \times 10^6$ erg cm^{-3}. To estimate the magnitude of the anisotropy energy density U_a for cobalt, we note that, if the magnetization is directed along the c-axis (the easy direction), then $\theta = 0$, and $U_a = 0$. If the magnetization is along a direction lying in the basal plane of the hexagonal structure, then $\theta = (\pi/2)$, and, from (10.53a), $U_a = 5.4 \times 10^6$ erg is the anisotropy energy per unit volume. As mentioned earlier, the difference in anisotropy energy density between hard and easy directions of magnetization in cobalt is reflected in the larger applied field necessary to magnetize cobalt to saturation in the hard direction.

We should also note that the magnitude of the anisotropy energy U_a in equations (10.53) depends on the values of the anisotropy constants K which are, of course, different for different materials. The larger the anisotropy constants in equations (10.53), the larger is the energy difference between the easy and hard directions of magnetization in the crystal. The magnitudes of the anisotropy constants cover a wide range. In addition to the values given above for hexagonal cobalt, values[65] for cubic iron at room temperature are $K_1 = 4.8 \times 10^5$ erg cm^{-3} and $K_2 = 1.2 \times 10^5$ erg cm^{-3}. For the hexagonal intermetallic compound $SmCo_5$, the anisotropy energy is described[65a] by the first term of (10.53a) with $K_1 = 1.7 \times 10^8$ erg cm^{-3}, a very large value.

We now turn to the region of transition between domains with different directions of magnetization. In Figures 10.2 and 10.3, this transition was shown schematically as if it were discontinuous, which is not the case. Consider two adjacent domains, such as those in Figure 10.2(b), in which the directions of magnetization are opposite. If the magnetization were to be changed discontinuously in direction by π radians between adjacent atoms, then equation (10.45) would give the total exchange energy between these adjacent spin magnetic moments as $[2JS^2 + (N-1)(-2JS^2)]$, where we are including $(N-1)$ pairs of parallel spins and the one pair of antiparallel spins. The total exchange energy is

$$U_x = 4JS^2 \qquad (10.54)$$

where, as before, we choose to set $(-2JS^2) \equiv 0$, and where S is the spin quantum number and J the exchange integral (assumed positive). Consider next a situation in which the magnetization in adjacent domains changes by π radians over a region which is $(N+1)$ atomic planes wide, as shown schematically

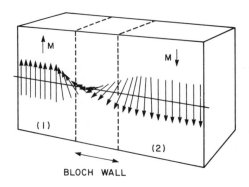

BLOCH WALL

Figure 10.6. Bloch wall (schematic) showing the reversal of the direction of magnetization M between domains (1) and (2) taking place over (N+1) atomic planes (after Kittel,[40a] Figure 23).

Figure 10.7. Schematic Bloch wall for a (hypothetical) ferromagnetic simple cubic crystal of lattice constant a. The Bloch wall contains $(N+1)$ atomic planes and has a width Na. (The drawing shows $N = 5$; typically N is several hundred.) Also indicated are lines of atoms normal to the plane of the wall and an area a^2 normal to one of the lines of atoms.

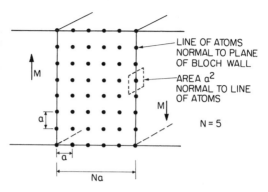

in Figure 10.6. The transition region is shown in Figure 10.7 for a (hypothetical) case in which $N = 5$; actually N is the order of 100. In this transition layer, called the Bloch wall (or domain wall), the angle φ between adjacent spin magnetic moment vectors is (π/N) radians and so is small. Now, using equation (10.49), the difference in exchange energy between adjacent spins is $JS^2(\pi/N)^2$. There are N pairs of adjacent spins in one line of $(N+1)$ atoms directed normal to the plane of the Bloch wall. The total exchange energy U_x of one such line of $(N+1)$ atoms is therefore

$$U_x = NJS^2(\pi/N)^2 = \pi^2 JS^2/N \tag{10.55}$$

where again, as in (10.49), we set $(-2JS^2) \equiv 0$. Comparing equations (10.55) and (10.54), we see that the exchange energy difference between the two domains is reduced by a factor of $(4N/\pi^2)$ when the transition between domains occurs over a line of $(N+1)$ atoms instead of taking place discontinuously. Equation (10.55) gives the exchange energy U_x of one line $(N+1)$ atoms of length Na, where a is the interatomic distance along the line. As shown in Figure 10.7, this line of atoms is in a direction normal to the plane of the Bloch wall. Since U_x is the exchange energy of one line of atoms, U_x times the number of lines of atoms per unit area of the wall is equal to the exchange energy per unit area of the Bloch wall, denoted by σ_x. From Figure 10.7, there are $(1/a^2)$ lines of atoms per unit area of wall, so we have

$$\sigma_x = (\pi^2 JS^2/Na^2) = (\pi^2 JS^2/a[Na]) \tag{10.56}$$

as the exchange energy density per unit area of the Bloch wall.

From equation (10.56), we see that the exchange energy density σ_x of the

Bloch wall varies inversely with the width Na of the wall. Thus σ_x decreases as the width of the wall increases. Based on the exchange energy density alone, then, the thickness of the wall would be expected to increase indefinitely in order to minimize the energy of the wall. However, as we will see, the anisotropy energy (per unit area) associated with the Bloch wall is directly proportional to the width Na of the wall. The anisotropy energy density of the wall thus increases with increasing wall width. Combined with the decreasing exchange energy density in equation (10.56), the result is a value of N (or, alternatively, Na) for which the total energy density, exchange plus anisotropy, of the wall is a minimum. We can calculate this value of N, denoted by N_m, as follows.

We need first to find the anisotropy energy per unit area, denoted by σ_a, of the Bloch wall. The anisotropy energy density (per unit volume) U_a is given by equation (10.53) for cobalt. Suppose we consider a model which is a (hypothetical) simple cubic ferromagnetic crystal of lattice constant a for which the anisotropy energy per unit volume U_a is given by a simpler form of (10.53), in which

$$U_a = K \sin^2 \theta \tag{10.57}$$

where K is the anisotropy constant and θ is the angle between the magnetization and the crystallographic easy direction of magnetization. Dimensional analysis suggests that the desired anisotropy energy per unit area is given by U_a times the number N of atoms times the volume a^3 per atom, divided by the area a^2 per atom in the plane of the Bloch wall. The result (albeit, hand-waving) is that σ_a is of the order [66] of KNa. One may calculate[67] that $\sigma_a = (KNa/2)$ for the model of a ferromagnet described above. In any event, we will take

$$\sigma_a \cong KNa \tag{10.58}$$

for our calculation of the width Na of the Bloch wall. We find (10.58) plausible by noting that most of the spins in the wall will point away from the easy direction, so it is reasonable to expect σ_a to increase as the number of spins in the wall increases.

Using equations (10.56) and (10.58), we find the total energy density σ per unit area of the Bloch wall to be

$$\sigma \cong (\pi^2 JS^2/Na^2) + KNa \tag{10.59}$$

To minimize the total energy density, we set

$$(\partial\sigma/\partial N) \cong -(\pi^2 JS^2/N^2 a^2) + Ka = 0 \qquad (10.60)$$

leading to

$$N_m \cong (\pi^2 JS^2/Ka^3)^{1/2} \qquad (10.61)$$

for the value N_m of N for which the wall energy density is a minimum. Rewriting (10.61) gives

$$W_m = N_m a \cong (\pi^2 JS^2/Ka)^{1/2} \qquad (10.62)$$

as the value W_m of the Bloch wall width for which σ is a minimum. Using equation (10.61), one may estimate[68] a representative value of N_m (for iron) to be about 150, corresponding to a wall thickness W_m of the order of several hundred to a thousand Ångstroms. We will take 1000 Å as a representative figure for the thickness of a Bloch wall.

In summary, the Bloch wall is a transition region separating domains with opposite directions of magnetization. From (10.56), the exchange energy density of the wall decreases with the increasing number N of spins over which the spin reversal of π radians is "spread." From (10.58), the anisotropy energy density increases with N, balancing the decrease in exchange energy density. The result is a Bloch wall containing an equilibrium number N_m of spins.

The Magnetization Curve and Hysteresis

If one has a piece of "unmagnetized" ferromagnetic material, for which the net magnetization **M** is zero, it may be "magnetized" by applying an external magnetic field **B**. One could then, in principle, observe the value of **M** for the sample as a function of **B**. The more usual procedure[69] is to observe **B** (rather than **M**) as a function of the magnetic intensity **H** inside the ferromagnet. Since **B**, **H**, and **M** are related by

$$\mathbf{B} = \mathbf{H} + 4\pi\mathbf{M} \qquad (10.63)$$

M can be calculated if **B** and **H** are known. A plot of B as a function of H is sometimes called the magnetization curve of the ferromagnetic material.

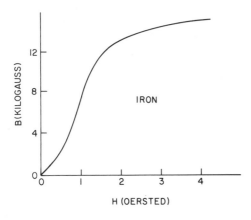

Figure 10.8. Magnetization curve for iron, showing the magnetic induction B (in gauss) as a function of the magnetic intensity H (in oersted) (after Purcell,[1] Figure 11.31).

Consider the application of an applied field H to an initially unmagnetized sample of a ferromagnet, for which therefore B and H are zero. As seen from the "typical" magnetization curve[70] (for iron) shown in Figure 10.8, increasing the magnitude of the magnetic intensity H causes an increase in the magnetic induction B in the sample. Writing a scalar form of equation (10.9) gives us

$$B = \mu H \tag{10.64}$$

as the relation between B and H. From (10.64), we define the permeability μ of the ferromagnetic material as

$$\mu \equiv B/H \tag{10.65}$$

The magnetization curves for ferromagnetic materials are nonlinear, so the permeability μ depends on the magnetic intensity H, and $\mu = \mu(H)$. For iron, as seen from Figure 10.8, the maximum values of the slope and of the permeability occur for a value of H about 1.0 oersted. With increasing values of H, the curve levels off and B approaches saturation as the magnetization of the iron increases and approaches[69] the saturation value M_s. Since B is much larger than H over almost all of the curve in Figure 10.8, equation (10.63), in scalar form, gives us

$$B \cong 4\pi M \tag{10.66}$$

and the difference between B and $4\pi M$ is negligible.[69]

What domain processes in the ferromagnet are taking place to give rise to the magnetization curve in Figure 10.8? Figure 10.9 shows a schematic plot of the magnetization M of the ferromagnet as a function of the applied magnetic field H. [Since, from (10.66), $M \cong (B/4\pi)$ for the ferromagnet, the curves in Figures 10.9 and 10.8 give us the same information.] As the applied magnetic field is increased, the magnetic moment per unit volume (the magnetization) of the ferromagnet increases until the saturation value M_s is reached. As indicated in Figure 10.9, there are two different kinds of processes[71] by which the magnetization of the ferromagnet increases. The first process, at lower values of the applied magnetic field, is motion, both reversible and irreversible, of domain boundary walls. The second process, occurring at higher applied fields, is rotation of the magnetization vector $\mathbf{M}$, within a domain, toward the direction of the applied field $\mathbf{H}$. We now discuss these processes in a qualitative manner.

Consider first the motion of a domain boundary (Bloch) wall in a single crystal sample. In zero applied magnetic field, we would expect[72] the unmagnetized sample to consist of several domains, each with its magnetization vector $\mathbf{M}$ oriented along one of the easy directions of the crystal structure.

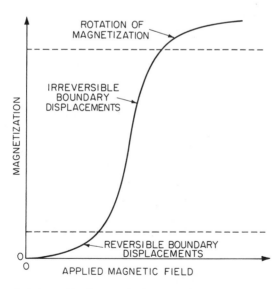

Figure 10.9. Graph (schematic) of magnetization as a function of applied magnetic field, showing different domain processes important at increasing values of the magnetic field (after Kittel and Galt[12]).

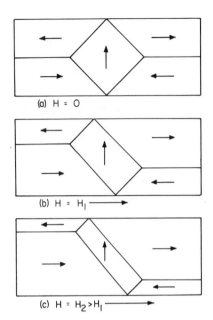

(a) H = O

(b) H = H_1 ⟶

(c) H = $H_2 > H_1$ ⟶

Figure 10.10. Motion of domain boundary walls in a ferromagnet in an applied magnetic field **H** directed along an "easy" direction. The volume of domains whose magnetization is parallel to the applied field increases by boundary wall displacements (after Kittel[73]).

The situation in zero applied field is thus as shown in the domain structure of Figure 10.10(a), which is adapted from a figure, given by Kittel,[73] of domains in an iron crystal, which has six equivalent cubic [100] easy directions.[74] Figure 10.10(b) shows the effect of applying a magnetic field of magnitude H_1 (directed to the right in the figure) along one of the easy directions of magnetization. There is a decrease in the volume of domains whose magnetization vector **M** is antiparallel to the applied field; this decrease is due to motion of the domain walls. The domain walls move in the presence of the applied field because the spins making up the wall can lower their energy by changing their orientation to become parallel to the applied field. This can be illustrated using Figure 10.6, in which we can consider the effect of applying a magnetic field **H** (not shown in the figure) directed parallel to the magnetization **M** in region (1), that is, "upward" in the plane of the diagram. In this situation, the spins in region (1), to the left of the Bloch wall, have a lower energy density, by an amount $2MH$, than do the oppositely-directed spins in region (2) to the right of the Bloch wall. The wall will move toward region (2), to the right in the figure, as the spins in region (2) lower their energy by aligning themselves parallel to the applied field. The energy of the sample decreases as the Bloch wall moves to the right since this wall motion increases the volume of the domain, region (1), whose energy density is lower because its magnetization vector is parallel to the applied magnetic field. The decrease in the magnetic energy of the domain with magnetization parallel to the applied field is converted[75] into an increase in the surface energy density of the Bloch wall. One may think of an effective pressure p, equal to $2MH$, exerted by the applied field on the wall,[75,76] causing the Bloch wall to move.

Returning to Figure 10.9, we consider the processes of Bloch wall displacement and magnetization rotation, mentioned earlier, which increase the magnetization of a ferromagnetic sample in an applied magnetic field. Consider first a single crystal sample which is "perfect" in the sense of being ideal and free of defects. In such a perfect crystal, with its complete translational periodicity, one would expect[77–79] that the energy of a domain wall would not depend on the position of the wall in the crystal. Under such circumstances, then, the Bloch wall between domains should be easily displaced by a very small applied field.[80]

However, in a real, nonideal single crystal, there will exist various kinds of imperfections of defects. Among these we might expect impurities, inclusions, cavities, internal strains, and so on. In such a nonideal sample, we would expect that the energy of the Bloch wall would no longer be independent of its location, but would instead be a function of the position of the wall. Figure 10.11 shows a (schematic) plot of boundary wall energy E as a function of distance x (in some direction in the crystal). In the absence of an applied magnetic field, the Bloch wall would be located at some equilibrium position $x = x_0$ (point A in Figure 10.11) of minimum energy E_A. One speaks of the domain wall being "pinned"[81] in an energy minimum, such as point A, introduced by defects in the crystal. One model[82] for such pinning represents a defect by a region of the crystal in which anisotropy, exchange,

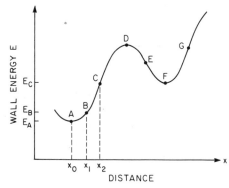

Figure 10.11. Graph (schematic) of boundary wall energy E as a function of distance x, in a crystal containing defects. Points A and F are relative minima, and points C and G are points of inflection at which (dE/dx), and the "restoring force", are. maxima.

and saturation magnetization have values different from those in the homogeneous bulk of the crystal.

Next, suppose than an external magnetic field H_1 is applied and is directed so that the Bloch wall moves toward increasing x in Figure 10.11. Consider a situation in which the magnitude of the applied field is small, so that the increase $(E_B - E_A)$ in the wall energy E corresponds to a small displacement $(x_1 - x_0)$ of the wall to point B. The effective pressure on the wall is $2MH_1$, where M is the magnetization of the domain. As the wall moves from A toward B, its energy increases and we may think of the increasing slope (dE/dx) as a "restoring force"[83] pushing the wall back toward point A. The point $x = x_1$ will be the value of x at which the restoring force (dE/dx) is equal to the force exerted by the pressure $2MH_1$ on the wall.

With the wall now located at point B, suppose that the magnitude H_1 of the applied field is decreased to zero, without, changing the direction of the field. The pressure on the wall due to the applied field is thus reduced to zero, and the restoring force pushes the wall toward the original point A at which (dE/dx) is zero, and at which the wall is in equilibrium. In the process of moving from A to B and back to A, the displacement of the wall is reversible, in that the wall returns to its original equilibrium position when the applied field is removed. Such reversible small displacements occur at low values of the applied magnetic field, as indicated on the magnetization curve in Figure 10.9.

We now consider the situation at values of the applied magnetic field larger than those producing reversible boundary wall displacements. Suppose that a magnetic field H_2, larger than H_1 but in the same direction, is applied to the boundary wall situated in an equilibrium position at point A of the curve in Figure 10.11. The wall moves from A to point B, and now, since the applied field is larger, the magnetic pressure $2MH_2$ is larger, and the wall moves on to point C. Point C is located at $x = x_2$, and the displacement $(x_2 - x_0)$ corresponds to a larger energy increase $(E_C - E_A)$. Point C is a point of inflection, at which (dE/dx) is maximum. At points of the curve beyond C, the restoring force (dE/dx) decreases in magnitude, and the wall moves[85,85] spontaneously to point G. Point G is a point with the same value of (dE/dx), and hence the same magnitude of the restoring force, as at point C. The motion of the wall from point A to point G is irreversible in the following sense. If the applied field is decreased to zero with the wall at point G, the wall moves from point G to the energy minimum at point F, and not back to its original position at point A. If the field is then reversed in direction, the wall will move toward point E. These irreversible boundary displacements thus take place, as shown on the magnetization curve in Figure 10.9, at higher values of the applied magnetic field than do the reversible displacements described earlier.

So far, we have been considering domain wall displacements, reversible and irreversible, in an unmagnetized sample placed in an applied field directed along one of the easy directions of magnetization as shown in Figure 10.10. We discuss next the situation when the applied field $\mathbf{H}$ is along a "hard" direction of magnetization of the unmagnetized crystal,[86] which is assumed to be composed mainly of domains whose magnetization is either parallel or antiparallel to an "easy" direction. The initial effect of the field $\mathbf{H}$, which makes some acute angle θ with the easy direction, is to produce domain wall displacements. As the field $\mathbf{H}$ is increased in magnitude, these wall displacements increase the volume of energetically favored domains whose magnetization makes an angle θ with $\mathbf{H}$, and decrease the volume of energetically unfavored domains whose magnetization makes an angle $(\pi + \theta)$ with $\mathbf{H}$. When the process of boundary displacement is complete, the sample will still not have achieved saturation of the magnetization because the net magnetization $\mathbf{M}$ of the sample is not parallel to $\mathbf{H}$, but makes an angle θ with the applied field. In order to obtain complete saturation of the magnetization, the net magnetization $\mathbf{M}$ must be made to rotate until it is parallel to $\mathbf{H}$. Such a rotation requires energy because the magnetization must be rotated toward a direction parallel to the hard direction of the crystal. This rotation requires the expenditure of energy[87] because of the difference in anisotropy energy between the easy and hard directions of magnetization. This process of magnetization rotation therefore takes place at applied magnetic fields larger than those that cause boundary displacements. This is indicated on the magnetization curve in Figure 10.9. Further, as mentioned earlier, the larger the anisotropy constants K in equation (10.53), or its equivalent, for the anisotropy energy density U_a, the larger is the energy difference between easy and hard directions of magnetization in the crystal, and the larger the energy which must be supplied by the external applied field to rotate the magnetization vector $\mathbf{M}$ toward the hard direction. We conclude that crystals with a larger anisotropy energy density will require higher applied magnetic fields to rotate $\mathbf{M}$ toward the hard direction in achieving saturation.

Finally, we discuss the magnetization of very small particles of a ferromagnetic material. It is found[88-90] that, for particles which are small enough (diameters smaller than about 10^{-5} cm), domain wall formation is energetically unfavorable. Such particles are composed of a single domain, in which there can be no boundary wall displacements. The magnetization of such particles must take place only by rotation of the magnetization vector within the single domain of the particle. We have seen that this process of magneti-

zation rotation generally requires higher applied magnetic fields than boundary wall displacements. We will find that this effect is important in materials
for permanent magnets.

Using these ideas about domain wall movement and magnetization rotation, we can now discuss the familiar hysteresis curve[91] of a ferromagnet.
Figure 10.12 shows, in schematic form, a representative hysteresis curve or
loop. (The portion OAA' of Figure 10.12 is often referred to as the magnetization curve.) The vertical axis is the magnetic induction B, which will be
approximately equal to $4\pi M$ since B is much larger than H over much of the
graph. We may therefore think of the hysteresis curve as a plot of magnetization M as a function of applied field H.

Consider an initially unmagnetized ($\mathbf{M} = 0$) ferromagnetic single crystal
sample in zero applied field, at point O of Figure 10.12. We assume, as before, that the sample is composed of a number of domains whose magneti-
zations are aligned parallel to easy directions of the crystal. Assume that the
applied field $\mathbf{H}$ is in a direction parallel to a hard direction and that the
magnitude H of $\mathbf{H}$ is increased from zero. As reversible, and then irreversible,
boundary wall displacements take place, and the volume of energetically
favored domains increases, the magnetization of the sample increases from
zero as the dashed curve OA is traced out. As H is increased further, domain
wall motion is complete and the magnetization increases slowly by rotation
of the vector $\mathbf{M}$ within domains. In this way, the magnetization approaches
its saturation value M_s, and the portion AA' of the curve is produced. The
magnetic induction at point A' is the saturation induction $B_s \cong 4\pi M_s$.

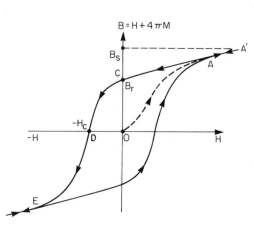

Figure 10.12. Plot of magnetic induction $B (= H + 4\pi M)$ as a function of applied field H for a ferromagnet, showing a typical (schematic) hysteresis curve. Since $B \gg H$
over most of the plot, we can take
$M \cong (B/4\pi)$, giving essentially a plot
of magnetization M as a function of
H. The point O is the initial unmagnetized state of the sample, taken to
be a single crystal. The quantity B_s is
the saturation induction and B_r is the
remanence, the induction remaining
when H is decreased to zero. The ar-
rowheads on the curves indicate an
increasing or decreasing magnitude
of H.

Suppose next that the magnitude H of the applied field is decreased while its direction is unchanged. The magnetization M decreases reversibly[92,93] along the curve $A'A$ as the magnetization $\mathbf{M}$ within domains rotates away from the "hard" direction of $\mathbf{H}$ as H is decreased. When point A is reached, the magnetization no longer decreases reversibly in that the original magnetization curve AO is not retraced. Instead, the magnetization decreases to a value called the remanent magnetization M_r at $H = 0$. The corresponding value B_r of the magnetic induction is called the remanence, and is shown at point C of Figure 10.12. The reason for this hysteresis (lack of reversibility) when the applied field is decreased is the irreversibility of the domain wall displacements (comprising most of section OA of the curve) which occurred as the sample was originally magnetized. The magnetization which was produced on going from point O to point A is largely "locked in" by the irreversible nature of the displacements of the domain walls. Thus, as H is decreased to zero, the domain structure remains in an arrangement giving the nonzero value M_r of the magnetization and the nonzero value B_r of the magnetic induction. The net result is that the sample has a spontaneous magnetic moment in the absence of an applied magnetic field, and so is what is called a "permanent magnet."

At point C, with $H = 0$, the direction of the applied field H is reversed (so $\mathbf{H}$ is now antiparallel to its former direction) and its magnitude H is increased from zero. As H increases, domain wall motion continues to take place, now decreasing the volume of the domains whose magnetization had been oriented favorably to the original direction of $\mathbf{H}$. The overall magnetization and magnetic induction of the sample therefore decrease with increasing H until the magnetic induction becomes zero. This is point D of the curve where $\mathbf{B}$ is zero and the reversed applied field has a magnitude H_c, called the coercivity. The coercivity H_c of the sample is thus the (reverse) applied field necessary to make the magnetic induction of the sample vanish. There is another definition,[94,95] in which the coercivity is defined as the reverse field necessary to reduce the magnetization (rather than the magnetic induction) to zero. This reverse field, usually called the intrinsic coercivity and denoted by H_{ci}, $_MH_c$, or $_iH_c$, is thus the field necessary to reverse half of the magnetization of the sample, thereby giving a net magnetization of zero. Further, one can show[96] that the coercivity of a small spherical single-domain particle is given by $(2K/M_s)$, where K is the anisotropy constant of the (uniaxial) crystal and M_s is the saturation magnetization. The quantity $(2K/M_s)$ is called the anisotropy field H_A and is often taken as a theoretical upper limit to the coercivity for the process of uniform magnetization rotation in a sample composed of spherical single-domain particles.[97,98]

In describing ferromagnetic materials for applications, the remanence B_r and the coercivity H_c are two of the most important parameters. These quantities are specified by discussing the first two quadrants (B, H positive; B positive, H negative) of the hysteresis curve in Figure 10.12. We will therefore not discuss the remainder of the curve shown there; the interested reader is referred to the literature.[99]

As a last point, we have so far been explicitly discussing single crystals of ferromagnetic materials. For a polycrystalline specimen, the situation[100] is more complicated than for a single crystal. Consider an unmagnetized polycrystalline sample which is composed of grains (small crystallites) with random crystalline orientations. An applied magnetic field will make different directions with the easy directions of magnetization of the grains. For some grains, the applied field will be parallel, or nearly so, to the easy direction; these grains will magnetize easily at relatively small applied fields. Other grains, for which the applied field is along a hard direction, will require larger values of the applied field to reach their saturation magnetization. The overall effect of this situation is the usual sort of hysteresis curve, such as Figure 10.12, seen for practical ferromagnetic polycrystalline materials. Since we are interested primarily in the qualitative physical picture of the domain processes giving rise to the magnetization and hysteresis curves, the differences between single- and polycrystalline materials will not be discussed further.

Materials for Permanent Magnets

We can use the ideas developed above to discuss the physics of materials for permanent magnets. For a permanent magnet, one wants a ferromagnetic material with a high value of the remanence B_r in order to have a "strong" magnet. For the permanent magnet to remain magnetized, one wants a high value of the coercitivity H_c. A material having high values of both B_r and H_c would be a good permanent magnet, so a figure of merit would be a large value of the product $B_r H_c$. Since B_r is usually expressed in gauss, and H_c in oersted, the units of this figure of merit are gauss-oersted, which have the units of energy density. Since the product $B_r H_c$ is the area of a rectangle of height B_r and base H_c inscribed in the second quadrant of the hysteresis curve of the ferromagnet, the usual figure of merit for permanent magnet materials is the area $(BH)_{max}$ of the largest rectangle that can be inscribed in the second quadrant of the hysteresis curve of the material. The quantity $(BH)_{max}$ is referred to as the maximum energy product[101] for the material and is usually expressed in megagauss-oersted (abbreviated MGOe). One can show[101] that, for a given volume of ferromagnetic material, the largest magnetic field is produced when $(BH)_{max}$ has its largest value, so $(BH)_{max}$ is a useful figure of merit.

Consider an "ideal" permanent magnet material[102,103] whose magnetization is constant and equal to its saturation value M_s, independent of the applied magnetic field. If $M = M_s$, independent of H, then equation (10.6) gives

$$B = H + 4\pi M_s \qquad (10.67)$$

as the relation between B and H. In the second quadrant of the hysteresis curve (B positive, H negative), equation (10.67) is a straight line defining[104] a triangle of area $8\pi^2 M_s^2$. Using the theorem[104] that the area of the largest rectangle that can be inscribed in a triangle of area A has an area $(A/2)$, one may show[104] that the maximum energy product for this ideal material has the value $(BH)_{max} = 4\pi^2 M_s^2 = (4\pi M_s)^2/4$. The value $(4\pi M_s)^2/4$ of the maximum energy product serves as an upper limit to $(BH)_{max}$ in an ideal situation of complete alignment of magnetic moments and idealized packing density for a given material. For example, using the value $M_s = 1700$ gauss for iron at room temperature,[13] we get $(4\pi M_s)^2/4 = 114$ MGOe, a value of $(BH)_{max}$ much larger than the largest value[105] observed to date. Representative observed valued of $(BH)_{max}$, which are structure-sensitive and vary from sample to sample, will be given for specific permanent magnet materials as they are discussed.

What characteristics of a ferromagnetic material will favor large values of the remanence B_r and coercitivity H_c, thereby leading to a large value of $(BH)_{max}$? Discussing the coercivity first, we recall H_c is the reverse applied field H necessary to reduce the magnetic induction B to zero, starting from the remanent value B_r at $H = 0$. When the direction of H is reversed, domains which had been oriented favorably (in the energetic sense) to $+H$, are now unfavorably oriented to $(-H)$. As the magnitude $(-H)$ of the reverse field is increased, this field increases the volume of domains in which the magnetization vector M is oriented parallel to $(-H)$. As this process proceeds, the magnitudes M and B decrease (tracing out the demagnetization curve CD in Figure 10.12) until $B = 0$ at the value $H = -H_c$, the coercivity. What characteristics of the ferromagnetic material will make this demagnetizing process more difficult and thus requiring a larger value of the coercivity to achieve $B = 0$? In general, characteristics which impede domain wall motion[106–108] will make the demagnetizing process more difficult and will increase the coercivity of the material. An example is furnished by the decreased coercivity of iron[109] with decreasing amounts of precipitated copper and with the removal of internal strains by annealing. The presence of the precipitates and the regions of strain introduce local energy minima, which pin domain walls and thereby impede magnetization reversal by domain wall motion. The result is an increased coercivity.

Another method of obtaining a high value of the coercivity is to use small particles of the ferromagnetic material. As mentioned earlier, sufficiently small particles are composed only of a single domain. The magnetization vector **M** of such a particle can reverse only by rotation, a process which generally requires higher applied fields than processes involving domain wall motion. If the ferromagnetic material (let us say uniaxial) has a large crystalline anisotropy, a large applied field will be required to supply the energy necessary to rotate **M** over the energy barrier of the hard direction of magnetization. We therefore expect high values of the anisotropy constant K to favor high values of the coercivity. This conclusion is reflected in the expression $H_A = (2K/M_s)$ for the anisotropy field H_A which, as mentioned earlier, serves as a theoretical upper limit to the coercivity.

In sum, an important way[110] to make a good permanent magnet material with high coercivity is to use a ferromagnetic material with a large uniaxial anisotropy in the form of small particles containing only a single domain.

We turn next to the magnitude of the remanence B_r. From the general shape of the portion $A'AC$ of the hysteresis curve in Figure 10.12, it seems reasonable to conclude that, all other things being equal, a larger value of the saturation magnetization M_s would imply a larger value of the remanence. We can consider the factors determining the magnitude of M_s by considering equation (10.12), written for 300 K as

$$M_s = n_B N \mu_B \qquad (10.68)$$

where n_B is the effective magneton number of the formula unit involved (such as Fe, SmCo$_5$, and so on), N is the number of formula units per unit volume, and μ_B is the Bohr magneton. The effective magneton number n_B is determined by the atomic magnetic moments involved, their relative orientations, and any band structure effects in the itinerant electron picture. The number N of formula units per unit volume will be a function of the crystal structure of the ferromagnetic material. In a perfect single crystal of a material characterized by particular values of n_B and N, these values will be determined by the inherent atomic, crystal, and electronic structures of the material, and cannot be readily altered. However, in a sample composed of small particles of ferromagnetic material dispersed in a possibly nonmagnetic matrix, one would want to be sure the ferromagnetic particles are concentrated enough to give a satisfactory value of M_s, and hence a large enough remanence B_r.

We will now apply these ideas to a discussion of several materials of current use for permanent magnets. Values of relevant parameters will be given, but it should be kept in mind that literature values of coercivity, remanence, and maximum energy product are strongly structure-sensitive, so the figures quoted here are representative of commercial material. Generally, only the values of the saturation magnetization, Curie temperature, and anisotropy constants found in the literature are characteristic of the material and not structure-sensitive.

While it is not a material used for permanent magnets (because of its limited availability), it is useful to list some of the magnetic properties of elemental cobalt for use as a "base line" relative to which modern materials can be compared. At room temperature, the saturation induction B_s (approximately equal to $4\pi M_s$) is 17600 gauss,[111] the remanence[112] $B_r = 4900$ gauss, and the coercivity[112] $H_c = 8.9$ oersted. The anisotropy constant $K_1 = 6.8 \times 10^6$ erg cm^{-3} for hexagonal (uniaxial) cobalt[113] at 0 K.

One of the first of the modern permanent magnet materials is the series of alloys known as "Alnico" and containing various percentages of iron, aluminium, nickel, cobalt, and other materials. One of the most widely used of these alloys is Alnico 5 (with a composition[114] of approximately 51% iron, 8% aluminium, 14% nickel, 24% cobalt, and 3% copper). For Alnico 5, important parameters[114] are $B_r = 12800$ gauss, $H_c = 630$ oersted, $4\pi M_s = 14000$ gauss, and the maximum energy product $(BH)_{max}$ typically has a value of 5.5×10^6 GOe. This alloy, developed in the 1930s, has a high remanence and a much larger coercivity than that of elemental cobalt. It is presently thought[116–119] that the high coercivity of Alnico 5 is due to a structure containing single-domain small particles, so the demagnetization process involves rotation of the magnetization vectors. The small particles have an elongated shape which contributes[120] to their anisotropy energy; the crystalline anisotropy energy appears to be relatively unimportant in this case. During preparation, the alloy is cooled in the presence of a magnetic field, which orients the elongated particles and increases the value of the maximum energy product. The commercial magnetic material is then mechanically compacted and bonded.

Newer materials for permanent magnets, developed in the 1960s, are the rare earth–cobalt compounds RCo_5, where R is a rare earth atom. Of this group, the most important[121–123] for applications is $SmCo_5$. This well-defined intermetallic compound can be prepared in single crystal form, has the hexagonal $CaCu_5$ structure,[124] and has a ferromagnetic Curie temperature of about 1000 K. At room temperature, important parameter values[124] are as follows. The saturation magnetization $4\pi M_s$ is 9700 gauss, less than that of Alnico 5, but $SmCo_5$ has the largest reported value, 1.7×10^8 erg cm^{-3}, of the anisotropy constant K_1. A typical value[126] of the maximum energy product $(BH)_{max}$ is 18 MGOe, more than a factor of three larger than the value for Alnico 5. Representative values[125] are $H_c = 8000$ oersted and $B_r = 8700$ gauss for the coercivity and the remanence. While the remanence of $SmCo_5$ is less than that of Alnico 5, its coercivity is more than an order of magnitude higher.

Considering $SmCo_5$ as composed of samarium and cobalt sublattices, it is found[126] that the magnetic moments of the Sm and Co atoms are aligned parallel to each other, conforming to the general observation[127] that light rare earth and transition metal moments couple ferromagnetically. This leads to the high value of the saturation magnetization observed for $SmCo_5$; the saturation magnetic moment[128] is more than $6\mu_B$ per formula unit. It has also been concluded[129] that $SmCo_5$ magnets do not contain small single-domain particles. The magnet particles thus reverse their magnetization by domain wall motion (including pinning) and/or nucleation. For this reason, it appears that the high coercivity of $SmCo_5$ cannot be ascribed to the difficulty of magnetization rotation in small particles of a strongly anisotropic ferromagnet. However, the general increase of coercivity with increasing crystalline anisotropy[130] suggests that the large value of the anisotropy constant of $SmCo_5$ plays a role in determining its high coercivity. The exact mechanism determining the coercivity is still under investigation.

The newest material[131,132] for permanent magnets, developed in the early 1980s, is $Nd_2Fe_{14}B$. The crystal structure is tetragonal,[133] with all magnetic moments parallel to the easy c-axis[134] of the unit cell (which contains four $Nd_2Fe_{14}B$ formula units). The Curie temperature T_c is 586 K (which is rather low) and the saturation magnetization $4\pi M_s$ is 16000 gauss at room temperature,[135] corresponding[135,136] to a saturation magnetic moment of about $33\mu_B$ per formula unit. The material is strongly anisotropic, as suggested by its crystal structure, with[136] an anisotropy constant $K = 4.5 \times 10^7$ erg cm^{-3} at room temperature, a value somewhat lower than that of $SmCo_5$. Data on the structure-sensitive properties are usually reported for compositions varying from $Nd_2Fe_{14}B$. As an example,[137] for a sample with the composition $Nd_{2.0}Fe_{12.6}B_{1.0}$, the remanence B_r is about 14000 gauss, and the coercivity H_c is about 9300 Oe. The maximum energy product $(BH)_{max}$ for this sample is slightly more than 50 MGOe, an extremely large value. While this sample is certainly better than most, these figures certainly indicate that $Nd_2Fe_{14}B$ is an exceptional material for permanent magnets, especially since the raw materials are more readily available than those used in $SmCo_5$. Finally, it appears that a detailed picture of the mechanism determining the coercivity has not yet appeared, and the field is one of currently active research.

Magnetic Materials for Other Applications

The previous section was concerned with "hard" magnetic materials for permanent magnets, in which one wishes to have high values of the coercivity (hence their name) and the remanence. For other applications, these qualities are not necessarily desirable. An example is furnished by the "soft" magnetic materials used, for example, in transformer cores. In these applications, one wishes to produce a high magnetic flux density B in some region of space.[138] This can be obtained by using a ferromagnetic material with a high initial permeability, the value approached by $\mu \equiv (B/H)$ as H goes to zero. This results in the saturation flux density B_s being achieved at a small value of the applied field H. From the magnetization curve in Figure 10.9, we recall that the saturation magnetization M_s (or flux density $B_s \cong 4\pi M_s$) is approached in the regime in which the domain magnetization vectors are rotated against the crystalline anisotropy by the applied field. A crystal with a low anisotropy would therefore be expected to reach its saturation magnetization at a lower applied field, and hence have a higher permeability.

For applications in an AC device, in which the applied field changes its direction, one also wants the coercivity to be small so that the magnetization of the material can reverse its direction readily at small applied fields. An example is the soft iron–nickel–molybdenum alloy "supermalloy," which has a corcivity of 0.005 oersted.[139] As we have seen, high coercivity can be achieved in a material for a permanent magnet by making domain wall displacements difficult. On the other hand, to obtain a low coercivity, one wishes to facilitate domain wall motion, which can be done[140] by using very pure, annealed, oriented materials. The fact that hard and soft magnetic materials exhibit different degrees of ease of domain wall motion is reflected in the observation[41] that materials with high coercivities usually have low permeabilities, and *vice versa*.

Further, in applications[138] of this kind, one also wants a low energy loss through dissipation when the magnetic flux is changed. If the soft ferromagnetic material has a low electrical resistivity (as does, for example,[139] supermalloy, with a resistivity of 6.5×10^{-5} ohm-cm), the changing magnetic flux will set up eddy currents, leading to large energy losses. This loss is called[142] the "core loss," and can be reduced by increasing the intrinsic resistivity of the material by alloying. Another approach is the use[138] of a different class of ferromagnetic materials called "soft ferrites." Ferrites are ferrimagnetic oxides[143] which have resistivities much higher than those of metals, thereby decreasing eddy-current losses. An example[139] is a manganese–zinc ferrite with a coercivity of 0.038 oersted and a resistivity of 5 ohm-cm; the latter value is five orders of magnitude larger than that of supermalloy.

A relatively new approach to soft magnetic materials is the use of amorphous alloys[144–146] formed by very rapid cooling from the melt. Since these amorphous materials do not have a crystal structure, they exhibit an almost complete absence[147] of the crystalline anisotropy found in crystalline ferromagnetic materials. This, in turn, lead to the low coercivity and high permeability desired in a soft magnetic material. For example, an amorphous $Fe_{80}B_{20}$ alloy has a coercivity,[148] of about 0.08 oersted.

In summary, the desirable properties of low coercivity and high permeability of soft magnetic materials are essentially the opposites of those required in a hard permanent magnet material. In the soft materials, these properties are obtained by facilitating domain wall motion in crystals of low anisotropy, again the opposite of what is done in hard materials. For recent discussions of soft ferromagnetic materials, the reader is referred to the literature.[139,146,149,150]

Finally, while a detailed discussion is outside the scope of this introductory book, mention should be made of one other application of ferromagnetic materials. This is the magnetic recording[151–156] of information, using either analog or digital signals. In the recording process, the electrical signal to be recorded produces a time-varying magnetic field in a region of space. Ferromagnetic material (say, on a disk or tape) is moved mechanically through this time-varying magnetic field, producing a spatially-varying magnetization in the moving ferromagnetic material. To "read" this magnetically stored information, the ferromagnetic material, with its spatially varying magnetization, is mechanically moved past a coil of wire. The moving, spatially-varying, magnetization produces a time-varying magnetic field in the coil. This magnetic field, via Faraday's Law of Induction, produces a voltage (which is the output) in the coil. If the direction of the spatially-varying magnetization is parallel to the plane of the ferromagnetic medium, one speaks of "longitudinal" recording. If the direction of the magnetization is normal to the plane of the medium, the process is called "perpendicular" recording.[156,157] The magnetic recording and storage process may be used with both analog and digital signals.

The ferromagnetic material used as a recording medium is functioning as a permanent magnet. In this case, however, it is not desirable to have the largest possible coercivity, because then the magnetization could not be readily reversed, as is necessary in the recording process. One wishes to have a value of the coercivity high enough to prevent demagnetization by external fields, but smaller than the field used in recording signals. A typical value[158] of the coercivity of a recording medium is 500 oersteds, with values of 300–700 oersteds generally being required.[159] However, higher coercivity leads to a sharper boundary[159] between the regions with opposite directions of magnetization[160] used in recording digital signals. This, in turn, increases the sharpness of the output electrical pulse marking the boundary and thus aids the signal-to-noise ratio obtained. The iron oxide particles commonly used as a recording medium now often have cobalt added to increase the coercivity. In addition to particulate ferromagnetic media (such as iron oxide), an area of active research is that of thin films. In these materials, a continuous thin metallic or alloy film, such as CoCr, is used (on a substrate) as the ferromagnetic medium. The interested reader is again referred to the literature.[153,157,161,162]

Problems

10.1. *Anisotropy Energy Density.* Consider a (hypothetical) model of a ferromagnetic material with a simple cubic crystal structure of lattice constant a, for which the anisotropy energy density $U_a = K \sin^2 \theta$ where K is an anisotropy constant (in erg cm^{-3}) and θ is the angle between the direction of the magnetization and an "easy" direction. (a) Calculate the anisotropy energy of a line of N atoms directed normal to the plane of the Bloch wall, which has a width $(N-1)a$. (b) Use your result in (a) to show that the anisotropy energy per unit area σ_a of the Bloch wall (in erg cm^{-2}) is given by $\sigma_a = (KNa/2)$. This calculation shows that the anisotropy energy per unit area of the Bloch wall increases as the wall width increases.

10.2. *Bloch Wall Thickness for Iron.* (a) Using equation (10.61) and values of S, a, J, and K from the text and from Kittel's book,[7] estimate N_m and W_m for iron. (b) Show that the value of N_m given by (10.61) is dimensionless, as it should be. (See Kittel and Galt,[12] page 476.)

10.3. *Minimum Value of Bloch Wall Energy Density.* (a) Using results derived in the text, calculate the minimum value σ_m of the total energy per unit area of the Bloch wall. This value σ_m obtains when the wall width has the value $(N_m a) = W_m$. (b) Check the units of your expression for σ_m for dimensional consistency. (c) Estimate σ_m for iron, using parameter values given in the text. (Note, from your answer to (a), that the exchange and anisotropy contributions to σ_m have the same magnitude. A more accurate calculation of σ_m for the "180° Bloch wall" shown in Figure 10.6, in a (100) plane of iron, gives $\sigma_m = 2(KA)^{1/2}$, where $A = 2JS^2/a$. [See Kittel and Galt,[12] page 478, equation (11.12), and page 461.]

10.4. *Effective Pressure on a Bloch Wall.* Consider a ferromagnetic specimen in an applied magnetic field **H**. Suppose a domain (of magnetization **M** parallel to **H**) increases its volume when its boundary Bloch wall moves by an amount dx. Assuming that the increase $d\sigma_w$ in the surface energy density (energy per unit area) of the wall is due to the decrease in the domain energy, show that $d\sigma_w$ equals $2MH\,dx$. Using this result, show that $(2MH)$ has the dimensions of pressure, and so is taken as the effective pressure on the wall. (See Kittel and Galt,[12] page 521).

10.5. *Maximum Energy Product for an Ideal Material.* Consider the ideal permanent magnet material mentioned in the text for which the relation between B and H is given by equation (10.67) as $B = H + 4\pi M_s$. (a) Show that the graph of B as a function of H in the second quadrant of the hysteresis curve defines a triangle of area $8\pi^2 M_s^2$; (b) Prove, using analytic geometry, the theorem that the area of the largest rectangle that can be inscribed in a triangle of area A is $(A/2)$; (c) Use results (a) and (b) to show that the maximum energy product $(BH)_{max} = 4\pi^2 M_s^2 = (4\pi M_s)^2/4$ for this ideal material.

References and Comments

1. E. M. Purcell, *Electricity and Magnetism*, Second Edition (Berkeley Physics Course, Volume II), McGraw-Hill, New York (1985), pages 431–437.
2. J. D. Jackson, *Classical Electrodynamics*, Second Edition, John Wiley, New York (1975), page 189.
3. A. Shadowitz, *The Electromagnetic Field*, McGraw-Hill, New York (1975), pages 315–321.
4. See E. M. Purcell, Reference 1, page 433 for a discussion of the nomenclature of B and H and page 422 for the units of M, etc.
5. For a discussion of the case of anisotropic materials for which the susceptibility is a second rank tensor, see, for example, J. F. Nye, *Physical Properties of Crystals*, Oxford University Press (1957), pages 53–67; see also J. D. Jackson, Reference 2, pages 13–17.
6. E. M. Purcell, Reference 1, page 435.
7. C. Kittel, *Introduction to Solid State Physics*, Sixth Edition, John Wiley, New York (1986), pages 397–399.
8. E. M. Purcell, Reference 1, pages 413–418.
9. J. R. Reitz, F. J. Milford, and R. W. Christy, *Foundations of Electromagnetic Theory*, Third Edition, Addison-Wesley, Reading, Mass. (1979), Chapter 10. This chapter is recommended for a brief introduction to magnetism, for readers who may not have covered the topic in a solid state physics course.
10. E. M. Purcell, Reference 1, page 423; C. Kittel, Reference 7, page 399.
11. C. Kittel, Reference 7, pages 400–416; E. M. Purcell, Reference 1, pages 418–421.
12. C. Kittel and J. K. Galt, *Solid State Physics*, F. Seitz and D. Turnbull (editors), Volume 3,

Academic Press, New York (1956), pages 457–458.

13. C. Kittel, reference 7, page 429, Table 2.

14. C. Kittel, Reference 7, pages 430–431.

15. J. M. Ziman, *Principles of the Theory of Solids*, Second Edition, Cambridge University Press (1972), pages 339–341.

16. N. W. Ashcroft and N. D. Mermin, *Solid State Physics*, Holt, Rinehart and Winston, New York (1976), pages 674–681.

17. C. Kittel, Reference 7, pages 423–426.

18. A. J. Dekker, *Solid State Physics*, Prentice-Hall, New York (1957), pages 464–482.

19. J. C. Slater, *Quantum Theory of Matter*, First Edition, McGraw-Hill, New York (1951), Section 14.5.

20. F. Seitz, *Modern Theory of Solids*, McGraw-Hill, New York (1940), pages 232, 613.

21. L. I. Schiff, *Quantum Mechanics*, Third Edition, McGraw-Hill, New York (1968), pages 374–375.

22. The problem is the same as the Heitler–London approach to the calculation of the energy of the H_2 molecule. See, for example, C. A. Coulson, *Valence*, Second Edition, Oxford University Press (1961), Section 5.2, for a discussion.

23. N. W. Ashcroft and N. D. Mermin, Reference 16, Page 674.

24. N. W. Ashcroft and N. D. Mermin, Reference 16, Page 679.

25. J. M. Ziman, Reference 15, page 338.

26. J. C. Slater, Reference 19, page 420.

27. J. S. Blakemore, *Solid State Physics*, Second Edition, Saunders, New York (1974), pages 444–446.

28. J. M. Ziman, Reference 15, pages 339–341; C. Kittel, Reference 7, pages 430–431; J. C. Slater, Reference 19, Chapter 14.

29. J. C. Slater, *Quantum Theory of Matter*, Second Edition, McGraw-Hill, New York (1968), Chapter 33.

30. N. W. Ashcroft and N. D. Mermin, Reference 16, pages 679–681.

31. F. Seitz, Reference 20, page 613.

32. See, for example, R. Eisberg, *Fundamentals of Modern Physics*, John Wiley, New York (1961), page 446, equation (13-54').

33. J. C. Slater, Reference 29, pages 205–206.

34. See, for example L. I. Schiff, Reference 21, Section 41; F. Seitz, Reference 20, pages 203–208, 613.

35. N. W. Ashcroft and N. D. Mermin, Reference 16, page 680; F. Seitz, Reference 20, page 613.

36. The reader should note that some authors define the energy difference $(E_s - E_t) \equiv J$, rather than $2J$ as done in (10.22), so the Heisenberg Hamiltonian then lacks the factor of 2 shown in equation (10.40). Among these authors are Ashcroft and Mermin, Reference 16, page 680, and Ziman, Reference 15, page 336. Note also that the factor of $\hbar^2$ is usually omitted from the Heisenberg Hamiltonian, as in Ashcroft and Mermin's equation (32.19).

37. C. Kittel, Reference 12, page 460.

38. N. W. Ashcroft and N. D. Mermin, Reference 16, page 680, Footnote 20; page 701, Footnote 8.

39. N. W. Ashcroft and N. D. Mermin, Reference 16, page 674.

40. C. Kittel, Reference 12, page 462.

40a. C. Kittel, "Physical Theory of Ferromagnetic Domains," *Rev. Mod. Phys.* **21**, 541–583

(1949).

41. C. Kittel, Reference 7, Chapter 15.

42. N. W. Ashcroft and N. D. Mermin, Reference 16, pages 718–722.

43. C. Kittel and J. K. Galt, Reference 12, pages 445–449.

44. J. C. Slater, Reference 19, pages 428–432.

45. C. Kittel and J. K. Galt, Reference 12, page 446; C. Kittel, Reference 7, page 454.

46. C. Kittel and J. K. Galt, Reference 12, page 448.

46a. C. Kittel and J. K. Galt, Reference 12, pages 481–483; 450.

47. N. W. Ashcroft and N. D. Mermin, Reference 16, pages 718–722, give a brief but physical discussion of domains.

48. C. Kittel and J. K. Galt, Reference 12, pages 457–480.

49. C. Kittel, Reference 7, pages 448–454.

50. J. R. Beam, *Electronics of Solids*, McGraw-Hill, New York (1965), Chapter 9.

51. C. Kittel and J. K. Galt, Reference 12, page 459.

52. C. Kittel and J. K. Galt, Reference 12, page 461, equation (5.6).

53. C. Kittel, Reference 7, Table 2, page 429.

54. C. Kittel and J. K. Galt, Reference 12, page 462; C. Kittel, Reference 7, page 424.

55. C. Kittel, *Introduction to Solid State Physics*, Second Edition, John Wiley, New York (1956), page 403.

56. C. Kittel and J. K. Galt, Reference 12, pages 463–467.

57. C. Kittel, Reference 7, pages 450–452.

58. N. W. Ashcroft and N. D. Mermin, Reference 16, pages 720–721.

59. J. R. Beam, Reference 50, pages 444–452.

60. See, for example, R. Eisberg, Reference 32, pages 338–346; L. I. Schiff, Reference 21, pages 433, 482.

61. C. Kittel and J. K. Galt, Reference 12, pages 448–449; C. Kittel, Reference 7, pages 450–452.

62. S. Kaya, *Sci. Rep. Tohoku Imp. Univ.*, **17**, 1157–1177 (1928). Data of Kaya's Table 2, pages 1161–1162, plotted by the author.

63. C. Kittel and J. K. Galt, Reference 12, page 463.

64. C. Kittel and J. K. Galt, Reference 12, pages 463–464, 557.

64a. A. H. Morrish, *The Physical Principles of Magnetism*, John Wiley, New York (1965), page 313.

65. E. P. Wohlfarth, "Iron, Cobalt and Nickel," in *Ferromagnetic Materials*, E. P. Wohlfarth (editor), North-Holland, Amsterdam (1980), Volume 1, page 39, Table 12.

65a. A. Menth, N. Nagel, and R. S. Perkins, "New High Performance Permanent Magnets Based on Rare Earth–Transition Metal Compounds", *Ann. Rev. Mat. Sci.*, **8**, 21–47 (1978), Table 1, page 22.

66. C. Kittel and J. K. Galt, Reference 12, page 476.

67. See Problem 10.1.

68. See Problem 10.2.

69. See, for example, E. M. Purcell, Reference 1, pages 441–442 for an experimental arrangement to measure B as a function of H for a ferromagnetic sample of toroidal shape.

70. Adapted from J. R. Reitz, F. J. Milford, and R. W. Christy, Reference 9, page 199, Figure 9.5.

71. R. M. Bozorth, *Ferromagnetism*, Van Nostrand, New York (1951), page 480, Figure 11-4.

72. J. C. Slater, Reference 19, page 429.

73. C. Kittel, *Introduction to Solid State Physics*, Fourth Edition, John Wiley, New York (1973), page 563, Figure 31a.
74. C. Kittel and J. K. Galt, Reference 12, page 448.
75. C. Kittel and J. K. Galt, Reference 12, page 521.
76. See Problem 10.4.
77. W. F. Brown, *Magnetostatic Principles in Ferromagnetism*, North-Holland, Amsterdam (1962), page 165.
78. S. Chikazumi, *Physics of Magnetism*, John Wiley, New York (1964), page 264.
79. D. J. Craik, *Structure and Properties of Magnetic Materials*, Pion Limited, London (1971), page 191.
80. C. Kittel and J. K. Galt, Reference 12, page 519.
81. R. M. White and T. H. Geballe, *Long Range Order in Solids*, Academic Press, New York (1979), pages 310–313.
82. R. Friedberg and D. I. Paul, "New Theory of Coercive Force of Ferromagnetic Materials," *Phys. Rev. Lett.*, **34**, 1234 (1975).
83. C. Kittel and J. K. Galt, Reference 12, page 545.
84. A. H. Morrish, *The Physical Principles of Magnetism*, John Wiley, New York (1965), pages 383–385.
85. C. W. Chen, *Magnetism and Metallurgy of Soft Magnetic Materials*, North-Holland, Amsterdam (1977), pages 128–129.
86. J. C. Slater, Reference 19, page 430.
87. C. Kittel and J. K. Galt, Reference 12, page 455.
88. C. Kittel, Reference 7, pages 455–456 and Problem 5, page 459.
89. C. Kittel and J. K. Galt, Reference 12, pages 502–519.
90. A. H. Morrish, Reference 84, pages 340–360.
91. See, for example, C. Kittel, Reference 7, page 450, Figure 29.
92. L. Solymar and D. Walsh, *Lecture on the Electrical Properties of Materials*, Second Edition, Oxford University Press (1979), page 289.
93. C. W. Chen, Reference 85, page 63; S. Chikazumi, Reference 78, page 246.
94. C. Kittel and J. K. Galt, Reference 12, page 452, Footnote 2.
95. J. D. Livingston, "Present Understanding of Coercivity in Cobalt–Rare Earths," in *Magnetism and Magnetic Materials, 1972* (A.I.P. Conference Proceedings, No. 10, Part 1), C. D. Graham and J. J. Rhyne (editors), American Institute of Physics (1973), pages 643 –657; page 643.
96. C. Kittel and J. K. Galt, Reference 12, pages 507–509; C. Kittel, Reference 7, page 459, Problem 5.
97. A. Menth, H. Nagel, and R. S. Perkins, Reference 65a, page 27.
98. J. D. Livingston, Reference 95, page 645.
99. See, for example, R. M. Bozorth, Reference 71, pages 507–518. More recent discussions are given by A. H. Morrish, Reference 84, Chapter 7, and C. W. Chen, Reference 85, Chapter 4.
100. J. C. Slater, Reference 19, pages 429–431; L. Solymar and D. Walsh, Reference 92, page 289.
101. A. H. Morrish, Reference 84, pages 363–367.
102. National Materials Advisory Board, *Report of the Committee on Magnetic Materials*, R. M. White (editor), Publication NMAB-426, National Academy Press, Washington, D. C. (1985), pages 11–12.

103. R. M. White, "Opportunities in Magnetic Materials," *Science*, **229**, 11–15 (1985), page 11.

104. See Problem 10.5.

105. The largest value of $(BH)_{max}$ reported as of late 1987 is about 45 MGOe for an FeNdB alloy. See R. M. White, Reference 103, Figure 1, page 12.

106. C. Kittel, Reference 55, pages 416 and 422–425.

107. R. M. White and T. H. Geballe, Reference 81, page 310.

108. C. Kittel and J. K. Galt, Reference 12, pages 452–455 and 519–533.

109. C. Kittel and J. K. Galt, Reference 12, page 454.

110. R. M. White, Reference 103, page 12.

111. Calculated from the value $M_s = 1400$ gauss given by C. Kittel, Reference 7, Table 2, page 429.

112. R. M. Bozorth, Reference 71, page 266.

113. A. Menth, H. Nagel, and R. S. Perkins, Reference 65a, Table 2, page 23.

114. G. Y. Chin, "New Magnetic Alloys," *Science*, **208**, 888–894 (1980), Table 2, page 889.

115. R. M. Bozorth, Reference 71, Figure 9-37, page 391, gives this value of $B_s \cong 4\pi M_s$ for an Alnico 5 type alloy of slightly different composition from the figures given in the text. The author is indebted to G. Y. Chin for pointing out this data.

116. G. Y. Chin and J. H. Wernick, "Magnetic Materials, Bulk," in *Encyclopedia of Chemical Technology*, John Wiley, New York (1981), Volume 14, pages 646–686, page 670.

117. S. Chikazumi, Reference 78, page 504.

118. A. H. Morrish, Reference 84, page 365.

119. C. Kittel and J. K. Galt, Reference 12, pages 513–519.

120. H. Zijlstra, "Permanent Magnets: Theory," in *Ferromagnetic Materials*, E. P. Wohlfarth (editor), North-Holland, Amsterdam (1980), Volume 3, pages 37–106, Sections 3.2, 5.1, 5.2; see also A. H. Morrish, Reference 84, page 345.

121. A. Menth, H. Nagel, and R. S. Perkins, Reference 65a.

122. E. A. Nesbitt and J. H. Wernick, *Rare Earth Permanent Magnets*, Academic Press, New York (1973).

123. G. Y. Chin and J. H. Wernick, Reference 116, pages 675–678.

124. G. Y. Chin, Reference 114, Table 1, page 889.

125. G. Y. Chin, Reference 114, Table 2, page 889.

126. G. Y. Chin, Reference 114, page 888.

127. J. F. Herbst, R. W. Lee, and F. E. Pinkerton, "Rare Earth–Iron–Boron Materials," *Ann. Rev. Materials Science*, **16**, 467–485 (1986), page 471.

128. K. H. J. Buschow, "Rare Earth Compounds," in *Ferromagnetic Materials*, E. P. Wohlfarth (editor), North-Holland, Amsterdam (1980), Volume 1, pages 297–414, Table A.3c, page 386.

129. R. M. White, Reference 102, page 13.

130. R. M. White and T. H. Geballe, Reference 81, page 310; C. Kittel and J. K. Galt, Reference 12, page 467.

131. M. Sagawa *et al.*, "Nd–Fe–B Permanent Magnet Materials," *Jap. J. Appl. Phys.*, **26**, 785–800 (1987).

132. J. F. Herbst *et al.*, Reference 127.

133. J. F. Herbst *et al.*, "Relationship between Crystal Structure and Magnetic Properties in $Nd_2Fe_{14}B$," *Phys, Rev. B*, **29**, 4176–4178 (1984).

134. J. F. Herbst *et al.*, "Structural and Magnetic Properties of $Nd_2Fe_{14}B$," *J. App. Phys.*, **57**, 4086–4090 (1985).

135. J. F. Herbst *et al.*, Reference 127, Table 1, page 472.
136. M. Sagawa, Reference 131, Table 1, page 787.
137. M. Sagawa, Reference 131, page 795, Figure 14.
138. R. M. White, Reference 103, page 13.
139. G. Y. Chin, "Magnetic Materials," in *Frontiers in Materials Technologies*, M. A. Meyers and O. T. Inal (editors), Elsevier, Amsterdam (1985), Chapter 13, page 421.
140. C. Kittel, Reference 55, pages 416 and 425–428.
141. C. Kittel, Reference 55, Figure 15.18, page 425.
142. G. Y. Chin, Reference 114, page 892.
143. C. Kittel, Reference 7, pages 438–439.
144. G. Y. Chin and J. H. Wernick, Reference 116, page 667.
145. F. E. Luborsky, "Amorphous Ferromagnets," in *Ferromagnetic Materials*, E. P. Wohlfarth (editor), North-Holland, Amsterdam (1980), Volume 1, pages 451–530.
146. G. Y. Chin, Reference 139, page 426.
147. F. E. Luborsky, Reference 145, pages 505–514.
148. F. E. Luborsky, Reference 145, Table 8, page 508.
149. G. Y. Chin and J. H. Wernick, Reference 116, pages 650–669.
150. C. W. Chen, Reference 85.
151. J. C. Mallinson, "Tutorial Review of Magnetic Recording," *Proceedings IEEE*, **64**, 196–208 (1976).
152. R. M. White (editor), *Introduction to Magnetic Recording*, IEEE Press, New York (1985).
153. M. H. Kryder and A. B. Bortz, "Magnetic Information Technology," *Physics Today* (December 1984), pages 20–28.
154. R. M. White, Reference 102, pages 51–68.
155. R. M. White, Reference 103, pages 13–15.
156. T. C. Arnoldussen and E.-M. Rossi, "Materials for Magnetics Recording," *Ann. Rev. Mat. Sci.*, R. A. Huggins (editor) **15**, 379–409 (1985).
157. R. M. White, "Magnetic Discs: Storage Densities on the Rise," *IEEE Spectrum* (August 1983), pages 32–38.
158. R. M. White, Reference 152, page 10.
159. R. M. White, Reference 157, page 33.
160. M. H. Kryder and A. B. Bortz, Reference 153, page 21.
161. T. C. Arnoldussen and E.-M. Rossi, Reference 156, pages 396–407.
162. M. H. Kryder, "Data Storage Technologies for Advanced Computing," *Scientific American*, **257**, 117–125 (October 1987).

Suggested Reading

C. KITTEL, *Introduction to Solid State Physics*, Sixth Edition, John Wiley, New York (1986), Chapter 15. This chapter is the basic background reference on ferromagnetism. Much of our treatment parallels Kittel's work.

C. KITTEL, *Introduction to Solid State Physics*, Second Edition, John Wiley, New York (1956), Chapter 15. This chapter in an earlier edition of Kittel's text contains additional material on ferromagnetism.

N. W. Ashcroft and N. D. Mermin, *Solid State Physics*, Holt, Rinehart, and Winston, New York (1976), Chapter 32 and 33. This text, at a level more advanced than that of Kittel's book, discusses, among other topics, magnetic interactions and structures, including ferromagnetism.

C. Kittel and J. K. Galt, "Ferromagnetic Domain Theory", in *Solid State Physics*, F. Seitz and D. Turnbull (editors), Academic Press, New York (1956), Volume 3, pages 437–564. This review discusses its subject at the advanced level and many of its discussions are followed in our treatment.

E. P. Wohlfarth, editor, *Ferromagnetic Materials*, North-Holland, Amsterdam (1980). A multivolume treatise on its subject at the advanced level, this collection of articles by different authors includes discussions of materials for applications (such as recording). Volume 4 (1988) includes articles on the newer materials for permanent magnets.

A. H. Morrish, *The Physical Principles of Magnetism*, John Wiley, New York (1965). An advanced treatise on many aspects of magnetism, this volume includes a detailed discussion of ferromagnetism.

J. C. Slater, *Quantum Theory of Matter*, First Edition, McGraw-Hill, New York (1951), pages 418–432. This part of Slater's text offers a brief discussion, with much physical insight, of the Heisenberg and energy band pictures of ferromagnetism, and of domain structure. The second edition (1968) of the book updates the energy band description.

E. A. Nesbitt and J. H. Wernick, *Rare Earth Permanent Magnets*, Academic Press, New York (1973); A. Menth, H. Nagel and R. S. Perkins, "New High Performance Permanent Magnets Based on Rare Earth Transition Metal Compounds," *Ann. Rev. Materials Sci.* **8**, 21–47 (1978). This monograph and this review article discuss permanent magnet materials such as $SmCo_5$.

J. F. Herbst, R. W. Lee, and F. E. Pinkerton, "Rare Earth–Iron–Boron Materials," *Ann. Rev. Materials Sci.* **16**, 467–485 (1986); M. Sagawa *et al.*, "Nd–Fe–B Permanent Magnet Materials," *Jap. J. Appl. Phys.* **26**, 785–800 (1987). These recent review articles discuss the newest class of permanent magnet materials.

Appendix:
References on Some Other Topics

The following is a brief list of references on some interesting topics which are not discussed in the text. For compound semiconductor devices and the ballistic transistor, I suggest "Compound Semiconductor Transistors," by L. F. Eastman, *Physics Today*, **39**, 77–83 (October, 1986), "The HEMT: A Superfast Transistor," by H. Morkoc and P. M. Solomon, *IEEE Spectrum*, **21**, 28-35 (February, 1984), and "The Quest for Ballistic Action," by T. E. Bell, *IEEE Spectrum*, **23**, 36–47 (February, 1986).

On superlattices and quantum well devices, I suggest as a brief introduction "Quantum Tailored Solid State Devices," by T. J. Drummond *et al.*, *IEEE Spectrum*, **25**, 33–37 (June, 1988). Additional information may be found in "Band Gap Engineering," by F. Capasso, *Science*, **235**, 172–176 (9 January 1987), "Quantum Wells for Photonics," by D. S. Chemla, *Physics Today*, **38**, 57–64 (May, 1985), and "Crystalline Semiconductor Heterostructures," by V. Narayanamurti, *Physics Today*, **37**, 24–32 (October, 1984). A volume devoted to these topics is "Application of Multiquantum Wells, Selective Doping, and Superlattices," edited by R. Dingle, Volume 24 of the series *Semiconductors and Semimetals*, edited by R. K. Willardson and A. C. Beer, Academic Press, San Diego (1987).

The February 1990 issue of *Physics Today* includes discussions of nanoscale and ultrafast devices, including quantum electron devices.

As an introduction to light wave communications and integrated optics, I suggest "Integrated Optics," by E. M. Conwell, *Physics Today* **29**, 48–56 (May, 1976), "Components for Optical Communications Systems: A Review," by D. Boetz and G. J. Hershkowitz, *Proc. IEEE*, **68**, 689–731 (June, 1980), and "Light Wave Telecommunication," by T. Li, *Physics Today*, **38**, 24–31 (May, 1985). For additional detail, Volume 22 of the series

Semiconductors and Semimetals, edited by R. K. Willardson and A. C. Beer, Academic Press, Orlando (1985), is a five-part treatise on all aspects of optical communication physics and technology.

Index